Transcultural Cities

Transcultural Cities uses a framework of transcultural placemaking, cross-disciplinary inquiry and transnational focus to examine a collection of case studies around the world, presented by a multidisciplinary group of scholars and activists in architecture, urban planning, urban studies, art, environmental psychology, geography, political science, and social work. The book addresses the intercultural exchanges as well as the cultural transformation that takes place in urban spaces. In doing so, it views cultures not in isolation from each other in today's diverse urban environments, but as mutually influenced, constituted and transformed.

In cities and regions around the globe, migrations of people have continued to shape the makeup and making of neighborhoods, districts, and communities. For instance, in North America, new immigrants have revitalized many of the decaying urban landscapes, creating renewed cultural ambiance and economic networks that transcend borders. In Richmond, BC, Canada, an Asian night market has become a major cultural event that draws visitors throughout the region and across the US and Canadian border. Across the Pacific, foreign domestic workers in Hong Kong transform the deserted office district in Central on weekends into a carnivalesque site. While contributing to the multicultural vibes in cities, migration and movements have also resulted in tensions, competition, and clashes of cultures between different ethnic communities, old-timers, newcomers, employees and employers, individuals and institutions.

In *Transcultural Cities* Jeffrey Hou and a cross-disciplinary team of authors argue for a more critical and open approach that sees today's cities, urban places, and placemaking as vehicles for cross-cultural understanding.

Jeffrey Hou is Associate Professor and Chair of Landscape Architecture at the University of Washington, USA.

Transcultural Cities

Border-Crossing and Placemaking

Edited by Jeffrey Hou

Routledge
Taylor & Francis Group
NEW YORK AND LONDON

First published 2013
by Routledge
711 Third Avenue, New York, NY 10017

Simultaneously published in the UK
by Routledge
2 Park Square, Milton Park, Abingdon, Oxon OX14 4RN

Routledge is an imprint of the Taylor & Francis Group, an informa business

Library of Congress Cataloging-in-Publication Data
Transcultural cities : border crossing & placemaking / edited by Jeffrey Hou.
p. cm.
Includes bibliographical references.
1. Cities and towns–Cross-cultural studies. 2. City planning–Cross-cultural studies. 3. Community development–Cross-cultural studies 4. Urbanization–Cross-cultural studies. I. Hou, Jeffrey, 1967–
HT151.T747 2014
307.76–dc23 2012025449

ISBN: 978–0–415–63141–9 (hbk)
ISBN: 978–0–415–63142–6 (pbk)
ISBN: 978–0–203–07577–7 (ebk)

Typeset in Humanist by
Keystroke, Station Road, Codsall, Wolverhampton

SFI Certified Sourcing
www.sfiprogram.org
SFI-00453

Printed and bound in the United States of America
by Edwards Brothers, Inc.

Contents

Figures

Tables

Contributors

Mallika Bose is Associate Professor of Landscape Architecture at Penn State and Director of the Hamer Center for Community Design. Motivated by her interest in understanding how social structures are spatially embedded, she pursues research on Built Environment and Active Living/Healthy Eating, Public Scholarship and Community-based Design.

Caitlin Cahill is Assistant Professor in the Public Science Project and Environmental Psychology Program at the CUNY Graduate Center. A community-based scholar, her research focuses on young people and global restructuring. She co-directed the Mestizo Arts & Activism Collective that engages young people as catalysts of change.

Felicity Hwee-Hwa Chan is a PhD candidate at the University of Southern California. She received her BA (Hons.) in Geography from the National University of Singapore and her Master of Urban Planning from Harvard University. Her research interests include multi-ethnic cities and intercultural relations, migration, urban design, and land-use planning.

Shenglin Elijah Chang is Associate Professor in the Graduate Institute of Building and Planning, National Taiwan University. She directs the New Ruralism Research and Development Center, focusing on culturally and ecologically sustainable ways to revitalize rural areas in East Asia. She is the author of *The Global Silicon Valley Home* (2006).

Hung-Ying Chen is an activist at the Taiwan Association for the Justice of Urban Renewal. She is currently pursuing her doctorate at the Graduate Institute of Building and Planning, National Taiwan University. Her research focuses mainly on migrant issues that intersect with spatial planning and cultural activism.

Kirk Dimond currently serves as the primary faculty academic adviser for undergraduate architecture and landscape architecture students at the Pennsylvania State University. His professional experience includes residential and commercial/institutional design. He continues to practice and research with a primary interest and focus on edible landscapes for individuals and communities.

Yenchew Foo is Associate Research Fellow at the Centre for Malaysian Chinese Studies in Kuala Lumpur. Her research focuses on the history of Malaysian Chinese in Kuala Lumpur and interactions between space and minority groups. She received her Master's degree from the Graduate Institute of Building and Planning, National Taiwan University.

José Edgardo Abaya Gomez, Jr. is Assistant Professor and training director of the University of the Philippines' School of Urban and Regional Planning. His research interests include street-level sociology, public space redevelopment, Southeast Asian urban history, and city cultures.

Hongguang He is Lecturer in the Department of Public Administration at Nanjing Audit University, China, and a PhD candidate in the Faculty of Arts and Social Sciences at the University of Sydney. His research focuses on spatialization and local government.

Jeffrey Hou is Associate Professor and Chair of Landscape Architecture at the University of Washington, Seattle. As the editor of *Insurgent Public Space: Guerrilla Urbanism and the Remaking of Contemporary Cities* (2010), his work focuses on public space, design activism, and engagement of marginalized communities in the making of place.

Jayde Lin Roberts is Lecturer in the School of Asian Languages and Studies at the University of Tasmania, Australia. Specializing in Chinese diaspora and Asian cities, she received her PhD in the Built Environment and MA in China Studies from the University of Washington and her BA in Architecture from University of California, Berkeley.

Duanfang Lu is Senior Lecturer in the Faculty of Architecture, Design and Planning at the University of Sydney, Australia. She has published widely on modern Chinese architectural and planning history. Her recent publications include *Remaking Chinese Urban Form* (2006, 2011) and *Third World Modernism* (2010).

Willow Lung-Amam is a postdoctoral scholar in the Department of City and Regional Planning at the University of North Carolina, Chapel Hill. She is currently writing a book on the politics of development, planning, and design in new immigrant suburbs in the Silicon Valley.

Trinh Mai is faculty with the College of Social Work at the University of Utah. At Hartland Partnership Center, a University Neighborhood Partners program, she supervises the Social Work Partnership where students and faculty collaborate with community and university partners to run programs building community capacity and promoting transcultural integration.

Lynne C. Manzo is an Environmental Psychologist and Associate Professor of Landscape Architecture at the University of Washington, Seattle. Her research focuses on place meaning, social justice, public housing, and cultural landscapes. She co-authored *Environmental Dilemmas: Ethical Decision-Making* (2008) and co-edited *Place Attachment: Advances in Theory, Methods and Applications* (2013).

Rachel Miller is an urban designer and planner with Makers Architecture and Urban Design in Seattle. She gained a holistic view of the built environment while obtaining a dual Master in Landscape Architecture and Urban Planning from the University of Washington and a BS Architecture and Geography minor from the Ohio State University.

Adam Prince earned his MA in Urban Studies from the San Francisco Art Institute in 2010. His research focuses primarily on social and political matters of concern in the context of prevailing food discourses. He coordinates a collaborative studio dedicated to the formation of substantive social change through creative practices.

Lorenzo Rinelli received his PhD in Political Science from the University of Hawai'i at Manoa in 2012. His thesis analyzed everyday practices of externalization of

African migration control beyond and within Europe's geopolitical limits. His research interests include border studies, migration theory, postcolonial theory and aesthetics in world politics.

Michael Rios is Associate Professor and Chair of the Community Development Graduate Group at University of California, Davis, where he directs the Sacramento Diasporas Project at the Center for Regional Change. His research intersects issues of marginality, urbanism, and public space. He is co-editor of *Diálogos: Placemaking in Latino Communities* (2012).

Clare Rishbeth is Lecturer in Landscape Architecture at the University of Sheffield. Her research focuses on perceptions of place by first generation migrants, particularly how memories are evoked, and how these inform everyday use of urban outdoor spaces. She also explores participatory approaches to produce creative on-site representations of place experience.

Kimberly Schmit is the Community Capacity Building Partnership Manager for University Neighborhood Partners at the University of Utah. As a community organizer, Kim spends her time developing sustainable and comprehensive partnerships with university stakeholders and community groups to ensure reciprocal learning, action, and benefit.

Arijit Sen is Associate Professor of Architecture at the University of Wisconsin Milwaukee. He is the co-coordinator of the Buildings–Landscapes–Cultures doctoral research area between the University of Wisconsin–Madison and Milwaukee. His research interests include physical and cultural landscapes of ethnicity and immigration in the United States.

Kin Wai Michael Siu is Professor and Lab Leader of the Public Design Lab, School of Design at the Hong Kong Polytechnic University. He was Fulbright Scholar at MIT and Visiting Scholar at University of California, Berkeley. His research interests are in public design, user reception, and design and culture.

Vera Zambonelli is a doctoral candidate in Urban and Regional Planning, University of Hawai'i, Manoa. Her research focuses on examining processes of placemaking, transcultural places, and everyday forms of cosmopolitanisms using videographic methodologies. Vera co-authored with Karen Umemoto "Cultural Diversity" for the *Oxford Handbook of Urban and Regional Planning* (2012).

Acknowledgements

This book represents the culmination of a three-year long collaboration among a multidisciplinary group of individuals located in four continents—Asia, Australia, Europe, and North America. The project began as a response to a request for proposals by the Worldwide Universities Network (WUN), a consortium of eighteen universities around the world. The request sought projects that would address one or more of the WUN Global Challenges—*Adapting to Climate Change, Furthering the Frontiers of Cultural Understanding,* and *Opportunities and Challenges of Globalization*. As I scrolled down the list of universities, a map of potential collaborators and possible research topics began to emerge. I am grateful to Mallika Bose (Pennsylvania State University), Shenglin Chang (National Taiwan University), Sam Dennis, Jr. (University of Wisconsin, Madison), Duanfang Lu (University of Sydney), Michael Rios (University of California, Davis), and Clare Rishbeth (University of Sheffield) for agreeing to join the project immediately and making it possible in the first place. We are indebted to the Selection Committee at WUN for choosing our project and to our respective institutions for providing the matching financial support. I am especially grateful to the former Office of Global Affairs at the University of Washington, specifically Steve Hanson and Lauren Jorelle for their unwavering support and assistance throughout the project.

While spending her sabbatical year in Seattle, Mallika Bose and I discussed how the project could be structured to produce more diverse and meaningful outcomes. Our conversation led to the development of a working symposium to engage a much broader network of researchers and individuals in the project. Working with the initial title of *Immigrants, Places and Cross-cultural Understanding*, we formulated the theme of *Transcultural Cities* for the two-day symposium, as a way to frame the often volatile, transitional, and transformative processes of migrations, place-making, and cross-cultural interactions. As the WUN project was in progress, I was fortunate to join another multidisciplinary team of faculty at the University of Washington in hosting the 2010–2011 John E. Sawyer Seminar titled "Now Urbanism: City Making in the 21st Century and Beyond" funded by the Andrew W. Mellon Foundation. Many thanks to Thaisa Way and Margaret O'Mara, the project's principal investigators, for supporting our contribution to the public lecture series, titled "Transcultural Urbanism." I am grateful to Anastasia Loukaitou-Sideris, Michael Rios, and Arijit Sen for speaking at the event that kicked off the two-day symposium and to Dan Abramson for moderating the discussion following the public lecture. We are equally grateful to our many colleagues and new friends who presented and enriched the two-day symposium at the University of Washington with their fascinating case studies and insights. During the symposium, and even as draft chapters were being revised repeatedly, we struggled with defining the notion of transculturation. I believe that in the end we have arrived at a nuanced, grounded, and productive understanding of the term and its relevance to placemaking and cross-cultural understanding in the contemporary cities. I salute your efforts and perseverance. As in all projects of this nature, in moving from the symposium proceedings into an edited book manuscript, we were not able to accommodate many compelling case studies. I would like to express my deep regret to the shadow contributors of this book.

In many ways, this project was as collaborative as it was personal. As an adolescent immigrant in New York City in the 1980s, my own experiences as a stranger in a new environment set the stage for my inquiry into relationships between migrations, placemaking, and cross-cultural understanding. This book explores not only the questions that are relevant to the changing conditions of our cities, but also my own curiosity into the dynamics of identity, design, and community building. For this, I would like to thank my parents for enabling me to pursue such a rich and diverse life experience in the United States, and the many friends, colleagues, and mentors whose lives have shaped mine over the years, from the East Coast to the West Coast and between the two coasts of the Pacific Ocean.

This project would not have been completed without the support of many capable and outstanding individuals. I would like to thank Jayde Lin Roberts for her immaculate organizational skills in staging the symposium and compiling the proceedings and the initial manuscript (and Lee Roberts for supporting Jayde), Joming Lau for his "timely" assistance during the symposium, Mackenzie Waller for providing graphic design for the symposium poster and the visual identity of the project, Kelley Pagano and Rachel Ward for their administrative support, and Audrey Maloney for her impeccable skills, confidence, and commitment in bringing the project to the finishing line. Lastly, but most importantly, I offer my deepest thanks to Emily Grigg-Saito who assisted me early on in the project but whose life ended unexpectedly to the shock and sorrow of many of us. Emily was an aspiring, kind, and hard-working student with her own transcultural life story between the USA and Japan. This book is dedicated in her memory.

Jeffrey Hou, Seattle

CHAPTER 1

Your Place and/or My Place?

Jeffrey Hou

In cities and regions around the world, movement and migration of people have continued to shape the makeup and making of neighborhoods, districts, and communities. In North America, new immigrants have played a critical role in revitalizing many decaying urban landscapes, creating new and renewed cultural ambiance and economic networks that transcend borders. In Seattle, for instance, Vietnamese businesses have transformed a declining urban area with vacant warehouses and storefronts into a vibrant commercial center at the edge of downtown. In Los Angeles, Latina/o immigrants turn front yards and streets into social gathering places and building façades into canvases for murals and cultural icons. In Richmond, BC, Canada, an Asian night market has become a major cultural event that draws visitors throughout the region and across the US/Canadian border. Across the Pacific, guest workers in Hong Kong transform the deserted office district in Central into a site of carnivalesque gatherings every Sunday. In Taipei, the Songkran Festival has become an annual event celebrated not only by immigrants, refugees, and foreign workers from Burma, Thailand, and Cambodia, but also by local residents and civic leaders.

While contributing to the multicultural vibes in cities, movement and migration have also resulted in tensions, competition, and clashes of cultures between old-timers and newcomers, employees and employers, and individuals and institutions. In Los Angeles, Korean and Bangladeshi immigrant communities competed over the official naming and recognition of their overlapping neighborhoods (Jang 2009). In Brazil, Chinese employers and local workers clash over work ethics and lifestyles (Brooks 2011). In Taipei, migrant workers are often harassed by police and portrayed negatively in the media as threats to law and order. In North London, the fatal shooting of an African–Caribbean by the police led to days of rioting across different cities in Britain in Summer 2011. These recurring conflicts illustrate the profound challenge of increasingly diverse populations in contemporary cities.

As cities continue to serve as the main destination of transnational and intra-national migrations, what role can they play in supporting the growing diverse populations? How can urban places function as vehicles for cross-cultural learning and understanding rather than just battlegrounds and turfs? How can cross-cultural interactions be constructed, enabled, or "staged" through social and spatial practices in the contemporary urban environment? As migration, diasporas, and translocality have further destabilized existing meanings and identities of places, how can we re-envision placemaking in the context of shifting cultural terrains? These are the questions and challenges that we intend to investigate in this book.

Figure 1.1
Grocery markets in Seattle's Little Saigon brought new economic vitality to an area that has long been dormant at the edge of downtown. Photograph by Jeffrey Hou.

Rise of New Multicultural Cities

Movement and migration has long been a central characteristic of civilization and human settlement (see Appadurai 1990; Hall 2000; Wood and Landry 2008). This was how Polynesians managed to settle in the countless islands and archipelagos throughout the Pacific, and how historic cities such as Rome and Xian, and megalopolises today such as New York and Tokyo, came to being. In the decades since World War II, advancement in transportation and communication coupled with economic globalization, along with conflicts and large-scale natural catastrophes, have accelerated movement and migration of people (Castles and Miller 2009). In

Figure 1.2 El Pedorrero Muffler Shop in East Los Angeles: a case of self-made neighborhood transformation. Photograph by Jeffrey Hou.

2010, the population of international migrants was estimated at over 213 million, or 3.1 percent of the world's population, compared with 76 million in 1960 (UNDESA 2011; UNDESA 2005, cited in Castles and Miller 2009). In the United States, the estimated number almost doubled from 23.3 million in 1990 to 42.8 million (13.5 percent of total population) in 2010 (UNDESA 2011).

Collectively, these movements and migrations have resulted in fundamental changes in global political, economic, and social and cultural systems (Roseman et al. 1996). Demographically, for example, international migration has increased ethnic diversity in immigrant-receiving countries (Castles and Miller 2009). Furthermore, in many North American and European cities, the new ethnic diversity is not limited just to the traditional sites of settlements in core downtown areas, but is instead widely dispersed to the suburbs, creating a complex patchwork of new ethnic enclaves (see Qadeer 1997; Krase 2002; Li 1999). Similarly, it was observed that the patterns of migration have diversified to include intra-regional migration within and between the Asia-Pacific, Latin America, and Middle East (Castles and Miller 2009). In terms of social and political consequence, Holston (1998: 51) argues "as new and more complex kinds of ethnic diversity dominate cities, the very notion of shared community becomes increasingly exhausted," and "the notions of formal citizenship" becoming increasingly problematic.

In recent writings on migration and urbanism, various terms and concepts have been proposed to illustrate these growing phenomena. Geographer Curtis Roseman and others (1996) used *EthniCities* to characterize cities with a variety of people having distinctive cultures and origins. Anthropologist Arjun Appadurai (1990: 7) suggests *ethnoscapes* to describe "landscapes of persons who constitute the shifting world in which we live." Urban designer Noha Nasser (2004b) uses "Kaleido-scapes" to describe the landscapes of migrant groups as "a hybrid urban morphology that combines local vernaculars with global (or imported) elements." Borrowing from Salman Rushdie, planning historian Leonie Sandercock (2003: 1) put forward *Mongrel Cities* to conceptualize the new urban condition "in which difference, otherness, fragmentation, splintering, multiplicity, heterogeneity, diversity, plurality prevail."

But beyond simply concepts and descriptors, how can we develop a framework for not only understanding but also for actions that facilitate active making of space and places to engender diversity, hybridity, and cross-cultural learning and understanding? How can we as individuals and as designers and planners actively support the making of *Mongrel Cities*, *ethnoscapes*, and *kaleido-scapes* to better serve diverse communities? Before we launch into our proposed framework, it is useful to revisit some of the past and current attempts under the rubric of multiculturalism.

Multiculturalism and the Planning Challenges

In the late twentieth century, growing migration and new settlements have given rise to multiculturalism as a new cultural and institutional paradigm. In countries including Canada, Australia, the Netherlands, and Sweden, multiculturalism has been enshrined into social policies that recognize identities and equal rights of different ethnic groups since the 1970s (Castles and Miller 2009; Parekh 2000; Sandercock 2003). Specifically, the Canadian Multiculturalism Act of 1988 defines multiculturalism as a policy designed to preserve and enhance the multicultural heritage of Canadians while working to achieve the equality of all (Qadeer 1997). Significant spending has been provided to support "maintenance of various cultures and languages" (Sandercock with Brock 2009). At the city level, municipal governments in Frankfurt, Rotterdam, Vancouver, and several Australian cities have adopted multicultural policies to serve their diverse citizens (Sandercock 2003; Thompson 2003). In the UK, developed in the wake of racial riots in 2001, the community cohesion agenda has promoted "neighborly mixing as a means of tackling racism and problematic race divides" (Wise and Velayutham, 2009: 5). In the United States, the rise of multiculturalism reflects shifts in cultural discourses, as minority groups (defined by race, ethnicity, gender, and sexuality) question the long-standing norm and assumption of *melting pot* and seek recognition for their distinct cultural identities and practices—in what Cornel West (1990) describes as "new cultural politics of difference."

In the field of urban planning and design, there has also been an awakening of sorts to the multicultural reality of today's cities (see Sandercock 1998a, 1998b, 2003; Burayidi 2000a; Qadeer 1997). Increasingly, planners and city administrators are being asked to develop cultural competency in working with diverse communities and constituents, as they are confronted with competing expectations about municipal services and patterns of urban growth (Pestieau and Wallace 2003). Sandercock (1998a: 16) argues that planners need to develop "a multicultural literacy more attuned to the cultural diversity, and to redefine and reposition planning according to these new understandings." She further suggests that an open and communicative planning process is needed, which requires planners to have "life experience, communicative skills and, in multicultural or multiethnic contexts, cross-cultural understanding" (Sandercock 2000: 23). Similarly, Umemoto (2001: 17) argues that designing and facilitating planning processes to accommodate cultural differences requires planners to "extend their thinking into other epistemological worlds." More than just an ethical responsibility, Burayidi (2000a) argues that multiculturalism has become practically necessary.

A number of planning approaches have been recognized or developed to meet the challenges of growing cultural diversity. For example, Sandercock (2000) sees Paul Davidoff's *Pluralist Planning* as a model to address the differential impacts of planning on race, gender, and class. Burayidi (2000b: 45) suggests *Holistic Planning* as "a means of social actions based on diversity, tolerance, and cooperation." Ameyaw (2000: 101) proposes *Appreciative Planning* as a model based on mutual respect, trust, and care-based action, and "to create contexts in which planners and multicultural groups can continuously learn and experiment, think systematically, engage in meaningful dialogue, and create visions that energize action and inclusion in city planning." In engaging the multicultural publics, Qadeer (1997: 493) notes that "the scope and procedure of citizen involvement in the planning process have to be modified to accommodate multicultural policies." He calls for more flexibility in planning norms and practices in recognition of ethnic and social diversity (ibid.). Similarly, Hou and Kinoshita (2007) suggest informal mechanisms as an equally important tool for bridging social and cultural differences in diverse communities.

As commonly understood, multiculturalism posits that all ethnic groups in society should be able to exercise equal rights without having to give up their own culture, religion, and language (Castles and Miller 2009). In practice, however, multiculturalism and multicultural planning is far from being fully realized. While the discourse and institutionalization of multiculturalism has been important in reversing the predominant practice of assimilation, the actual outcomes in terms of embracing diversity, identities, and cultural differences remain highly contested. Planning scholar Anastasia Loukaitou-Sideris (2002: 335) observes that ethnic minority neighborhoods are often described in negative terms by authorities "that wish to eliminate, control, or regulate them." Similarly, Jacobs and Fincher (1998: 1) argue that "city governance has vacillated between celebrating and enhancing such diversity, on the one hand, and regulating and repressing it, on the other." Sandercock (1998a: 164) notes that the multicultural city "is perceived by many to be much more of a threat than an opportunity."

In many cities today, despite growing cultural diversity, the social and spatial barriers between different ethnic groups have persisted. In North America, as immigrants increasingly move to the suburbs, so have the racial and ethnic lines that commonly delineate the inner city (Talen 2006; Krase 2002). Using the Chinese takeaway in London as an example, Parker (2000) wrote about the unequal terms of exchange between European and Asian in both past and present. In cities such as Tokyo, Taipei, and Seoul, migrant communities are largely invisible and unrecognized in the everyday landscapes and the cities' social policy. In many European countries, immigrants and refugees are largely confined to the outskirts of cities, creating pockets of concentrated poverty and joblessness, setting the context for negative

Figure 1.3 Except for a few businesses and weekend gathering, the Filipino guest workers remain largely invisible in Taipei. Photograph by Jeffrey Hou.

perception of these communities. In Paris and other French cities, these pockets of migrant settlements became sites of a series of civil unrest in 2005.

As institutionalized social policy, multiculturalism has been challenged by critics of different ideological orientations. Hall (2000) notes that multiculturalism has been contested by the conservative Right as well as the liberals, for threatening the cultural purity of the nation on the one hand and the universalism and neutrality of the liberal state on the other. Expressing a libertarian view, Welsh (2008: 16) argues that "racism and multiculturalism are forms of tribalism that promote separation of individuals based on racial, ethnic, and/or linguistic characteristics." He states that, in the context of the United States "the sharp differentiation of ethno-racial blocs in culture and politics may function to reproduce elements of racist social patterns" (ibid.: 2). In place of multiculturalism, Hollinger (1995) suggests a "post-ethnic" perspective that favors voluntary over involuntary affiliations and promotes solidarity of people with different ethnic and racial backgrounds.

Facing the complexity of today's migration, movements, and social change, the model of multiculturalism seems no longer sufficient or adequate. Scholars such as Martin and Midgley (2003) argue that both assimilationists and pluralists have failed to grasp the dynamic reality of immigrants and immigration. Furthermore, as the boundaries of nation-states become more porous, Nasser (2004a) argues that the presence of transnational cultures presents a challenge to the traditional claims of citizenship. Sandercock with Brock (2009: 22) suggests "identities are multiple and cannot be subsumed under an ethnic umbrella." As reflected in actual planning practice, Wood and Landry (2008: 252) note that the commonly deployed strategies of consultation and participation assuming communities as defined by ethnicity and consulted in isolation are "disconnected from the complex intercultural relations that actually exist between people."

Towards Transcultural Placemaking

The challenges facing multiculturalism and multicultural planning today suggest a need for new discourses, frameworks, and a point of departure based on the complexity of today's cities and societies. Specifically, such discourse needs to address the dynamic processes of cultural change, overlays, and cross-cultural interactions in the context of migration and diversity. It also needs to address the politics of intercultural struggles and the agency of space and placemaking for cross-cultural learning, dialogues, engagement, and political empowerment.

This book is an attempt to formulate such a framework, one that we call *transcultural placemaking*. The framework embodies a set of ideas that recognizes the instability of culture(s) and the emergent nature of cultural formation and reconstitution in the shifting terrains of today's cities. Distinct from *interculturalism* (Sandercock 2004; Sandercock with Brock 2009), it addresses not only the intercultural exchanges but also the cultural *trans*-formation that takes place in urban places and through urban placemaking. It sees cultures not as isolated from each other in today's urban environments but able to be mutually influenced, constituted, and transformed. It recognizes the importance of *trans*-locality and *trans*-location in the processes of identity formation and placemaking, in which our affiliation with, and attachment to, place may not be tied to a fixed or singular location. It focuses on the capacity and agency of individuals and communities to adapt and transform the cultural landscapes of today's cities. Furthermore, the concept of *transcultural placemaking* addresses transcultural processes and understanding as a building block for a more inclusive democracy and critical embrace of diversity. Most importantly, it highlights the instrumentality of placemaking as a vehicle for cross-cultural learning, individual agency, and collective actions. Taken together, transcultural placemaking is a framework that can guide the current practice of planning, design, and community development in the context of diverse cities and communities.

The term *transculturation* has been used by ethnographers to describe "how subordinated or marginal groups select and invent from materials transmitted to them by a dominant or metropolitan culture" (Pratt 1992: 6). While the underlying context here is somewhat different, *transcultural placemaking* implies similar subjectivity on the part of migrants and migrant communities while addressing the full complexity and scope of change faced by cities and their occupants. It recognizes that cultural boundaries can be porous, and cultural practices can be reinvented and, at best, self-determined.

Recently, discourses of intercultural exchange and dialogue have begun to overtake multiculturalism as the preferred model for addressing the demographic complexity in today's cities. Sandercock (2004; with Brock 2009) argues for re-theorizing multiculturalism in twenty-first-century cities and suggests re-naming it as *interculturalism* to address the shortcomings of twentieth-century multiculturalism. Amin (2002) suggests the term *intercultured* to stress cultural dialogue, vis-à-vis multiculturalism that emphasizes differences solely. Van Leeuwen (2010: 640) further suggests a form of "intercultural citizenship" not as a set of legal rights but as "competences and attitudes that are required for everyday dealing with cultural diversity."

This volume shares and builds upon these critical reflections and propositions. In addition, it draws from recent attempts to re-conceptualize culture, place, and placemaking, as well as identity, citizenship, and marginality. Specifically, Massey's (1994) articulation of place as particular moments of intersecting social relations, not necessarily tied to any particular locality, provides a starting point that enables us to address notions of agency, subjectivity, change, and flow. This is echoed by Gupta and Ferguson (1997) who argue that the processes of placemaking are always contested and unstable, and the relations between places are continuously shifting as a result of the political and economic reorganization of space in the world system. Similar to a more dynamic understanding of place and placemaking, the discourse of transcultural placemaking also builds upon the idea of culture and identity as an unstable product and process (see Appadurai 1990; Friedman 1994; Parekh 2000; Gupta and Ferguson 1997; Hall 2003; Sandercock 2004; Sandercock with Brock 2009; Nasser 2004b; Hesse 2000). Specifically, Nasser (2004b: 77) notes that "migration in today's globalized world has served to unsettle older territorial notions of 'community', 'culture', 'group identity' within fixed and bounded geographies." We are also informed by Hall (2003) who argues that cultural identity is a matter of "becoming" as well as "being" and undergoes constant transformation.

This Book

This book consists of a collection of case studies which have been developed to inform understanding of transcultural placemaking. The studies are presented by a multidisciplinary group of scholars and activists with backgrounds in architecture, art, environmental psychology, geography, landscape architecture, political science, social work, urban planning, and urban studies, who came together through a shared interest in the cross-cultural complexity of today's cities. Together, the chapters offer twenty snapshots of sites, communities, social groups, and their interactions with one another. The variety of settings, actors, and stories, covering eighteen cities across five continents, offers a glimpse into the bewildering complexity and richness of today's intersecting cities and communities.

The volume is organized in five parts, corresponding to the different contexts, opportunities, and challenges for transcultural placemaking. Part I: Placemaking at the Margins addresses the liminal sites in which transcultural placemaking often occur. Located in Chicago, Tokyo, Seattle, and Yangon, the sites include restaurants and shops, as well as new and existing ethnic enclaves, where cultural practices of immigrants and residents find new expressions in the given settings. In some cases, these places have become islands of exchange between different cultural groups. Part II: Placemaking in the Space of Flows focuses on spaces of transition and diaspora as created by transnational migrants that include Korean in the Philippines, Chinese in the suburbs of Sydney, "Trans-Asian migrants" in the rural townships of Malaysia, and African immigrants in Sheffield, UK, who carry with them multiple knowledge and frames of references as they negotiate their presence in the new environment.

Part III: Bridging Spaces of Difference examines how cultural differences are negotiated and sometimes bridged through the process of placemaking. The case studies focus on social ties forged among multicultural residents in a public housing complex in Seattle, contestation of territorial claims and place identities in Central Long Beach, California, collaboration between multiple ethnic groups in San Diego, and negotiation of racial and ethnic relations among suburban high school students in Fremont, California. Part IV: Building Communities across Cultures focuses on organized efforts in engaging communities across cultural barriers. It begins with two case studies from Utah—a participatory action research project focusing on immigration and education through the eyes of undocumented youths, and a university outreach program to support immigrant transition and integration. This is followed by a case study of efforts to engage immigrant residents in neighborhood design in Seattle's International District, and another that examines community gardens as a site of everyday transcultural learning and community building.

Finally, in Part V: Struggles for Transcultural Cities, the book closes by examining the continuing barriers and injustice faced by migrants and migrant communities. These range from institutionalized practices of open space planning and urban governance, to long-standing historical and cultural biases in immigrant-receiving societies. Four cases are presented here, including studies of Latina/o experience in Pennsylvania, Little Indonesia in Taipei, newly arrived mainland Chinese residents in Hong Kong, and internal borders within European cities as experienced by minority immigrants in Rome.

Learning from Transcultural Cities

In the end, what have we learned from these places and stories? How can the lessons contribute to the practice of transcultural placemaking as we seek to better engage and empower diverse communities in the making of place, and by extension the city? The following highlights a few collective lessons.

Supporting everyday sites of interactions. We learned from the case studies that transcultural encounters and interactions often occur in the everyday landscapes, in seemingly mundane locations such as restaurants, shops, streets, schools, libraries, parks, and community gardens—places that Amin (2002) describes as the *sites of banal transgression* and *micropublics* of everyday social contact and encounter. They are also what Wood and Landry (2008) describe as the modern *zone of encounter* and spaces of day-to-day exchange, as well as what Wise (2009) calls *quotidian transversality*—everyday intercultural modes and spaces that facilitate sociality across difference. The omnipresence of transcultural processes in places such as streetscapes and restaurants in Chicago, a library in Long Beach, California, and public spaces in Hong Kong suggests the need for greater attention to how precisely these everyday landscapes support or impede transcultural processes. They also suggest opportunities and needs for greater investment in planning and design for participation (both formal and informal) by migrants and marginalized populations.

Making safer space and time. Related to the sites of everyday encounter is the concept of *safe space* (or *safer space*, as no urban space is absolutely immune from conflicts and contestation). Here, they exist in the form of third places in the case of Brazilian restaurants in Tokyo and a community garden in Oakland, and as places of social support in the case of the Hartland Partnership Center in Salt Lake City. They also exist as urban refuges that protect the migrant communities against outside hostility, as in the case of Devon Street in Chicago where Muslim migrants practice

Figure 1.4 Through daily activities and events, Chinatown residents in New York City claim the Columbus Park as a community space and activate it as a multi-functional stage. Photograph by Jeffrey Hou.

prayers inside ethnic restaurants sheltered from external gaze and scrutiny. The concept of safeness applies to the temporal dimension as well, as events and activities can also function as safe spaces in which cultural barriers, perceived or real, are temporarily suspended, enabling boundaries to become porous between different cultural groups and practices. Safer space and time both lower barriers or thresholds for transcultural interactions and reduce fear of perceived danger and intimidation. As John Forester (2000: 151) observes in his interviews with professional facilitators who deal with contentious cross-cultural negotiations, "safe spaces are not found; they are created." As such, the making of "safer space and time" suggests another important agenda for transcultural placemaking. It also calls for attention to what otherwise can make a space/time unsafe for immigrants and other social groups.

Developing medium for understanding. The cases of Easy Targets in Salt Lake City and Walking Voices in Sheffield demonstrate that cross-cultural learning can be mediated and supported through specific use of media. Similar to spaces and events, recording media in the case of Easy Targets provided a safe space/time in which undocumented youths spoke to state legislators and the public about their plight without having their identities disclosed. In the case of Walking Voices, the participants were able to articulate and record their experiences and reflections in the newly settled urban landscapes in their own daily routines and in their own voices. The mediated interactions here are socially indirect and temporally flexible. They enable the subjects to participate without the intimidating presence of an unfamiliar or unsympathetic audience. Similar to video production and audio recording, structured or facilitated interactions can also function as a medium, as in the case of planning for the Village of Market Creek Plaza in San Diego and community planning in Seattle's International District, in which community differences were mediated, leading to better cross-cultural understanding.

Learning from in-between-ness. Transcultural placemaking is by nature in a state of in-between, on the move, and yet to settle. It is engendered often by individuals who occupy multiple worlds and carry with them multiple forms of knowledge and frames of reference. As culture and identities evolve with new interactions and influences, so do their spatial expressions and manifestations. The phenomenon of in-between-ness is exemplified here by the adaptation of public space by newly arrived residents in Hong Kong, the formation of Little Shanghai in Sydney, the day-to-day operation of the Hartland Partnership Center in Salt Lake City, the evolution of Devon Street in Chicago, and the multiple relocations of Little Indonesia in Taipei, where the space of transcultural processes are in constant flux and transition. Evolving spaces like these often appear to be hybrid and messy. As a result, they easily escape full understanding and sometimes invite fear. As such, their situation can be precarious, threatened by institutional forces that are indifferent if not hostile to their existence. Supporting transcultural placemaking in these scenarios requires greater patience and empathy. It also requires greater attention to the process vis-à-vis product of placemaking. It calls for a process-driven focus that supports continuing adaptation and change. It calls on planners and designers to question the static and reifying notion of culture and place.

Working with transcultural agents. As demonstrated in the cases here, transcultural placemaking does not exist in a social vacuum. It is often facilitated by individuals and organizational actors who are able to traverse different cultural traditions and multiple *epistemological worlds* (Umemoto 2001). The transcultural agents here, or *transversal enablers* (Wise 2009), are those who can negotiate different knowledge and value systems, engage in iterative, cross-cultural translation, and facilitate co-adaptation of culture and place. In doing so, they help forge new place meanings, identities, and transcultural understanding. In this volume, they include owners and managers of Brazilian restaurants in Tokyo, Indonesian businesses in Taipei, shops in Sydney's Little Shanghai, and language schools in Manila. They are the community gardeners in Oakland, as well as designers and facilitators in San Diego and Seattle and social workers in Salt Lake City. In addition, the title of transcultural agents can be further extended to customers, shoppers, park users, and public housing residents who are also active participants in the collective making of place and production of transcultural relationships and understanding. The role of transcultural agents suggests the importance of supporting individuals and organizations as active players in the process of transcultural placemaking. This includes capacity building on the part of the participants as well as training of professional staff and organizers for transcultural competency. It also involves opening of the planning process to non-professional actors who may be better equipped in handling cross-cultural translation and transactions.

Turning conflicts into opportunities. When different cultures and cultural groups interact with each other, it is not surprising that intercultural conflicts can often arise. Rather than disguising conflicts or avoiding them at all cost, the case

Figure 1.5
Interpretation at a workshop in Little Saigon played a critical role in helping residents express their thoughts in informing a neighborhood design project. Photograph by Jeffrey Hou.

studies here tell us that it is important to recognize these conflicts as opportunities for critical reflections and transcultural engagement. Wood and Landry (2008) argue that social conflict as a result of heterogeneity can foster individual and social maturity. They argue that an intercultural city is "an engaged, argumentative and an essentially political space," and that "conflict can be ... a site of creativity" (Wood and Landry 2008: 280 and 321). As demonstrated in the cases of Cambodia Town in Central Long Beach, Sydney's Little Shanghai, and the Hartland Partnership Center in Salt Lake City, conflicts and disagreements are instances where differences in cultural values and practices are illuminated, providing opportunities for transcultural learning and exchange. Turning conflicts into learning experiences rather than further entrenchment of social and intercultural tensions should be on the agenda of transcultural placemaking.

Border-Crossing and Placemaking

As movement and migration become more widespread and increasingly a routine aspect of the contemporary life, border-crossing emerges as a critical part of our everyday urban experience. Crossing of borders in this sense is no longer solely about traveling from one nation to another, but rather a part of the everyday encounters in the city. As urban dwellers, we experience it in streets, schools, markets, gardens, restaurants, and parks, and participate regularly in the negotiation of space, identities, values, and rights in our encounters with others. These border-crossing activities and experiences have unsettled prescribed notions of culture, identity, place, and placemaking. As challenges to the institutionalized practice of planning and design, the encounters and exchanges also offer opportunities for a richer understanding of culture and place in a diverse society, as well as making a more inclusive and dynamic cityscape.

Returning to our questions at the outset—*how can urban places serve as a vehicle for cross-cultural learning and understanding? How can cross-cultural interactions be constructed, enabled, or "staged" through social and spatial practices in the contemporary urban environment? How can we re-envision the process of placemaking in the context of shifting cultural terrains?*—we believe that the stories in this volume offer some concrete answers to these questions. Specifically, the case studies here show how everyday places provide opportunities for encounters, and in some cases function as "safe space/time" for interactions. To varying extents, the everyday places provide opportunities for participation and civic engagement and by extension the practice and expression of a renewed notion of citizenship. They provide settings in which culture and identities can continue to be negotiated and reconstituted, and how our broader understanding of place and justice can be enriched. In the stories presented here, we see how immigrants, guest workers, refugees, business people, and other forms of transnational beings act as transcultural agents in facilitating the remaking of urban landscapes imbued with new identities and meanings. We learn how landscapes and places can serve as vehicles for cultural adaptation and transformation.

Transcultural placemaking is a framework for both inquiries and actions, as well as a point of departure for reconceptualizing and repurposing placemaking as a medium for cross-cultural learning and understanding. It recognizes the unsettling characteristics of today's culture and place, and their continuing co-adaptation as an important part of citymaking in the twenty-first century. With the notion of transcultural agency, it recognizes the rights and abilities of individuals and groups to engage actively in the making of our urban spaces while finding and founding their own place in the changing society. Placemaking in this context is about not only staking and claiming but also learning and positioning in the shifting terrains of today's society. As a vehicle for activism and citizenship, it can serve as a building block for democracy in our ever more complex, unwieldy, and intersecting worlds.

References

Ameyaw, S. (2000) 'Appreciative Planning: An Approach to Planning with Diverse Ethnic and Cultural Groups', in Burayidi, M. A. (Ed.), *Urban Planning in a Multicultural Society*. Westport, CT: Praeger, pp. 101–114.

Amin, A. (2002) 'Ethnicity and the Multicultural City: Living with Diversity', *Environment and Planning A,* 34: 959–980.

Appadurai, A. (1990) 'Disjuncture and Difference in the Global Cultural Economy', *Public Culture,* (2)2: 1–24.

Brooks, B. (2011) 'To Do Things the Chinese Way or the Brazilian Way?' *Northwest Asian Weekly*, 30(23): 5, 13. June 4–10.

Burayidi, M. A. (Ed.) (2000a) *Urban Planning in a Multicultural Society*. Westport, CT: Praeger.

Burayidi, M. A. (2000b) 'Tracking the Planning Profession: From Monistic Planning to Holistic Planning for a Multicultural Society', in Burayidi, M. A. (Ed.), *Urban Planning in a Multicultural Society*. Westport, CT: Praeger, pp. 37–51.

Castles, S. & Miller, M. J. (2009) *The Age of Migration: International Population Movements in the Modern World*, Fourth Edition. New York: Guilford Press.

Forester, J. (2000) 'Multicultural Planning in Deed: Lessons from the Mediation Practice of Shirley Solomon and Larry Sherman', in Burayidi, M. A. (Ed.), *Urban Planning in a Multicultural Society*. Westport, CT: Praeger, pp. 147–168.

Friedman, J. (1994) *Cultural Identity and Global Process*. London: Sage.

Gupta, A. & Ferguson, J. (1997) 'Culture, Power, Place: Ethnography at the End of an Era', in Gupta, A. & Ferguson, J. (Eds), *Culture, Power, Place: Explorations in Critical Anthropology*. Durham, NC: Duke University Press, pp. 1–29.

Hall, S. (2000) 'Conclusion: The Multi-cultural Question', in Hesse, B. (Ed.), *Un/settled Multiculturalisms*. New York: Zed Books.

Hall, S. (2003) 'Cultural Identity and Diaspora', in Braziel, J. E. & Mannur, A. (Eds), *Theorizing Diaspora: A Reader*. Malden, MA: Blackwell, pp. 233–246.

Hesse, B. (Ed.) (2000) *Un/settled Multiculturalisms*. New York: Zed Books.

Hollinger, D. A. (1995) *Postethnic America: Beyond Multiculturalism*. New York: Basic Books.

Holston, J. (1998) 'Spaces of Insurgent Citizenship', in Sandercock, L. (Ed.), *Making the Invisible Visible: A Multicultural Planning History*. Berkeley: University of California Press.

Hou, J. & Kinoshita, I. (2007) 'Bridging Community Differences through Informal Processes: Reexamining Participatory Planning in Seattle and Matsudo', *Journal of Planning Education and Research,* 26(3): 301–313.

Jacobs, J. M. & Fincher, R. (1998) 'Introduction', in Jacobs, J. M. & Fincher, R. (Eds), *Cities of Difference*. New York: Guilford Press.

Jang, M. (2009) 'Koreans and Bangladeshis Vie in Los Angeles District', *New York Times*. Online, available: http://www.nytimes.com/2009/04/07/us/07koreatown.html?sq=bangladesh%20koreatown%20los%20angeles&st=cse&scp=1&pagewanted=print (accessed on May 17, 2011).

Krase, J. (2002) 'Navigating Ethnic Vernacular Landscapes Then and Now', *Journal of Architectural and Planning Research*, 19(4): 271–281.

Li, W. (1999) 'Building Ethnoburbia: The Emergence and Manifestation of the Chinese Ethnoburb in Los Angeles' San Gabriel Valley', *Journal of Asian American Studies,* 2.1: 1–28.

Loukaitou-Sideris, A. (2002) 'Regeneration of Urban Commercial Strips: Ethnicity and Space in Three Los Angeles Neighborhoods', *Journal of Architectural and Planning Research*, 19(4): 334–350.

Martin, P. & Midgley, E. (2003). 'Immigration: Shaping and Reshaping America', *Population Bulletin*, 58(2): 5–42.

Massey, D. (1994) *Space, Place, and Gender*. Minneapolis: University of Minnesota Press.

Nasser, N. (2004a) 'From Multicultural Urbanities to the Postmodern Cosmopolis: A Praxis for Urban Democracy', *Built Environment,* 30(1): 5–16.

Nasser, N. (2004b) 'Southall's Kaleido-scape: A Study in the Changing Morphology of a West London Suburb', *Built Environment,* 30(1): 76–103.

Parekh, B. (2000) *Rethinking Multiculturalism: Cultural Diversity and Political Theory*. Cambridge, MA: Harvard University Press.

Parker, D. (2000) 'The Chinese Takeaway and the Diasporic Habitus: Space, Time and Power Geometries', in Hesse, B. (Ed.), *Un/settled Multiculturalisms*. New York: Zed Books.

Pestieau, K. & Wallace, M. (2003) 'Challenges and Opportunities for Planning in the Ethno-culturally Diverse City: A Collection of Paper—Introduction', *Planning Theory & Practice,* 4(3): 253–258.

Pratt, M. L. (1992) *Imperial Eyes: Travel Writing and Transculturation*. London: Routledge.

Qadeer, M. A. (1997) 'Pluralistic Planning for Multicultural Cities: The Canadian Practice', *Journal of the American Planning Association,* 63(4): 481–494.

Roseman, C. C., Laux, H. D., & Thieme, G. (1996) 'Introduction: Modern EthniCities', in Roseman, C. C., Laux, H. D., & Thieme, G., (Eds), *EthniCity: Geographic Perspectives on Ethnic Change in Modern Cities.* Lanham, MD: Rowman & Littlefield.

Sandercock, L. (1998a) *Towards Cosmopolis: Planning for Multicultural Cities*. New York: John Wiley.

Sandercock, L. (Ed.) (1998b) *Making the Invisible Visible: A Multicultural Planning History*. Berkeley: University of California Press.

Sandercock, L. (2000) 'When Strangers Become Neighbors: Managing Cities of Difference', *Planning, Theory & Practice,* 1(1): 13–30.

Sandercock, L. (2003) *Cosmopolis II: Mongrel Cities in the 21st Century*. London: Continuum.

Sandercock, L. (2004) 'Reconsidering Multiculturalism: Towards an Intercultural Project', in Wood, P. (Ed.), *Intercultural City Reader*. Stroud, UK: Comedia.

Sandercock, L., with Brock, S. (2009) 'Inventing a Multicultural State: Canada's Evolving Story', in Sandercock, L. & Attili, G. (Eds), *When Strangers Become Neighbours: Integrating Immigrants in Vancouver, Canada*. Dordrecht: Springer.

Talen, E. (2006) 'Design that Enables Diversity: The Complications of a Planning Ideal', *Journal of Planning Literature,* 20(3): 233–249.

Thompson, S. (2003) 'Planning and Multiculturalism: A Reflection on Australian Local Practice', *Planning Theory & Practice*, 4(3): 275–293.

Umemoto, K. (2001) 'Walking in Another's Shoes: Epistemological Challenges in Participatory Planning', *Journal of Planning Education and Research,* 21: 17–31.

UNDESA (United Nations Department of Economic and Social Affairs). (2011) International Migration Stock. Online, available: http://esa.un.org/migration/p2k0data.asp (accessed on September 16, 2011).

Van Leeuwen, B. (2010) 'Dealing with Urban Diversity: Promises and Challenges of City Life for Intercultural Citizenship', *Political Theory,* 38(5): 631–657.

Welsh, J. F. (2008) *After Multiculturalism: The Politics of Race and the Dialectics of Liberty*. Lanham, MD: Lexington Books.

West, C. (1990) 'The New Cultural Politics of Difference', in Russell, F., Gever, M., Minh-ha, T. T., & West, C., (Eds), *Out There: Marginalization and Contemporary Cultures*. Cambridge, MA: MIT Press.

Wise, A. (2009) 'Everyday Multiculturalism: Transversal Crossings and Working Class Cosmopolitans', in Wise, A. & Velayutham, S. (Eds), *Everyday Multiculturalism*. New York: Palgrave Macmillan.

Wise, A. & Velayutham, S. (2009) 'Introduction: Multiculturalism and Everyday Life', in Wise, A. & Velayutham, S. (Eds), *Everyday Multiculturalism*. New York: Palgrave Macmillan.

Wood, P. & Landry, C. (2008) *The Intercultural City: Planning for Diversity Advantage*. London: Earthscan.

PART I

PLACEMAKING AT THE MARGINS

CHAPTER 2

Transcultural Placemaking

Intertwined Spaces of Sacred and Secular on Devon Avenue, Chicago

Arijit Sen

Devon Avenue in Chicago is an ethnic and immigrant retail strip home to South Asians, Jews, Russians, Latin Americans, Koreans, and native-born white Americans of various ethnic backgrounds.[1] This cultural diversity belies any attempts to categorize this street as a site of a single community; rather, landscapes of different ethnic groups are interwoven into a complex tapestry. The architectural character of Devon Avenue is similar to many multiethnic retail streets in US cities. Typically, such streets include dense rows of multi-storied mixed-use buildings occupied by sequential waves of new immigrants, for whom the street becomes the initial launching pad into America. Narrow storefronts with eye-catching signage distinguish the buildings along such streets. Examples of such streets include New York's Lower East Side and Chicago's Lower West Side. However, the highly visual architectural quality of these spaces makes it difficult for us to see them as lived environments of a diverse group of people. The ephemeral and experiential interpretations of these places by different stakeholders may escape the gaze of a material culture scholar who studies building types. Architectural historians focus on who built a building, the construction technologies, and material details. But what happens when the users of the buildings change and new tenants occupy old buildings? In Devon Avenue we see instances where users temporarily transform parts of a building into very specific kinds of places which, after those users leave, then revert to what they were before. We encounter situations where one group of individuals uses a space in a certain way while another group simultaneously uses the same space differently.

The geography of Devon Avenue has a long multicultural past (Archer & Santoro, 2007; Jones, 1995; Koval et al., 2006). Jewish families moved into the area after World War II in the neighborhoods between Damen and Kedzie Avenues; by 1963, there were around 48,000 Jews in the West Rogers Park area (Cutler, 1996; Jones, 1995). In the 1980s, the now-aging Jewish inhabitants and their children began moving into suburban locations such as Skokie, Buffalo Grove, Highland Park, and Deerfield, and newer immigrants moved in. The remnants of the old community can still be seen in the remaining Orthodox neighborhoods that surround local synagogues on the western end of the avenue and in the northern neighborhoods of West Rogers Park, Peterson Park, and all of the immediately surrounding areas. Since the fall of the Soviet Union, Russian Jews have repopulated much of the west end of Devon between California and Kedzie Avenues. By the late 1970s newcomers of South Asian origin set up businesses that catered to a new ethnic clientele.[2] Initial stores carried spices and clothes; soon stores selling electronic items, bags, and jewelry appeared (Kalayil, 2009). They were interspersed amidst pre-existing Jewish and non-ethnic stores, creating a checkered multicultural street fabric. Today, Devon Avenue serves as a point of entry for many less-skilled or poorer immigrants. A large population of South Asian elderly lives in the immediate vicinity of West Ridge and Rogers Park neighborhoods. Since the late 1990s, Muslim immigrants from Bangladesh, Pakistan and India have settled between North California Avenue and North Ridge Avenue.[3]

This trend was reflected in the Greater Chicago region. For instance, Kniss and Numrich (2007) documented the increase of mosques in the Greater Chicago area from pre-1965, when only a handful served various groups and locations, to over 60 in 1997. In 2008 there were over 100 religious centers, many newly established ones serving the second and third generations of the community (M. Hermansen, personal communication, March 2009; Numrich, 1997, 2005; Schmidt, 2004). In proximity to Devon Avenue, there also developed centers of Islamic worship to cater to the new residents. These worship spaces ranged from the Jama Masjid (mosque) with a capacity of thousands for regular Friday services, to a dozen basement gathering places that attract as many as several hundred male worshippers per location on Fridays, while simultaneously functioning as schools and community centers throughout the rest of the week (M. Hermansen, personal communication, 2008, March 2009). Other services catering to local Muslim residents include legal firms, Islamic bookstores, Islamic schools, halal produce stores, women's and children's services, and restaurants with prayer spaces.

Unlike Herbert Gans's (1982) famous urban villages where immigrants live and interact almost exclusively with members of their in-group within the safe boundaries of their ethnic neighborhood, the situation on Devon Avenue is quite different. For instance, if you step out of the IQRA book center (2749 W. Devon Avenue), an Islamic bookstore, you cross paths with non-Muslim patients emerging from the Prism Medical Center (2744 W. Devon Avenue) or newly arrived Bangladeshis and Russians entering the nearby Reliable Driving School. While the customers in the Tiger International Video store (2750 W. Devon Avenue) opposite may be members of the immigrant in-group, those entering the next door Elita Wholesale Video, Audio and CD store (2753 W. Devon Avenue) or the Russian Book Store (2746 W. Devon Avenue) on the opposite side of the street are predominantly Russian and Jewish. In such an intertwined multiethnic neighborhood, drawing clear spatial and social boundaries is difficult, yet necessary to maintain community turf. In order to understand the culture of this street we may turn to recent theories of transculturation.[4] Theories of transculturation suggest that immigrants switch codes or translate between different contexts and cultures while retaining multiple cultural identities. Transcultural spaces are "contact zones" of the twenty-first century, or "social spaces where cultures meet, clash, and grapple with each other, often in contexts of highly asymmetrical relations of power" (Pratt, 2002, p. 607).

In 2009 I began a detailed architectural, observational, and ethnographic study of Devon Avenue to document how local South Asian residents (mostly poor, recently arrived, and Muslim), made this retail street their lived environment. I planned collaborations with Dr. Marcia Hermansen on the cultural practices of diverse South Asian Islamic sub-communities in this neighborhood. I was specifically interested in exploring prayer spaces intertwined within the ethnic marketplace along Devon Avenue. These spaces include mosques and prayer spaces located in store basements, back rooms of restaurants, and makeshift spaces in residential units. This chapter focuses on two restaurants that simultaneously worked as marketplaces and prayer spaces.

Transcultural Public Realm

Hannah Arendt (1989), in her discussion of public realm, identifies two interrelated phenomena that are necessary for the sustenance of a public. First is the need for visibility, or as Arendt calls it, "appearance." According to her, our understanding of ourselves and our identity is intensely related to the existence and visibility of the "other" (pp. 50–54). Second, is the importance of the material world that is shared by many and that acts as something that unites disparate people (p. 52).

Arendt's public realm exists not simply because of the many perspectives and differences embedded within that realm, but also because these differences and perspectival positions are visible and experienced by all. When Arendt mentions visibility, she is not referring to a uniform ability to scrutinize, see, and discuss the same reality. She explains that:

> Being seen and being heard by others derive their significance from the fact that everybody sees and hears from a different position. (p. 57)

Thus the very act of seeing and interpreting the other is dependent on the viewer and his or her point of view. This "relational visibility" is interactively reproduced when individuals encounter each other in specific locations. A visceral yet differential engagement between people and the public realm ensues. The notion of public space from the point of view of Arendt is very different from one that assumes commonality and unanimous identification from all. Arendt's public space takes the form of a ground for political exchange where people sharing a common world also share this common "space of appearance" and where public concerns and issues are articulated and debated from multiple perspectives. Political engagements of the public realm in the case of Arendt are contextual, performative, interactional, and visceral.

In a 2004 article titled "Muslims in the Performative Mode: A Reflection on Muslim–Christian Dialogue," Hermansen proposes a theoretical framework for studying cross-cultural interaction and inter-religious dialogue between Muslims and other religious groups in the United States that produces engaged transcultural publics. Hermansen uses the metaphor of performance and her work draws from the work of performative modes of language and speech acts (Austin, 1975). She argues that an "emphasis on performance in dialogue among Muslims in America is partly due to their need not only to assert some sort of authority over an imagined Muslim space but to claim both the authority and the space" (Hermansen, 2004,

p. 392). We confirmed in our study that such engagement allows immigrants to take charge of their environment (often temporarily) and claim authority and space in ways mentioned by Hermansen. This is not the same as ownership of land or property or distinct territorial claims and power exerted by state and institutions of power. Rather, these moments of empowerment involve taking over a street corner with friends every evening, marking one's personal space of praying in a restaurant, using a restaurant as a place to meet, arranging child care, or engaging for a moment with a fellow shopper over the purchase of a common cultural artifact. These moments of interaction are important because they afford opportunities for imagining and experiencing cultural similarities, common interests, and solidarities. It allows people to identify neighbors, co-ethnics, citizens, friends, colleagues, and outsiders. Old stereotypes are broken and new ones are formed. Repeated over time these experiences coalesce and solidify, sustaining friendships, community membership, and interest groups. Yet, at any given moment these interactions are neither binding nor invested. The form, function, and layout of this shopping street encourage such performative affordances of belonging and peoplehood.

As we focused on stores on Devon Avenue that doubled as prayer spaces, we used Hermansen's emphasis on performance, Arendt's definition of public, and added Thomas Tweed's (2008) notion of embodied experience of religion in everyday life as a way to frame our study. Tweed's spatial metaphors of "crossing and dwelling" include mobility as a central mode of experience in contemporary society. According to Tweed, individuals "dwell" by maintaining boundaries between in-group and out-group. Crossing boundaries is central to the way individuals demarcate places, locate spatial experiences, and maintain (multiple) identities, allegiances, and subjectivities (p. 54). Boundaries, real and imagined, may be ambiguous to an outsider but are easily recognized by an insider. Tweed calls these acts of crossing (real or cognitive) boundaries "terrestrial crossings" (p. 59) and his spatial metaphor rings true for diasporic populations living around Devon Avenue and using these stores.

These embodied acts of crossing domains are made meaningful by what historian Paul Connerton (1989) calls "incorporating practices"—affective responses to the environment generated from internalized values, accepted maxims, and customs that are deeply cultural in nature. My use of the word "affective" refers to embodied, emotional and pre-cognitive responses to the built world by human users. I refer to Merleau-Ponty's 1995 work on the role of habit memory and body memory in our perceptions and engagements with the material world, a theme that is continued in contemporary scholarship on "affect theory" (Gregg & Seigworth, 2010, p. 6). According to this line of thought, examining the visceral interactions between humans and their material world is essential for an understanding of transcultural public spaces because those interactions are inflected by our cultural biases and racial, gender, class, national, or occupational backgrounds. Indeed, these everyday, mundane, taken-for-granted embodied acts—opening a door, bending while entering a room, taking off one's slippers at the threshold, shifting one's gaze, shunning stigmatized spaces—frame bodily and performative engagement of terrestrial crossings; these acts became central to our reading of place-making on Devon Avenue.

Embodied Placemaking: Case Study of Two Restaurants

Ghareeb Nawaz restaurant (2032 W. Devon Avenue) and Tahoora Sweets and Bakery (2345 W. Devon Avenue) are Muslim immigrant-owned restaurants located east of Western Avenue on Muhammad Ali Jinnah Way (2000 W to 2400 W blocks) and Sheikh Mujib Way (1800 W to 2000 W blocks).[5] The owner of Tahoora, Salim Shelia, is from Mumbai, India; Bashir Bozai, of Ghareeb Nawaz, has South Asian roots. Both Tahoora and Ghareeb Nawaz have spaces within the stores that are used for prayers and religious activities.

Crossing the First Boundary: The Façade

Our first encounter with these stores is visual. Even before we decide to enter these stores, their façades communicate to us. Elsewhere I have written about the plethora of signs, advertisements, posters, store names, and images on the ethnic storefronts along similar ethnic retail streets (Sen, 2009). There are many kinds of storefront signage. Tall pylon signs are freestanding signs. Monument signs are shorter free-standing signs, rare on Devon Street. Projecting signs, pylon and monument signs are most permanent, indicating a business that has been there for a long time or a more permanent kind of enterprise.

The most popular signs on Devon are wall signs. Wall signs include awning signs, wall posters, and window signs, flat acrylic store names and board signs, and projecting signs. Their location, size, and height target particular audiences—passengers and drivers in fast-moving vehicles, local residents on the sidewalk, insiders and outsiders, and linguistic subgroups. The wall signs along Devon can be categorized by their permanence. The most transient are the wall and window signs that often take the form of posters and painted boards (for example, the façades of Ghareeb Nawaz). By tracking how these signs change we can follow the rhythms of shopping seasons, seasonal merchandise advertisements, and festival sales that animate the street. During Ramadan, the holy month of prayers, fasting and post sundown feasting, the storefront signs advertise Ramadan sales and special events. The two case study restaurants are open late, special dishes are advertised, and local residents attend *Suhoor* meals before dawn. Participating in these communal festivities reproduces a sense of community and belonging.

For most non-South Asian shoppers the store names and storefront visual imagery advertise the nature of merchandise available in these stores. Names such as Medina, Tahoora, Ghareeb Nawaz, Mehrab, Laxmi, or Devi that have religious connotations to Hindus and Muslims may not have the same relevance to out-group customers. However all these names are uniformly exotic for members of the out-group—subtle regional, religious, cultural, language, and historic references are lost to the latter.

The name Ghareeb Nawaz refers to Hazrat Shaikh Khwaja Syed Muhammad Mu'īnuddīn Chishtī (1141–1230), a Sufi saint of the Chisti order who was known for his charity towards the poor. Chisti's following is large and includes Hindus and Muslims from the subcontinent. Bashir Bozai, 55, at Ghareeb Nawaz, expanded on this reference to the saint: "Our goal is to continue to provide low-cost Indian food to our customers . . . When people want Indian food, they usually go to Devon Avenue. The food there is very expensive. We want to provide quality, inexpensive food everyone can enjoy on Devon" ("Ghareeb Nawaz"). In reality, Ghareeb Nawaz caters to a wide range of customers—students, cab drivers, locals, non-South Asians, and takeout customers. Polyvalent signage is central to the success of their business. The storefront of Ghareeb Nawaz communicates low prices, choice, and variety—Indo-Pak dishes, sandwiches, Mediterranean recipes, regional cuisine (*Andhra thali*), and Greek gyros are advertised with prices noted (Figure 2.1). The Mediterranean sandwiches and gyros are aimed at cab drivers from the Middle East who prefer a quick, easy pick-up lunch. Much of the signage has come down due to city regulations and an awareness campaign spearheaded by the local Chamber of Commerce and South Asian business leaders. However, Bozai has plans to further extend the roof parapet as an advertising space (B. Bozai, interview, February 5, 2011).

Tahoora, the second case, communicates a different image and its name suggests a different genealogy. Its clean exterior façades with minimal advertisements communicate purity. Tahoora means pious and pure in Arabic (or the quality of people and place in paradise or *Daar-us-Salaam*). The sweetshop and deli are on the street level; on the basement level is a prayer room that also serves as a mosque. Its austere façades stand in stark contrast to that of Ghareeb Nawaz, with the former communicating the purity associated with piety and prayer while the latter displays the cacophony of a marketplace.

Traversing Interior Boundaries: Fronts and Backs

As customers enter and navigate the interior of the stores they experience the space in a processional manner. Erving Goffman's theatrical metaphor of space helps explain such an experience. Goffman (1959) distinguishes between fronts and

Figure 2.1 Ghareeb Nawaz, storefront. Photograph by Arijit Sen.

backs. Fronts are formal spaces where diverse individuals and groups interact while back refers to a domain of an in-group where formal manners are replaced with informal in-group interactions and behavior. Fronts and backs are reproduced by setting up physical boundary features like doors, curtains, and walls that may allow or hinder access into certain parts of the building. One may also encounter symbolic boundaries such as careful placement of images, subtle signs, changes in lighting intensity, sounds, and smells that suggest more subtle forms of spatial distinctions. The latter may be obvious and visible to members of the in-group while totally invisible to others who cannot read the esoteric and culturally determined cues. For instance, in the stores discussed above, the darker, interior spaces, accessible at the end of a procession, are always reserved for prayer, private events, and families. Both symbolic and physical boundaries are employed in order to cross over from one socio-spatial domain and dwell in another.

The entrance sequence begins a haptic processional experience. The double entry doors of Tahoora and Ghareeb Nawaz are typical of Chicago. These doors keep the cold away and help with heating during the bitter winters. But despite this architectural similarity the experiences of the double door entry sequence in the two restaurants are different. Tahoora's entrance is spotlessly clean. There are no advertisements and posters on the entrance door and store windows. The sales counter, visible from the entrance, is set way back at the interior of the store (Figure 2.2). Since customers waiting to buy or order food crowd around the sales counter in the deep interior, the front zone of the store remains relatively less crowded.

The elongated front zone between the sales counter and the entrance, 20′ long, serves yet another function that is not immediately obvious to the occasional customer. The front zone, occupying two structural bays, serves as a space for entry and transition for activities in the basement (Figure 2.3). The stairway has a chain

Figure 2.2 Tahoora, interior view from entrance. Photograph by Arijit Sen.

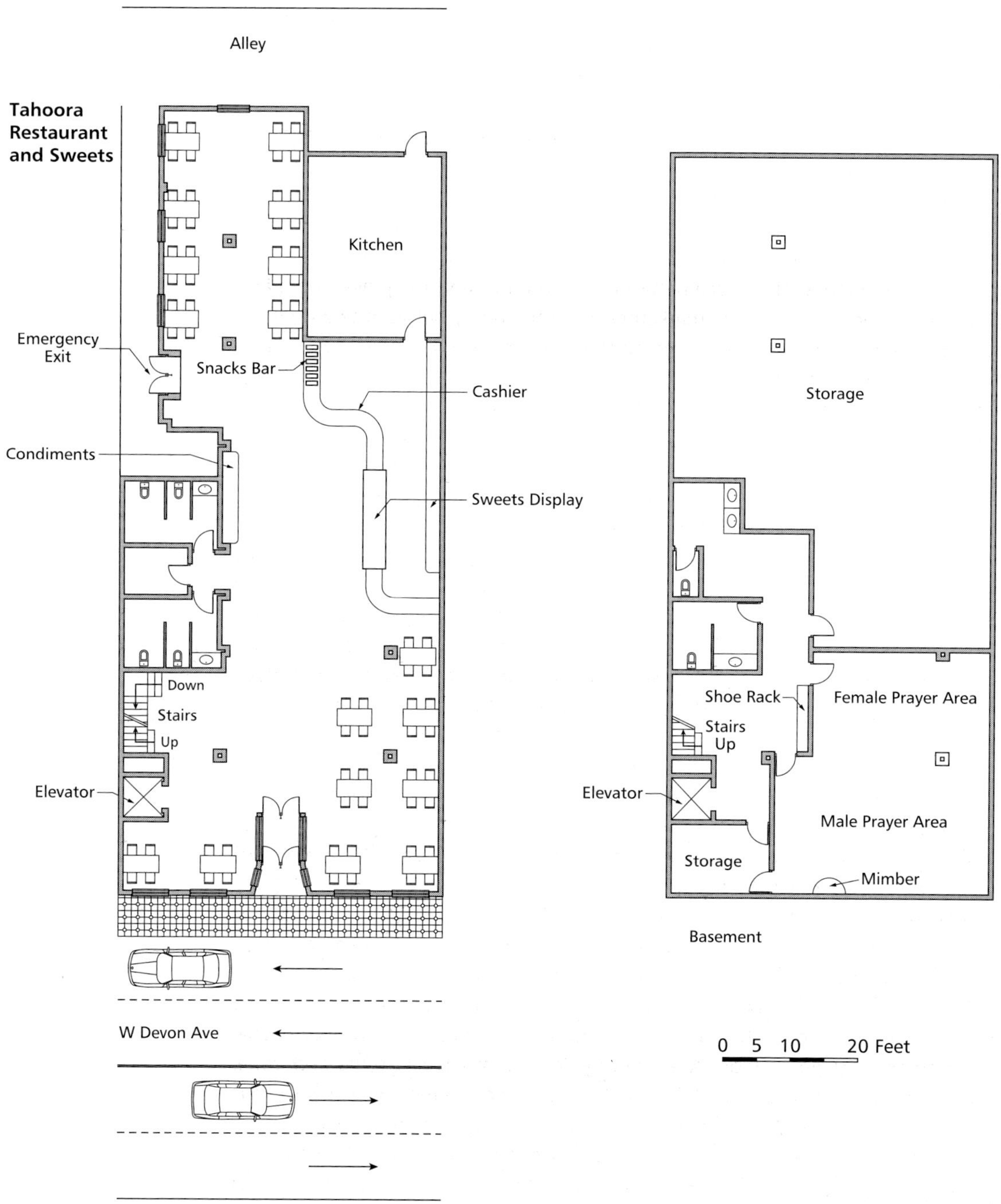

Figure 2.3 Tahoora—left: street level interior layout; right: basement level interior layout. Drawings by Sohail Khurram.

with a sign indicating that the lower level is private and out of bounds. In the basement of Tahoora is a "community hall" that is used as a basement mosque (*Masjid-e-Tahoora*) during prayer hours. A separate room serves women worshippers and community and educational activities. Yet shoppers who know about the basement mosque "find their way to the lower level when the congregation meets" (Anonymous customer, interview, February 5, 2011).

The doorway, entrance transitional zone, and withdrawn sales counter set up an important choreography of "crossing and dwelling." A brightly lit transition, signs, and a staircase separate the experience of being in the marketplace or in the prayer space inside Tahoora. This space between the entrance and the prayer room and between the entrance and the marketplace is a boundary between two domains rendered as a "thick edge" that "emerge[s] as not a plane but a zone, not physical but socio-spatial, not a division of things but a negotiation of flows" (Borden, 2000). Consider how the embodied act of entering into Tahoora is mediated by this thick edge. Customers entering the store see the sales counter and the food display. The long path between the entry and the destination food counters produces habitual and culturally choreographed responses from these users. They make a beeline for the sales counter. Moving straight towards the counter is not a discretely considered act—that is exactly how one behaves in a food deli. However, for the members of the mosque congregation, the very act of veering off midway, towards the basement mosque, is a break from the latter trajectory. It is an act of religiosity, an intention to pray, and a reiteration of one's religious identity, and a path taken by only those who are *actively seeking* to get to the basement spaces. The seemingly innocuous act of walking across an entrance zone and deciding to turn towards the stairs in Tahoora is an embodied act reiterating one's identity and allegiances. As one walks down the staircase, acts of ritual cleaning and taking off shoes precede entry into the prayer room. These sequential experiences enacted over a period of time reiterate a feeling of sacredness and enhance a perceptual distance from the marketplace above, preparing the devout for prayers. It also reproduces a sense of congregational and community identity.

Ghareeb Nawaz is located in an old corner building that is not as deep as the lot line. At the back of the store is a parking lot. The entrance to Ghareeb Nawaz is cluttered with signs, newspapers, advertisements, and brochures. One enters into the front room, brightly lit by fluorescent lights. The front room seems open and empty during lax hours, but during busy hours, the room fills up with people ordering and picking up food. A large portal leads customers into the back zone, also known as the "family room" (Figure 2.4). According to the owner, the room is so called to cater to parties with women and children. However, I have never observed strict gender divisions being observed in this tiny restaurant. Instead the primary difference between the front and back is haptic. The brightly lit front room is characterized by fast-paced activities and constantly moving customers. The tempo in the less bright back room is slow-paced as people take their food and settle down at the tables/booths. A prayer nook is located at the far end of the family room next to the toilets. One has to enter, cross the front room, and then cross over to the side room in order to reach the prayer room. The back zone is visually inaccessible, so that a praying individual remains unseen from the front pick-up area.

Figure 2.4 Ghareeb Nawaz, interior view. Photograph by Arijit Sen.

In December 2010, after this chapter was written, Mohammad Bozai, the American-born son of the owner, helped redesign the interiors. According to his father, the new and updated interiors have "modernized" the old place. This "modernization" project included a new coat of paint, a new flat screen TV, and new counters. But the two most important changes include relocation of the counter and partial removal of the party wall between the two rooms. The new position of the counter gives the person at the counter visual surveillance over the entire store. The wall between the two rooms is now a half wall. The new interior creates a sense of separation while allowing clear lines of sight from the counter. Although the new changes make the interior seem more open and lighted, the back room remains visually separated and the half wall retains the experiential boundary between front and back.

If Tahoora's sequential arrangement induces a slow unfolding and transformation from the marketplace to the prayer space, Ghareeb Nawaz's spatial transformation resembles the quick efficiency of a drive-in or takeout. Cab drivers rush in during prayer hours for *salah* or *namaz* (prayers), entering the backspace from the front room along a narrow passage. After prayers, some eat food but others rush out. Unlike Tahoora where the prayer space is delineated and separated from the commons by a barred stairway, Ghareeb Nawaz's prayer space is integrated within the store interior and its sequentially processional layout. The room, approximately 9′ × 9′, is enough to hold three praying individuals. A rug, floor patterns, and framed images of the *kaaba* distinguish this room from the other spaces (Figure 2.5). The processional order of the store interiors and the back location of the prayer room seclude it from the hubbub in the front.

The interior spatial characteristics reflect its non-congregational and temporary use (unlike Tahoora's congregational gathering space). Ghareeb Nawaz's interior prayer spaces belong to a quotidian landscape of "lived religion" that is distinctly different from sanctioned places of worship such as the local Jama Masjid located

Figure 2.5 Ghareeb Nawaz, prayer room. Photograph by Arijit Sen.

on 6340 N. Campbell Avenue, three blocks west of Western Avenue, or the less formalized but nevertheless formal congregational space in Tahoora Masjid. In their research on New York cab drivers who are Muslims, Courtney Bender and Elta Smith (2004) found a network of "free-standing prayer spaces" located across different locations in New York City. In Bender and Smith's study we find that practicing Muslim cab drivers in New York City cannot drive to a mosque to conduct their daily prayers while driving passengers around. They have to find alternative spaces to perform their rituals and prayers. As a result, prayer spaces have appeared across the secular domain allowing Muslim immigrants to integrate their religious and spiritual needs within the public and semi-public regions of urban life and to carry out their religious practices even while participating in the mainstream public realm. Bender and Smith argue that these spaces represent "an organizational innovation within the existing field of American mosques and complicate the analysis of immigrant religious life that focuses solely on congregational participation" (2004, p. 76). Calling them spaces of everyday "lived religion," Bender and Smith show the creative role that "immigrants' activities play in reconstructing the boundaries of public and private, ethnic and religious identities" while using such spaces (ibid., p. 77). Such an interweaving of spatial domains becomes necessary for Muslims since the public realm in the United States is not set up to support the daily religious and spiritual needs of practicing members of this community.

Whether in Tahoora or in Ghareeb Nawaz, the experience of traversing elongated processional spaces in order to access the prayer space engages the entire body of the user. Getting to back spaces is an intentional and embodied act since it not only leads to spaces for necessary bodily cleansing rituals associated with Islamic prayers but also involves walking across and beyond the restaurant space. The linoleum pattern, framed pictures, and lighting add to the changed ambience of the prayer zones, subconsciously conditioning the user into this temporarily private and sanctified space.

According to members of the Tahoora congregation, those people at the sales counter during daily prayer hours are marked either as non-practicing, or as people who are not members of their particular congregation. In both places, in order to effortlessly cross into the back zone one needs to know the exact location of this space and have a good knowledge of the layout of the restaurant interiors. In both places, the act of crossing over into the prayer zone is an intentional reiteration of religious practice and an embodied act of belonging in an ethno-religious community.

A salient feature of the prayer spaces in the restaurants described above is its relative illegibility to those who are not practicing members of the congregation (in case of Tahoora) or practicing Muslims (Ghareeb Nawaz). To an outsider or a tourist, conversant with the exoticized hypervisible ethnic spaces and ethnic enclaves in North America (Chinatown, Little Italy, and even Little India along Devon), the interweaving of domains (marketplace and prayer space in this case) is ideologically incomprehensible. The ontological invisibility of these intertwined spaces is important in the case of Muslims whose legibility and legitimacy in the American public realm has been under increased scrutiny, surveillance, and misrepresentation since September 11, 2001. During my interviews with the storeowners there was a clear hesitation on the part of the employees to give me their names and identifying details. Their reticence emerges from the negative media images, fear of intrusive government intelligence bodies, and a general feeling of vulnerability that recent violent events in Chicago have exacerbated among the local Muslim community. Thus the spatial illegibility provides an opportunity for the immigrants to carve out a safe zone for their community within the larger public realm.

Intertwined Spaces and Relational Visibility: Reading Ethnicity in the Built Environment

Reading authorship of the built environment in transcultural public spaces requires us to go beyond an understanding of origins, styles, and technologies of building and include a discussion of effects and performances of placemaking. Most South Asian storeowners along Devon Avenue occupy buildings that were built in the past by German and Jewish immigrants. Their buildings are neither exceptional nor exotic. Yet these new immigrants use these spaces in deeply embodied and culturally inflected ways, thereby recreating a new world in an old setting. They mark their stores and communicate with their customers via signs, banners, pamphlets, and posters (Sen, 2012). They transform the interiors, sometimes momentarily, through visual markings, transient behavior, protean performances, and momentary activities. As in linguistic code-switching, by merely changing the signage or altering behavior one transforms the nature and character of these spaces.

Boundaries, both physical and symbolic, allow individuals to circumscribe a safe in-group area. Boundaries allow users to cross over and experiment with multiple senses of belonging and allow immigrants to manipulate conditions of liminality to their own advantage. Thick edges maintain the sanctity of the interweaving domains while at the same time they provide flexible conditions where boundaries can be transgressed when necessary. Multiple domains within transcultural spaces need not be bounded, policed, and controlled. They need not be physically separated and zoned. Rather their visual and spatial co-presence within a larger public location provides opportunities for exchange and negotiations. Because these spaces are not sequestered behind the walls of residences or specially designed private institutional building types (e.g., mosques), restaurants such as Ghareeb Nawaz and Tahoora are places where the choice of being Muslim in America can be practiced beyond the absolute, private domain.

Four major points that have implications for future studies of urban transcultural public spaces emerged from this study. First, transcultural public spaces are ephemeral, protean, and transient sites where placemaking is more a momentary performance than a planned strategy. That means that a researcher has to go beyond the traditional analysis of building type and form that is common in material culture and architecture scholarship. Instead we will need to look at the built site as a complex theater stage where meaning and use of space depends on the event, activities, and actors.

Second, different people use and interpret transcultural places differently over time. Different users share the same space in different ways, make different territorial claims over time, and engage each other using varying proxemics standards. Complex overlapping and transmuting domains of insides and outsides, fronts and backs, in-group and out-group spaces, and cores and peripheries, maintained by real, imaginary, or experiential boundaries, sustain these public spaces. These multivalent experiences and interpretations of such places are phenomenological, and spatial orders reproduce sensory cues that allow individuals from different backgrounds to recognize and interpret this material world in inflected ways. A variety of responses may be generated by interior layouts (plan), façades decorations, architectural style, the processional and sequential experience of the interiors, and volumetric qualities of space. Uncovering the experiential order of transcultural space is the first step towards understanding place and placemaking in such locations.

Third, it becomes necessary to uncover the temporal rhythms and overlapping domains that bring life to transcultural spaces. One way of deciphering the multiple rhythms of transcultural spaces is to explore how spatial boundaries are viscerally experienced in these stores. Many of the boundaries that sustain multiple spatial domains inside these stores are time-based, protean, and experiential. They exist temporarily and then very quickly disappear. In the preceding discussion we saw how in a matter of minutes, store interiors turn into daily prayer spots and basements convert into congregational worship spaces and religious schools.

Finally, transcultural spaces are characterized by a relational visibility that sustains a very different kind of public space than the erstwhile agora and plaza. Here, on the one hand, these spaces sustain multiple publics—various subgroups within the Muslim immigrant community, non-Muslims, South Asians, non-South Asians, etc.—to participate with and see each other. On the other hand, it allows each individual to interpret and recreate these spaces in different (personalized) ways. The configurations of these spaces promote moving in, across, and out of them in multiple ways. These movements and experiences reproduce a variety of embodied forms of *being* American, Muslim, consumer, and ethnic. The civic, cultural, secular, and sacred domains are performatively bridged and crossed.

Intertwined transcultural spaces change the way we describe ethnic space in the United States. No longer is an ethnic space a site of difference. Instead it is a *choreographed experience* of difference and boundary crossing. Consequently, this research project hopes to inform material culture studies of the built environment in two ways. By making experience the center of our inquiry we deconstruct the built object and study parts of buildings. That means we write, compare, and catalog the production of sequential experiences of moving between exterior and interior spaces rather than only comparing the cultural production of building types. It means that we consider a notion of authorship that is not merely about craftsmanship or patronage. Micro-histories of individual users inhabiting, using, and producing experiential meanings explain everyday placemaking. Granted, our idea needs methodological fine-tuning, yet, as this chapter argues, reading embodied experiences of placemaking may help us decipher the many invisible worlds that lie behind the brick and mortar of our buildings.

Notes

1 The area of this study is very diverse. According to the 2000 census, this area has 49.7 percent white residents, 6.78 percent black, 15.5 percent Hispanic, 22.3 percent Asians, and 5.65 percent counted as others.
2 In 1965, the passage of a new Immigration Act made it possible for highly skilled South Asians to enter the United States. Although the 1965 law lifted all geographical and racial quotas and let in only skilled immigrants it was only in the 1980s that the family reunification clauses brought in less-skilled South Asian immigrants families to join their more educated and skilled compatriots (Minocha, 1987).
3 Many new and growing South Asian Muslim residents of this area work in low-paid and unstable jobs; but not all are unskilled. For instance, I interviewed skilled plumbers who due to their low English language skills and/or unstable immigration status cannot find regular and well-paid jobs. Because the US census does not collect information on religious groups the exact numbers of Muslim immigrants are not documented (except in the membership rolls of local cultural and religious bodies that in the recent political climate are difficult to obtain). Interview with GB/5/09; AZ/2/09, MH/2/09.
4 In the past, theories of assimilation assumed that immigrants enter ethnic ghettos on arrival, but lose their old habits and customs then move into suburbs as they are absorbed into American mainstream culture. Later, alternate theories of hybridity and diffusion emerged to explain that old and new customs mixed in order to produce new creolized forms.
5 Since 1991 parts of Devon Avenue have been symbolically renamed "Gandhi Marg," "Muhammad Ali Jinnah Way," and "Sheikh Mujib Road" to represent the Indian, Pakistani, and Bangladeshi communities respectively. These honorary street names, common practice in Chicago, pay homage to the founding fathers of India, Pakistan, and Bangladesh respectively and acknowledge the large Pakistani and Bangladeshi population living and doing business on this street.

References

Archer, J. D. & Santoro, J. W. (2007). *Images of America: Rogers Park.* Charleston, SC: Arcadia Publishing.
Arendt, H. (1989). *The Human Condition*. Chicago, IL: University of Chicago Press.
Austin, J. L. (1975). *How To Do Things with Words.* Cambridge, MA: Harvard University Press.
Bender, C. & Smith, E. (2004). Religious innovations among New York's Muslim taxi drivers. In T. Carnes & F. Yang (Eds), *Asian American Religions: The making and*

remaking of borders and boundaries (pp. 76–97). New York: New York University Press.

Borden, I. (2000). Thick edge: Architectural boundaries in the postmodern metropolis. In I. Borden & J. Rendell (Eds.), *InterSections: Architectural histories and critical theories* (pp. 221–246). New York: Routledge.

Connerton, P. (1989). *How Societies Remember.* New York: Cambridge University Press.

Cutler, I. (1996). *The Jews of Chicago: From shtetl to suburb.* Chicago: University of Illinois Press.

Gans, H. (1982). *The Urban Villagers: Group and class in the life of Italian-Americans.* New York: Free Press.

"Ghareeb Nawaz," (n.d.). Retrieved from: http://www.angelfire.com/ill/ghareebnawaz/

Goffman, E. (1959). *The Presentation of Self in Everyday Life*. New York: Anchor.

Gregg, M. & Seigworth, G. J. (2010). *The Affect Theory Reader,* Durham, NC: Duke University Press.

Hermansen, M. (2004). Muslims in the performative mode: A reflection on Muslim–Christian dialogue. *Muslim World* 94(3), 387–396.

Jones, P. d'A. (1995). *Ethnic Chicago: A multicultural portrait.* Grand Rapids, MI: Wm. B. Eerdmans Publishing.

Kalayil, Ann (March 2009). Interview. Devon Bank, 6445 North Western Avenue, Chicago, IL.

Kniss, F. & Numrich, P. D. (2007). *Sacred Assemblies and Civic Engagement: How religion matters for America's newest immigrants*. New Brunswick, NJ: Rutgers University Press.

Koval, J. P., Bennett, L., Bennett, M. I. J., Demissie, F., Garner, R., & Kim, K. (Eds). (2006). *The New Chicago: A social and cultural analysis*. Philadelphia, PA: Temple University Press.

Merleau-Ponty, M. (1995). *Phenomenology of Perception* (C. Smith, Trans.). New York: Routledge.

Minocha, U. (1987). South Asian immigrants: Trends and impacts on the sending and receiving societies. In J. T. Fawcett & B. V. Carino (Eds), *Pacific Bridges: The new immigration from Asia and the Pacific Islands* (pp. 347–374). Staten Island, NY: Center for Migration Studies.

Numrich, P. D. (1997). Recent immigrant religions in a restructuring metropolis: New religious landscapes in Chicago. *Journal of Cultural Geography* 17, 55–77.

Numrich, P. D. (2005). Muslims. In *Encyclopedia of Chicago*, Chicago Historical Society. Available online: http://www.encyclopedia.chicagohistory.org (accessed July 2, 2010).

Pratt, M. L. (2002). Arts of the contact zone. In D. Bartholomae and A. Petrosky (Eds), *Ways of Reading* (pp. 605–618). Boston, MA: Bedford/St. Martin's.

Schmidt, G. (2004). *Islam in Urban America: Sunni Muslims in Chicago*. Philadelphia, PA: Temple University Press.

Sen, A. (2009). Creative dissonance: Performance of ethnicity in banal space. *InTensions* 2. Available online: http://www.yorku.ca/intent/pastissues.html (accessed March 12, 2012).

Sen, A. (2012). From Curry Mahals to Chaat Cafes: Spatialities of the South Asian culinary landscape. In Tulasi Srinivasan and Krishnendu Ray (Eds), *Curried Cultures* (pp. 196–218). Berkeley: University of California Press.

Tweed, T. A. (2008). *Crossing and Dwelling: A theory of religion.* Boston, MA: Harvard University Press.

CHAPTER 3

Brazilian Restaurants and the Transcultural Making of Place in Tokyo, Japan

Vera Zambonelli

This chapter describes processes of placemaking as they occur in Brazilian restaurants in Tokyo, Japan. Placemaking, in its simplest definition, identifies the processes that transform an abstract, unfamiliar space into a familiar and meaningful place (Tuan, 2001). In the following analysis, placemaking is used to identify two different types of activities: place as product and place as process. Place as product is activities purposefully aimed at the production of physical, tangible, and designed sites; place as process is activities that occur in existing locations and transform those material settings into places of significance, memory, and attachment. The former entails the purposeful envisioning, designing, and constructing of the place and reveals an underlining agency. The latter, which may be influenced by the design place, results from repeated actions such as inhabitance (Lefebvre, 1996; Friedmann, 2007), naming and discourse practices (Carter, Donald, & Squires, 1994; Stokowski, 2002), walking (Wunderlich, 2008), and everyday routines (Feuchtwang, 2004). In other words, through seemingly mundane activities and reiteration of practices, the making of a place emerges (Cresswell, 2004).

Brazilian restaurants were chosen as informative sites for the analysis of placemaking because Japan exhibits internal contradictions in dealing with immigration and immigrant populations. In fact, Japan is known as a society hostile to immigrants where national policies discourage permanent settlement of immigrants and are aimed at controlling foreigners (Chung, 2006). Japan has no active policy to "incorporate [immigrants] into society or [for immigrants] to participate in Japan's political life" (Shipper, 2008, p. 4). Yet, the need for foreign workers is a structural aspect of the Japanese economy (Douglass & Roberts, 2003) and side mechanisms have arisen to deal with the demand for unskilled labor. One such default strategy is the granting of special visas to Brazilian *Nikkeijin* (Japanese descendants born and raised outside of Japan).

Relations between Brazil and Japan date back to 1908 when Japanese people first migrated to Brazil to work in coffee plantations. The flow of Japanese to Brazil increased in the aftermath of World War II and grew further with large joint venture projects in the 1960s and 1970s as the two countries' economies increasingly complemented each other.[1] However, in the early 1980s, the first instances of temporary return migration from Brazil to Japan appeared. The phenomenon went almost unnoticed, as the people who migrated were still first-generation Japanese, fluent in the language, and familiar with mainstream Japanese society. Their presence, rather undetectable and intentionally kept so to avoid the "loser" stigma (Higuchi, 2003), played a crucial role in establishing the initial connections for the subsequent flows of return migration.

In fact, in the 1980s, return migration from Brazil to Japan continued to grow as the Brazilian economy was deep in crisis and Japan experienced a mounting labor shortage. As the 1990 immigration policy reforms granted a special status for *Nikkeijin* and their family to live and work in Japan without restriction (Komai, 2001; Yamanaka, 2003; Tsuda, 2009; Morris-Suzuki, 2010), the flow of Japanese–Brazilian *degaseki* (temporary workers) from Brazil to Japan began to increase exponentially. Japan "admitted *Nikkeijin* to not undermine its official ban on unskilled immigrant workers or disrupt its ethnic homogeneity" (Tsuda, 2009, p. 208). The number of *Nikkeijin* increased from 1,955 in 1985 to a 2009 estimate of 267,456—representing the third largest group of foreigners living in Japan, after Chinese and Koreans.[2]

The Japanese–Brazilian population is concentrated in a few cities where many factories cluster such as Oizumi, Ota, and Tatebayashi in the Tomo area of Gunma Prefecture, Hamamatsu, Toyota, and Toyohashi in Aichi Prefecture, Nagoya, and Kawasaki, Tsurumi, and Fujisawa cities in Kanagawa Prefecture (Tsuda, 2009, p. 209). In Tokyo, the Japanese–Brazilian population is much smaller than in other areas (4,454 according to the Tokyo Statistical Yearbook 2007)[3] and consists of many non-Japanese–Brazilians as well.[4]

Despite the smaller population, I focused my research on Brazilian restaurants in Tokyo because I am not interested in understanding these places from a typical ethnic enclave perspective. Linger (2001) provides such a study in his analysis of Restaurant 51 in Nagoya. Restaurant 51 is a restaurant established by Brazilians for Brazilians. In the metropolitan area of Tokyo, some restaurants are operated as sites to maintain in-group ethnic contacts. However, in a city where there are not enough Brazilians to sustain a more ethnically exclusive establishment (as in the case of 51), Brazilian restaurants in Tokyo have become a particular type of public space where cultural diversity is negotiated (Zukin, 1995).

Restaurants as Transcultural Places

Restaurants, by being part of a wider socioeconomic landscape, are fascinating places to examine the intersections between ethnicity and place. They afford us opportunities to study how the images and traditions of a culture can be used to appropriate space and how these practices are located in larger networks and sociocultural practices (Beriss & Sutton, 2007). In brief, restaurants do more, much more, than simply serve food. In this chapter, I consider Brazilian restaurants both as (1) tangible sites where particular types of places are intentionally manufactured through creative design and even self-commodification, and (2) intangible practices that give rise to meaningful processes of placemaking.

To distinguish between the production (tangible) and reception (intangible) of the restaurant as a place, I adapted *rapports de production* (the relationships of production) from the field of artistic experience studies. *Rapports de production* questions both how and why art is produced, and how and why it is received

(Monaco, 2009, p. 34). An analysis of the relationship between the maker of the place and the place itself yields an understanding of the placemaking activities directed towards the production of the restaurant as a physical place. An analysis of the relationship between the place and its recipients generates an understanding of how the place is received, experienced, and given meaning.

Using these relationships as the basis of my analysis, I conducted interviews with both the producers of the restaurants—the managers, staff, and performers—and the recipients of the place—the customers. In addition, I observed the activities in the restaurants and tracked blogs and other digital media sources about these establishments. This research was conducted over four years during four different fieldwork periods, totaling about five months in Tokyo. Within the limited space of this chapter, I will focus on the relationship between the producers of place and the place itself while weaving in some observations and remarks from the clientele to illustrate the reception of these restaurants.

Through my fieldwork, it became clear that Brazilian restaurants are spread out all over Tokyo but have higher concentrations in three areas. The first is in the central part of Tokyo where a substantial concentration is in Omotesando/Aoyamadori. Another cluster is located around Tokyo's western Suginami Ward, while the last cluster is in the Kanagawa (Kawasaki and Tsurumi) Prefecture, southern part of the Greater Tokyo Area.

Despite similarities, these restaurants differ in several ways. The first and foremost difference concerns who opened the restaurant and the reason for starting the establishment. They can be divided into the following categories of makers, which also hint at the motivation for opening the restaurant aside from making profits. They are the: (1) traditional ethnic entrepreneur; (2) "Breaking Away" individual(s); (3) not-only-business promoter; and (4) Japanese returnee. In the following section, I discuss the restaurants that most significantly epitomize these categories.[5]

The Traditional Ethnic Entrepreneur

Traditional ethnic entrepreneurs are individuals who are Brazilians (most likely Japanese–Brazilians) and dedicate their businesses to pleasing a Brazilian clientele. Out of the 19 restaurants where I interviewed the owners/managers, only six belong to this category. These are Praça 11, Tucano, and Carioca, located in central Tokyo, and Chega Mais, Cabana, and Segredo, South of Tokyo in the Greater Tokyo Area. The latter three restaurants are in Kanagawa, an industrial zone with many factories that hire Brazilians as factory workers. Of these three, Chega Mais and Cabana are located in an area where many Brazilians actually live and understandably the restaurants have a stronger focus on serving Brazilians.

The restaurants in this area have a homier ambience than the ones in central Tokyo. As stated by the owner of Chega Mais, he wanted to create a *casa* (home) for his guests, which also resonates with the descriptions of Segredo, a *katei ryori* (home cooking) shop, and Cabana, *uma casinha simples, aconchegante e hospitaleira* (a simple, welcoming and hospitable small house). Indeed, the owners of these establishments are successful in creating a homey atmosphere, as evident in the remarks posted by their clients in various food blog entries such as: "doesn't feel like a restaurant, homey"; "you can imagine a Brazilian mom slow cooking the dish"; and "a home cooked meal taste."[6]

These places closely resemble the typical image of a restaurant in their layout and focus on food production and menus. People go there mostly to eat. However, the decorations within these places also create an environment in which one could temporarily journey to another country, Brazil, while consuming Brazilian food. From the highly decorated—almost cluttered—Segredo (Figure 3.1) to a moderately decorated Chega Mais and minimally decorated Cabana, clients, by entering the place, cross the threshold towards a place which takes them somewhere else. They feel they are offered "a different wonderland"[7] that may be intimidating at times "even for a person who lives in a melting pot of Tsurumi for four years, it takes courage"[8] but it is worthwhile because eating that food can make a customer feel like he was flying "over the carnival of Rio, in ecstasy, while out of the window it is still Japan."[9]

This Brazilian experience is nurtured through Brazilian related events. Cabana hosts monthly music events and invites local musicians (not necessarily Brazilian) to enhance the Brazilian experience. They provide a space for Japanese enthusiasts of Brazilian music, especially Bossa Nova, to gather. Segredo offers samba shows and karaoke. Chega Mais, which opened only few months before I interviewed its owners, has yet to establish planned events. However, several instruments are on display and accessible. I was told that on a few occasions customers have spontaneously asked to play the instruments and several other customers joined in. By offering these types of opportunities, these places insert themselves in a network of activities that promote Brazilian culture and around which people with interest in Brazil gravitate.

Given the demographics of Tokyo, the clientele in these restaurants is necessarily mixed. However, even with the declining number of Brazilians due to the economic downturn, the adaptation of the menu to Japanese tastes is met with resistance. When menus were adapted in the past, there was a loss in Brazilian patronage. This is what is remembered by the managers/owners, but in addition, the Brazilian menu seems to represent a site of proud resistance and stands as a mark of authenticity.[10]

In comparison to the restaurants in Kanagawa, the ones in central Tokyo are more flexible in their spatial organization. Tucano, in particular, can be easily transformed into a dance floor, music stage, and place for private parties. Tucano, Carioca, and Praça 11 have mostly a Japanese–Brazilian face. These establishments want to recruit

Figure 3.1 Interior decoration at Segredo. Photograph by Vera Zambonelli.

more Brazilians in their workforce but their clientele is mainly Japanese. Carioca's owner is Japanese, but Ramos, a famous soccer player from Rio (the reason for the name) is the manager. In fact, the main motif in the decorations is a soccer ball. Praça 11 was opened in 1981 because no one knew about Brazil or Brazilian music at that time; it remains one of the main venues to hear Brazilian music in Tokyo (Figure 3.2).[11] However, even these more centrally located restaurants are at the margins of central Tokyo.[12]

The presence of these restaurants at the margins can be explained by the limited access that non-Japanese have to the urban space and how the space at the margin seems to be more open to the cultural *other*. Moreover, the absence of any notion of permanence for Brazilian migrants in Japan makes them unwilling to invest in places that they will not be able to call home for long. Though officially it is possible for non-Japanese to own and run businesses in Japan, systemic barriers make the process difficult. For example, migrants encounter significant obstacles in obtaining loans and experience discrimination in renting places. This explains why so many of these activities are operated in partnership with Japanese citizens in order to ease dealings with the bureaucracy. Lastly, the physical marginality speaks to the fact that the majority of Brazilians have situated themselves as blue-collar workers in the local economy with very few venturing into the service sector.

The "Breaking Away" Individual(s)

The majority of Brazilian restaurants are entirely run by Japanese. The kind of self-immersion in or adoption of another culture portrayed by the character Dave Stoller in the movie *Breaking Away* is evident in Mr. Ishii, the owner of Alvorada. "Alvorada" means "dawn" in Portuguese. Mr. Ishii remarked how this name seemed so full of hope, promises, and possibilities. Before opening Alvorada, he experienced some

Figure 3.2 A live performance at Praça 11. Photograph by Vera Zambonelli.

difficulties. He was once a photographer who traveled a lot but at one point had to quit. With the help of a friend who was the previous owner of the property where Alvorada is located, he decided to open a Brazilian restaurant. The restaurant opened on September 7, 2004, and to Mr. Ishii the date is significant. September 7 is the day that Brazil celebrates its independence, and also served to launch Mr. Ishii's independence. In his words, "It has double meaning, Brazil's independence and my independence."[13] Yi-Fu Tuan writes, "Naming is power—the creative power to call something into being, to render the invisible visible, to impart a certain character to things" (Tuan, 1991 in Cresswell, 2004, p. 98). This is exactly what Mr. Ishii did in naming his restaurant Alvorada.

Furthermore, Mr. Ishii has taken on Brazilian culture and endeavored to create an authentic establishment through the decorations, information, and staffing in his restaurant. He has traveled extensively in Brazil and his shop is decorated with all sorts of trinkets that he personally brought back from his trips (Figure 3.3). The shop is also a hub of information on current and past events on and about Brazilian culture. When asked how he recruits his staff, Mr. Ishii stated that Portuguese language ability and knowledge about Brazilian culture are key. Through these actions, he holds himself accountable for creating a Brazilian experience in Alvorada.

A similar personal investment was found in the makers of Ilha Branca, Aparecida, and Copo do Dia. The makers of these places all have personal and sometimes very emotional stories of what made them open a Brazilian restaurant. Some, such as the owner of Aparecida, encountered Brazilian culture through Brazilian music, and offer a remarkable Brazilian music collection, not only the well-known Bossa Nova in the restaurant. Others, such as the owner of Ilha Branca, are friends with Brazilian musicians and performers.

Copo do Dia, run by a young couple that decided to do something different with their lives, opened up their restaurant to reconnect with their family history. The husband is an ex-salary man and the wife, Hayako, comes from a family with direct connections to Brazil. Hayako's mother and uncle migrated to Brazil after World War II but returned to Japan. Hayako remarked how little her mother talked about her experience in Brazil, only a few words here and there, and how she used to prepare some Brazilian food as snacks for Hayako. This heightened Hayako's curiosity about Brazil and led the couple on a trip to find the house where Hayako's mother once lived. The couple shared the anxieties associated with running a business but they were very happy to have created a place where people interested in Brazilian culture could visit. Copo do Dia hosts a wide range of events ranging from music performances to informational seminars on the relations between Japan and Brazil to slideshows of recent Brazilian trips presented by customers willing to share their experiences. The ambience in this place is welcoming and friendly, partly because of the small size of the restaurant.

In fact, most of these Brazilian restaurants are small with very few tables and customers usually have to share their tables with others. This physical proximity between customers encourages conversations between them even though they entered the restaurant as strangers. When asked what was unique about their Brazilian restaurants, a few owners, such as those

Figure 3.3 Brazilian memorabilia at Alvorada. Photograph by Vera Zambonelli.

of Ilha, Alvorada, and Favela, answered using the word "communication" in English to indicate how the customers were friendly and talkative among themselves, opening up to others sitting near them.[14] This is reminiscent of the notion of staging the self by performing the other (Soares, 1998). In a Brazilian restaurant, people behave as they think a Brazilian would behave and perform the other, thereby freeing a Japanese person to act in ways that are rarely entertained in Japanese culture. By being in a Brazilian space, they are temporarily able to behave differently, as they imagine a Brazilian person would behave.

As these Brazilian restaurants allow Japanese people to perform the Brazilian other, they also enable the Brazilian other to stage him or herself. These places provide a space/place for Brazilians to gather and to perform music and dance. In a conversation with a local journalist and filmmaker, Rolando, I was told the existence of this type of place has given him and other people the opportunity to organize a lot of events on Brazil such as film screenings. A place that stands out above the others for the opportunities that it grants to Brazilians for placemaking is Que Bom.

The Not-only-Business Promoter

Que Bom as well as Espeto Brazil belong to this category because of their underlying business orientation and because there are no Brazilians in a managerial position. However, their staff is predominantly Brazilian and the place is open to Brazilians to make a place while the Japanese clientele can embark upon or continue a transformative intercultural journey by frequenting the restaurant.

Que Bom is located in Asakusa, an area famous for its samba schools[15] and Brazilian carnival in late August. In Que Bom, a beautiful mural covers all of the walls in the stairwell that leads you down to the restaurant. The colorful and lively mural includes all sorts of Brazilian symbols such as Afro samba dancers, tucano birds, and soccer players. At the bottom of the stairs, there is a little area with a stand where books, pamphlets, and fliers for events are displayed. Then comes a wall with Athleta soccer shirts signed by famous people, and another piece of furniture used to display music CDs and T-shirts for sale. Many obviously Brazilian trinkets decorate the bar area and there are pictures of past events pinned here and there. A glass wall separates the seating and eating area from the entrance and bar area. On this glass wall, the Brazilian national anthem is engraved (Figure 3.4).

Mr. Kobe manages Que Bom and everybody at the restaurant calls him Marcio. He has never been to Brazil and has no special connections to the country. He had experience in managing restaurants before Athleta, a sportswear company with Brazilian origins, recruited him to run Que Bom. The cooks are also not Brazilian although they have learnt to cook Brazilian food. Half of the remaining staff is Brazilian while the other half is Japanese and uses Brazilian names (the importance of naming not only a place, but its people).

During the interview, Marcio mentioned he runs the place, but Athleta, a sportswear company, is the owner. Athleta was originally a Brazilian company that produced sportswear and was bought out by a Japanese company that maintained the original name and inventory. Athleta had initially opened this Brazilian restaurant to host its clients in order to expose them to a little bit of Brazil.

Figure 3.4
Glass wall at Que Bom engraved with Brazil's national anthem. Photograph by Vera Zambonelli.

When I asked who decorated the restaurant, Marcio told me that Athleta assigned the task to a Japanese interior design company. The corporate office made all the decisions concerning the opening, the name, and the design of the place and did so quickly. Therefore, there was little of the personal touch in the beginning. However, through the years, the staff and the Brazilian performers who work there have managed to make it one of the central places to disseminate Brazilian culture. Marcia, who waits at Que Bom twice or three times a week, stressed how she enjoyed working in this restaurant because of its atmosphere. She is granted permission to host events that Rabadas, a Brazilian cultural group, organizes to present different images of Brazil that diverges from the stereotypical perceptions existing in Japan.[16]

The images used in Que Bom to represent Brazil are quite repetitive and stereotypical: flag of Brazil, the colors green and yellow, the tucano, samba dancers, and soccer shirts. These are not necessarily perceived as bad by Brazilians, though. In response to an anonymous survey I circulated among some Brazilians I befriended during fieldwork, remarks such as the following were gathered. "Every time that I see a Brazilian flag or any kind of poster that reminds me of Brazil I get quite happy."[17] Another Brazilian answered "[I feel] praised by Japanese when they represent Brazil in their imagery. Although Brazil is seen as a violent country by Japanese media, people who love Brazil usually represent the country through its happy side, such as soccer, carnival, and so on."[18] Yet the constant use of women and samba dancers can be upsetting. As another Brazilian wrote, "when they use images of women dressed for the carnival, it means they are almost naked, and I feel quite bad."[19] Although the globally known carnival in Rio is Brazilian and represents Brazil and a Brazilian spirit, it is also a limited portrayal of Brazilian culture and reinforces stereotypes.

While Que Bom is the outcome of a professional design project, Espeto Brazil, located in Minami Otsuka, seems to be decorated with posters obtained from local travel agencies. In the stairwell down into the restaurant, there is a board with several pictures probably taken during previous music and dance performances at Espeto Brazil. On the walls of the performance area, the flag of Brazil is in a dominant position with "Espeto Brazil" written in katakana above it. The overall ambience is very cozy and not too pretentious. When there are performances, the place becomes crowded very easily, as the physical space available is relatively small.

In Portuguese, an "espeto" is a stick onto which one pierces meat to be grilled. This name creates a very vivid image of getting through the place called Brazil through the implied bodily and sensory experience. Grilled and barbequed, Brazil is served to the customers, who feast on it in small pieces, performance after performance. The menu includes some main dishes, but snacks and drinks predominate. At Espeto, I met some individuals who are passionate about Brazil and things Brazilian: Chihiro and Matsuo. As will be described below, their strong feelings for Brazilian culture and patronage of Espeto contribute to the placemaking process.

During a performance by Ms. Shindo, a Japanese singer who goes by the stage name Linda, I began chatting with a man who was carried away by Linda's music. He was singing all the songs and even had the words of a song printed out. He kindly offered to share the lyrics with me, which opened up our conversation. His name is Matsuo and he is a monk. He confided to me that he loves Brazil and has traveled extensively in the country. The first time Matsuo visited Brazil was during the carnival and it left a very strong impression. "Brazil? Shock desu! (Brazil? It's shocking!)," he said several times. He mentioned how he was so shocked by the place and felt so attracted to it, that he had to go back many times. He enjoyed coming to Espeto as a way to sustain his passion for Brazil.

Chihiro, whom I met at a samba performance at Espeto, expressed a similar feeling of having been overwhelmed. Chihiro said that she has been dancing samba for over fifteen years. When I asked what moved her to start dancing, she answered, "One day, while I was visiting New York City, I was reading this magazine editorial. It was about samba and the love this person had for samba. It struck me and I felt so strongly in my heart that I had to start dancing samba it was like a call."[20] Chihiro dances regularly at Espeto Brazil and other venues. She embraced her love for samba and does her best to live it out. Hence, Espeto, as other venues, is a part of a network of places that nurtures interests in Brazil.

The actions of the above two patrons give prominence to the role played by the customers/recipients in the making of these places. They were there because of a shared passion for Brazilian culture and a common need for an expressive outlet. As noted by Griswold (2008) in her study of the production of cultural objects and the recipients, "far from being a passive audience, cultural receivers are active meaning makers" (Griswold, 2008, p. 15). In pursuing their own interests and passions, such as Brazilian music and/or samba dance, Chihiro, Matsuo, and others carve a niche for themselves in their everyday life in Tokyo. Espeto would not exist in the way it does if these customers were not present. Customers such as Chihiro and Matsuo are the receivers of the physical restaurant and the food served, but they are also the creators of a shared meaningful experience who take advantage of a place that allows them to express their passions. The customers participate in place production in an ongoing process.

The consumers/recipients, no longer passive recipients, become active placemakers by not only arriving at the place but by also participating in its activities, in effect contributing to its existence. When a place is produced and not simply consumed but also granted the possibility to be used and remade, it activates processes of placemaking that can be understood as a transformational experience. By embracing and acting upon new learnt cultural practices through food, music, dance, and other forms of interaction, the consumers/clientele become placemakers and key transcultural agents.

The Japanese Returnees

Japanese returnees are people from the first wave of Japanese to migrate or live in Brazil. Upon their return to Japan some of them established Brazilian restaurants, during a time when very few in Japan understood anything about Brazil. These places target an almost exclusively Japanese clientele and are frequented by patrons of Brazilian arts. This is the case of Saci Perere, the first Brazilian restaurant in Tokyo, located in Yotsuya (Figure 3.5). Mr. Honda, the owner of Saci Perere, is now over 80 years old, and is seen from time to time at the restaurant. One of his two daughters runs the business and during our conversation shared how the business was not going too well due to the economic recession.

Making Transcultural Places

In concluding this chapter, I want to refer to the film *Do the Right Thing* (1989) by Spike Lee. In the film, Sal's pizzeria plays a central role. Sal's pizzeria was a focal place for the entire neighborhood. The pizzeria and, more specifically, the walls of the pizzeria were the main point of contention in the story, the place of conflict. For one of the leading characters, Buggin' Out, those walls were not simply walls but a contentious terrain where an individual could assert the right to represent and be represented. Sal denied Buggin' Out's request to put a black celebrity on the wall of fame. This denial signified to Buggin' Out his disempowerment, because he was not able to claim a place in his own space.

Figure 3.5 Exterior façade of Saci Perere, the first Brazilian restaurant in Tokyo. Photograph by Vera Zambonelli.

Buggin' Out was not allowed to appropriate what he thought was already his own. Ultimately he was denied the power to be present and represented. Although he supported the pizzeria daily, the walls of the restaurant were not open to him as he had no right to them. He was just a consumer, a paying customer and nothing else. Put simply, buying slices of pizza did not buy him the right to the pizzeria. But this does not have to be the case, and the restaurants I described above are situated at the opposite end of the spectrum compared to Sal's Pizzeria.

The Brazilian restaurants openly provide opportunities to individuals to navigate different cultures and their respective systems of meanings. These restaurants ask for pictures from their customers to put on their walls. The place belongs to both the makers and receivers and all actively contribute to its making. It would be hard to argue that these restaurants affect all of the clientele in the same way, but these various venues and the events housed within them emerge as "microspaces for intercultural contact where the foreign is made familiar and the global miniaturized" (Turgeon & Pastinelli, 2002, cited in Beriss & Sutton, 2007, p. 8). These places and the interactions within them are small yet significant steps that move us closer to the cultural other (Sandercock, 2003). In these restaurants, everyday experiences of cosmopolitanism can be localized as lived and lively embraced.

In sum, the stories I have told speak to the extraordinary in the ordinary. It is the story of so-called ordinary people who create places out of the ordinary; places we could term liminal as the individuals can be "removed from the concerns of their daily, practical lives, open themselves to a different quality of experience" (Duncan, 1995, p. 20). However, these places should not be understood as mere spaces of escapism. In these restaurants, images of a culture other than the Japanese mainstream are to be served and hence consumed, but also, another culture is imaged and imagined, adapted and negotiated. They are zones of contact. They are places where an initial consumption may lead to an enactment that ignites journeys of personal as well as social transformation, especially when established by traditional ethnic entrepreneurs and "Breaking Away" personalities. In a sense, they are extraordinary and subversive places, islands of intercultural exchange, precious and important in Japan, where competence in living and working with the cultural other has a long way to go before being fully appreciated and developed.

Notes

1 http://www.keidanren.or.jp/english/policy/2000/053.html (accessed July 23, 2011).

2 From the Japanese Ministry of Foreign Affairs website, http://www.mofa.go.jp/region/latin/brazil/index.html (accessed July 23, 2011).

3 From 2007, Brazilians disappear from the registered foreigner statistics of the Tokyo Statistical Yearbooks, http://www.toukei.metro.tokyo.jp/tnenkan/2007/tn07qyte0510b.htm

4 The dynamics of the relations between Japanese–Brazilians and non-Japanese–Brazilians in Tokyo represent a study in themselves, which I plan to make part of the larger study to which this chapter belongs.

5 There are also the in-between, such as Copa, Café do Centro, and Saudade, which resemble the not-only-business promoter and are dedicated to the Japanese clientele.

6 All the customers' comments are from Tabelog, http://alturl.com/m76vt (accessed July 15, 2010).

7 Ma Ra, 3/31/10, food blog entry. See: http://r.tabelog.com/kanagawa/A1402/A140210/14005062/dtlrvwlst/1069747/ (accessed July 2010).

8 Polako, 8/22/07, food blog entry. See: http://r.tabelog.com/kanagawa/A1402/A140210/14005062/dtlrvwlst/300777/ (accessed July 2010).

9 Tsururu, 3/18/08, food blog entry. See: http://r.tabelog.com/kanagawa/ (accessed July 2010).
10 Personal communication with the owner of Cabana.
11 Asada, 7/24/10, personal communication.
12 The only exception is Barbacoa which stands a few meters away from Omotesando. Though I visited and observed the place a few times, my calls and emails to set an interview were never returned. Similarly, Ricardo's owner did not appear for a meeting on three occasions.
13 8/6/10, personal communication.
14 7/13/10; 7/13/10; 8/6/10; personal communication. (The real names of interviewees for this study have been replaced with pseudo names to protect their identity.)
15 Samba schools don't really occupy a place of their own. As in the case of Capoeira, they usually rent other gym or municipal sports spaces. Capoeira Zoador is the only one which has its own space.
16 From their Facebook page: "Because Brazilian culture cannot be reduced to samba and the Asakusa carnival, members of the group 'the Rabadas Cultura Clube' wish to bring another Brazil to Tokyo." http://www.facebook.com/group.php?gid=17204817361
17 12/9/10, anonymous response number 2.
18 12/7/2010, anonymous response number 4.
19 12/8/10, anonymous response number 3.
20 11/13/2009, personal communication.

References

Beriss, D. & Sutton, D. E. (2007). *The Restaurants Book: Ethnographies of Where We Eat* (English ed.). Oxford: Berg.

Carter, E., Donald, J., & Squires, J. (1994). *Space and Place.* London: Lawrence & Wishart.

Chung, E. (2006). *Immigration and Citizenship in Japan*. Cambridge: Cambridge University Press.

Cresswell, T. (2004). *Place: A Short Introduction*. Oxford: Wiley-Blackwell

Douglass, M. & Roberts, G. S. (2003). *Japan and Global Migration.* Honolulu: University of Hawaii Press.

Duncan, C. (1995). The Art Museum as Ritual. In C. Duncan, *Civilizing Rituals: Inside Public Art Museums* (pp. 7–20). London: Routledge.

Feuchtwang, S. (2004). Theorizing Place. In S. Feuchtwang (Ed.), *Making Place: State Projects, Globalisation and Local Responses in China*. London: UCL Press.

Friedmann, J. (2007). Reflections on Place and Place-making in the Cities of China. *International Journal of Urban and Regional Research,* 31(2), 257–279.

Griswold, W. (2008). Culture and the Cultural Diamond. In W. Griswold, *Cultures and Societies in a Changing World* (pp. 1–19). 3rd edition. Thousand Oaks, CA: Pine Forge Press.

Higuchi, N. (2003). *Migration Process of Nikkei Brazilians*. CAS Symposium Series, Osaka: Japan Center for Area Studies.

Komai, H. (2001). *Foreign Migrants in Contemporary Japan*. Melbourne: Trans Pacific Press.

Lefebvre, H. (1996). *Writings on Cities*, Oxford: Blackwell.

Linger, D. T. (2001). Eating Brazil. In D. T. Linger, *No One Home: Brazilian Selves Remade in Japan*. (pp. 74–92). Stanford, CA: Stanford University Press.

Monaco, J. (2009). Film as an Art. In J. Monaco, *How to Read a Film* (pp. 24–75). Oxford: Oxford University Press.

Morris-Suzuki, T. (2010). *Borderline Japan: Foreigners and Frontier Controls in the Postwar Era*. Cambridge: Cambridge University Press.

Sandercock, L. (2003). *Cosmopolis II*. London: Continuum.

Shipper, A. (2008). Parables of Citizenship: Morality, Personhood, and Immigration Politics in Japan, the United States, and Sweden. Paper presented at the Annual Meeting of the American Political Science Association, Boston, August 28–31. Retrieved on December 2009, from: http://www.allacademic.com/meta/p278557_index.html

Soares, Luiz E. (1998). Staging the Self by Performing the Other: Global Fantasies and the Migration of the Projective Imagination. *Journal for Cultural Research*, 2(2), 288–304.

Stokowski, P. A. (2002). Languages of Place and Discourses of Power. *Journal of Leisure Research*, 34(4), 368–382.

Tsuda, T. (2009). Japanese–Brazilian Ethnic Return Migration and the Making of Japan's Newest Immigrant Minority. In M. Weiner (Ed.), *Japan's Minorities: The Illusion of Homogeneity* (pp. 206–227). London: Routledge.

Tuan, Y.-F. (1991). Language and the Making of Place. *Annals of the Association of American Geographers,* 81, 4,684–4,696.

Tuan, Y.-F. (2001). *Space and Place.* Minneapolis: University of Minnesota Press.

Turgeon, L., & Pastinelli, M. (2002). "Eat the World": Postcolonial Encounters in Quebec City's Ethnic Restaurants. *Journal of American Folklore,* 115(456), 247–268.

Wunderlich, F. M. (2008). Walking and Rhythmicity: Sensing Urban Space. *Journal of Urban Design*, 13(1), 125–139.

Yamanaka, K. (2003). "I Will Go Home, but When?" Labor Migration and Circular Diaspora Formation by Japanese Brazilians in Japan. In M. Douglass, & G. S. Roberts (Eds), *Japan and Global Migration: Foreign Workers and the Advent of a Multicultural Society* (pp. 123–152). Honolulu: University of Hawaii Press.

Zukin, S. (1995). *The Cultures of Cities*. Oxford: Blackwell.

Films

Breaking Away (1979) dir. Peter Yates. TCF.

Do the Right Thing (1989) dir. Spike Lee. UIP/Forty Acres And A Mule Filmworks.

CHAPTER 4

West African Immigrants' Hybrid Spaces and Identities in Rainier Valley, Seattle

Rachel Miller

> There are so many things that [some West African immigrants] are ignoring here. And we still [are] attached to some of our traditions. And actually we should feel like, having the world . . . perception. We have to mix now . . . to be competent in terms of culture. We live in the Western world. I don't say that we have to . . . quit how our ancestors [lived]. But actually being able to do both is the most important, and the helpful thing.
>
> (Fifi Doumbia, Gambian Malinke woman, Co-owner of West African Market, Seattle)

Diverse West African immigrants have created businesses and a mosque in the Rainier Valley of Seattle that provide culturally specific amenities and gathering places while facilitating transcultural interaction. Drawn to key elements of an already diverse neighborhood, they have adapted existing commercial buildings to make them their own, but also balance an expression of difference with a fluid exchange among multiple cultural identities. Using sixteen in-depth interviews, numerous informal conversations, and field observation, this chapter traces the story of their appropriation of space. It asks what places they have made, how they use the spaces as manifestations of their multiple cultural identities, what qualities draw people there and facilitate cultural exchange, and how customers and business owners may be negotiating multiple cultures through these spaces.

Hybrid Identities and Spaces

At the root of this story is the perception of self and cultural identity. People have embedded cultural identities (Sandercock, 2003) and seek to construct difference or represent their diverse identities (Hall, 2003). As migrants adapt to new homes, they can form hybrid identities, not bound solely to their original culture, but housing fluid, intercultural identities (Kim, 1988). The cultural competencies that grow out of shifting between two (or more) cultures allow a person to bridge cultural differences in a dynamic, interactive way that minimizes dichotomous conflicts (Byram, 2003). This dual insider/outsider perspective "has the potential to benefit individuals and communities alike through relativizing values, probing ignorance and testing assumptions that are held in relation to others and oneself" (De Korne et al., 2007, p. 303). Essentially, people who can identify as neither this nor that, but both simultaneously, have the tools to bridge sociocultural gaps.

Not leaving a culture behind is an important part of hybridity. The addition of cultures, without subtracting or losing a sense of belonging in the original culture, can relieve stressful transitions into a new society (Sussman, 2002). Regular trans-cultural interaction also eases that process (Manzo et al., 2008; Sonn, 2002). West African immigrants in the Seattle area are part of the new African diaspora, adding US identities to their African ones. The places on Rainier Avenue South help to keep the diaspora alive by offering amenities such as West African food, media, clothing, and health and beauty products which remind them of their homelands and fortify cultural identities; spaces for visiting each other and maintaining social ties; religious spaces for sustaining core beliefs and fostering fellowship; and ways of communicating with and transferring money to members in their homelands.

Understanding *places* that support the hybridity of cultures is especially important for sustaining and enhancing quality cross-cultural interaction. Places where different cultures can meet, redefine relationships to each other, and reach new, shared meanings have been described as "intermediate space" (Winnicott, 1971), "third space" (Bhabha, 1994), and "hybrid space" (AlSayyad, 2001). Hybrid places are sites of resistance, where people encounter and transform each other rather than one group being subverted by the other (Bhabha, 1994). They are also an expression of an individual's hybrid identity, as in the cases of the shopkeepers in this study.

Increasing migration complicates the relationship between place and culture, and reading urban form as a spatial manifestation of human behavior is becoming more difficult (AlSayyad, 2001). This is exemplified in the Rainier Valley, where the built environment provides few clues to the culturally specific uses inside the buildings. Flows of culture over boundaries and interconnectedness are "the prime and typical spaces of the global encounter" and "no longer simply occupy the margins of the periphery, for they have moved firmly to the center of the core" (AlSayyad, 2001, p. 16). This cluster of West African-run businesses is one of AlSayyad's hybrid places, firmly at the core of Seattle's global culture.

This chapter builds on the work of theorists who observe hybrid spaces formed by immigrant communities in the United States. For example, Hadjiyanni (2007) examines Somali immigrant women's adaptation of and cultural limitations to American housing in Minnesota. Dearborn (2008) explores two Hmong enclaves in a historically German Milwaukee neighborhood. In the first, traditional socio-spatial patterns are replicated to produce a predictable and comfortable reinvention of a Hmong village. People in the second enclave use the first as an anchor, but straddle Hmong and non-Hmong worlds to a greater extent. A third case is Davis's (1997) observation of immigrant groups' reuse and revitalization of existing auto-oriented commercial spaces. Rainier Valley businesses often fit Davis's descriptions. For example, a former gas station on Rainier Avenue South was used as a dry cleaner by Asian immigrants before becoming a West African-owned auto body shop. These places exemplify the negotiation of hybrid identities in an increasingly transcultural world.

West African Immigrants in the Seattle Area

Diversity provisions in federal immigration policies since the 1960s and particularly in the 1980s and 1990s, which were intended to allow more people to emigrate from underrepresented countries, have immensely benefited African immigrants (Takyi, 2009; Arthur, 2000). Most gained H-1B visas, given for advanced degrees and specialty occupations such as nursing, or were chosen through the Diversity Lottery Program, which requires a high school degree or work experience. These programs "resulted in a significant pool of human capital from Africa consisting of highly skilled and upwardly mobile professionals" (Margai, 2009, p. 166). Since 1996, two immigration policies have closed some doors for African immigrants and complicated their legal status in the United States.[1]

The population of people born in West Africa comprises the largest group of African immigrants in the USA, and nearly doubled in the Seattle area between 2000 and 2008 (US Census 2000, ACS 2006–2008). Although they are a quickly growing group, West Africans are a fledgling immigrant community in the Seattle area with only 3,817 documented West African immigrants in King, Pierce, and Snohomish Counties, as compared to 23,943 East Africans in these counties (ACS 2006–2010). Interviewees chose Seattle for work and education opportunities or to join family. The Alaskan fishing industry was often mentioned for bringing West African migrant workers to Seattle.

West Africa encompasses a diverse and complex population where nationality and ethnicity do not necessarily align. For example, the sixteen interviewees in this study represent six West African nations and eight ethnicities (Figure 4.1). Most are from Guinea, Mali, and Gambia, while the greatest ethnicity represented is Malinke, followed by Bambara and Wolof.

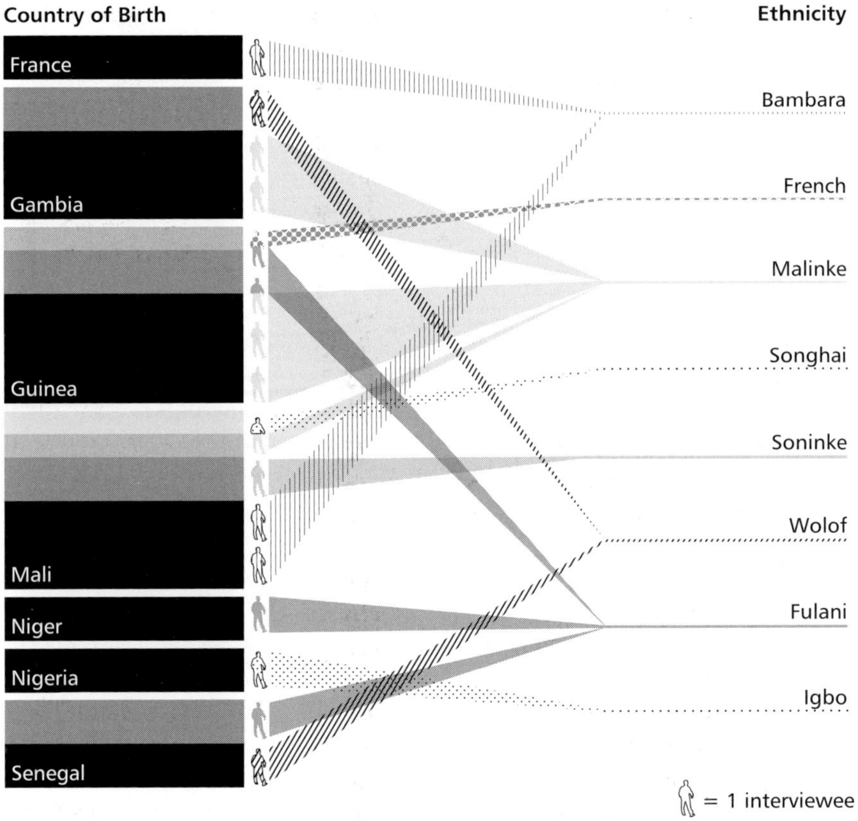

Figure 4.1 Country of birth and ethnicity of sixteen interviewees. Diagram by Rachel Miller.

West African residents are distributed throughout the Puget Sound, with no definitive residential clustering (US Census 2000). When West African immigrants move to the Seattle area, they are more likely to move to outlying cities such as Federal Way, SeaTac, or Lynnwood, rather than Seattle itself. Only four of the interviewees live within three miles of the cluster of West African-owned businesses.

Rainier Valley Cluster

Although their residences are not concentrated in any neighborhood, West African immigrants formed a cluster of businesses over the past ten years on a thirteen-block stretch of Rainier Avenue South in the Rainier Valley (Figure 4.2). The Rainier Valley has a long history of hosting immigrants, and today, over two-thirds of Rainier Valley residents are non-white, over 40 percent were born in foreign countries, and at least 59 languages are spoken (US Census 2000). It is known for offering an "authentic cultural experience" and as a business incubator for ethnically diverse entrepreneurs (CLUE Group, 2009, p. 88).

West African immigrants' businesses are found on Rainier Avenue South between Columbia City, a hip and gentrifying neighborhood, and Rainier Beach, a neighborhood seeing slower economic growth but also investment in a new community center and high school. Rainier Avenue South is a major arterial running north–south through the neighborhood, and this particular segment feels more like a throughway than a destination. Lacking a clear and cohesive pedestrian environment, it is reminiscent of Garde's (1999) leftover, marginal spaces.

The first five businesses—Rainier Beauty Supply, Kumba's African Hair Braiding, Rose's and Siga African Hair Braiding, West African Market, and Brighton Beach Autobody and Sale—opened between 2000 and 2006 and individually chose the area independent of the others' presence. Characteristics of the neighborhood—geographic centrality to a dispersed West African population, ease of access on a major arterial, a large, non-immigrant African–American population (users of hair-braiding services), and a prevalence of people with low incomes (the auto body shop's target clientele)—influenced their decisions to locate in the Rainier Valley. Once the cluster was underway, the next five establishments, Libra African Gifts and Fashion and Beauty Salon, Gambia International, Inc., Afrikando Afrikando, the mosque, and Baraka Information Technology, opening from 2008 to early 2010, chose the area primarily based on the presence of the existing West African businesses.[2]

The Spaces

I will focus on three buildings to demonstrate ways West Africans are using and adapting space to transculturally make places. Although all ten West African establishments and surrounding public space were studied, the three buildings chosen for this chapter exemplify the hybrid uses and identities of the users. These were the most loved spaces among interviewees and supported the most meaningful interaction among cultures, creating hybrid, shared spaces. The first building houses a Senegalese restaurant, a Gambian mini-mart, and a mosque; the second,

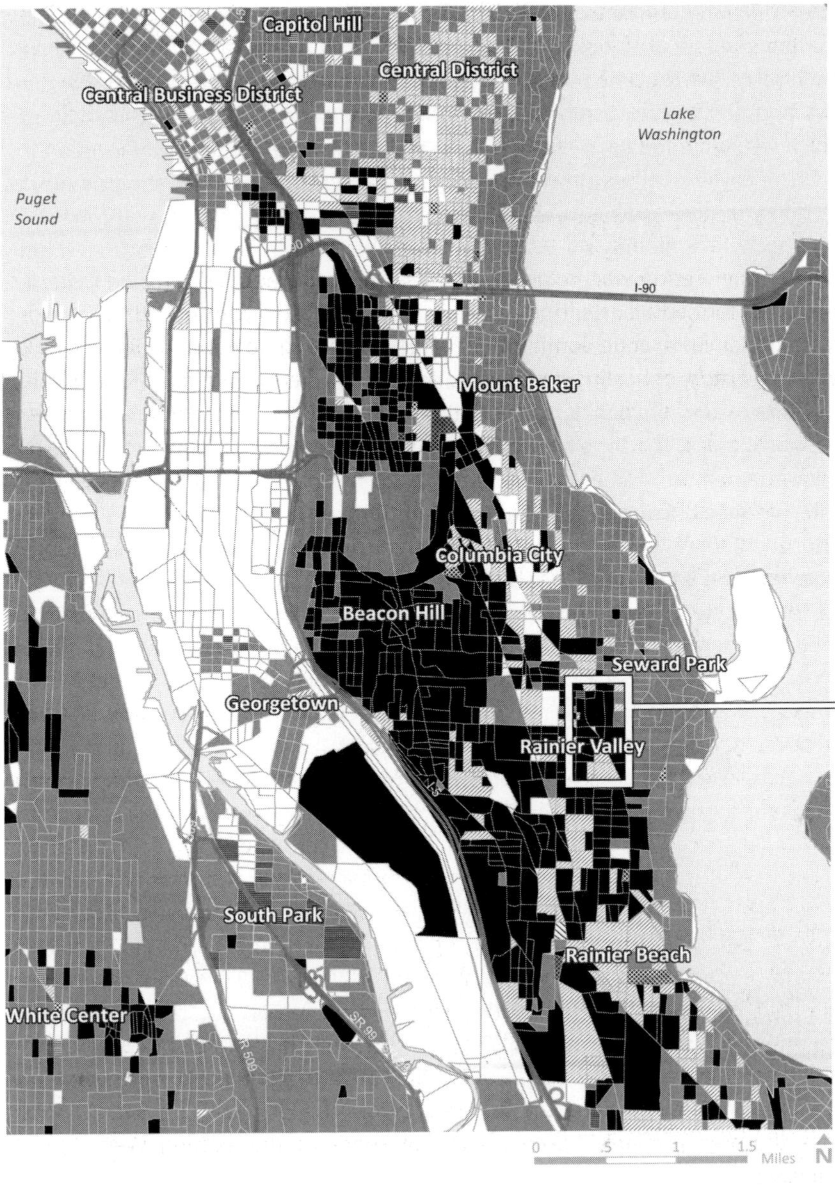

Figure 4.2 Map of race/ethnicity in the Rainier Valley: greatest proportion of block group. Map by Rachel Miller. (Data: US Census 2000)

a Guinean grocery store, a Guinean computer repair office, and a Filipino market; and the third, a Guinean auto body shop.

Afrikando Afrikando, Gambia International, Inc., and Masjid Ibadur Rahman

In 2008, Afrikando Afrikando, Gambia International, and a mosque, Masjid Ibadur Rahman, simultaneously opened in a building that housed an ice cream factory in the 1940s and most recently a nightclub. Mohammed Jacques Saar, a Senegalese Wolof, owns Afrikando Afrikando, a restaurant featuring West African cuisine. Gambia International is a mini-mart and Islamic book store owned by Gambian Malinkes. Gambia International also manages the mosque in the back of the building. The mini-mart and mosque relocated here after a post–9/11 federal Terror Task Force raid (which resulted in no convictions on terrorism charges) on their former location.

These three uses occupy distinct spaces and, even within their portions of the building, finer-grained uses occur in physically segregated areas (Figure 4.3). Afrikando Afrikando has a main dining area (used by infrequent patrons), a slightly separated semi-private dining room (usually used by frequent visitors), and a kitchen. The business owners built a wall to separate the restaurant from Gambia International. Gambia International displays Islamic books and DVDs, food, clothing, and gifts in the front and has a small gathering area in the back. The mosque has a prayer room and physically separate men's and women's entrances and washing (ablutions) areas. The shopkeepers constructed ablutions fountains for the mosque and formalized the women's entrance with paving. The three uses intertwine in the center of the building where restaurant patrons use the restrooms, mosque visitors perform ablutions, and regulars traverse between Gambia International, Afrikando Afrikando, and the mosque.

More people sit and hang out in Afrikando Afrikando on a regular basis than any other place in this study. Their visits often correlate with prayer times, and communal eating is important, particularly after the Friday afternoon prayer (*jumu'ah)*. The owners' and sometimes their friends' children play in the restaurant. Malinke, Bambara, Wolof, and French are commonly heard.

Jacques describes social and emotional refreshment as reasons for the popularity of Afrikando Afrikando's gathering place:

Figure 4.3 Activities in the Afrikando Afrikando, Gambia International, and mosque building. Photographs and diagram by Rachel Miller.

> So there's a place for them to refresh their heart, their *mémoire*, their everything. It's too much stress in this country. So African people, they hate to be lonely. Because we believe, loneliness don't bring nothing in the heart but disease. . . . So when you're around people, you talking, joking, everything you have before that day, you leave there. So then they have problem with their family, problem from the world, problem of whatever, as soon as they come here they forget about everything.

This gathering takes place in a room behind the main dining area. To an infrequent patron of the restaurant, the room appears to be a spillover seating area; it accommodates the path to the restrooms.

In an adjoining room in the back of Gambia International, people gather in a small space slightly separated from the retail area. A bench and a freezer box provide seating while tea is made on a hotplate. Regulars here seem to retreat the furthest from the outside world, presumably to find comfort in a place that is uniquely their own. Past experiences with the Terrorism Task Force and neighborhood complaints to the City about the mosque (prior to it receiving a land use permit), probably contribute to the desire to be hidden. Regulars walk between Afrikando Afrikando and Gambia International, especially to share tea, but most tend toward one gathering space or the other.

On one level of transcultural interaction, the frequent visitors are a group of West African immigrants who are diverse in country of origin, ethnicity, language, age, and duration in the USA. Baram Jeng, a young Gambian man, describes his fondness for meeting people from different places when he visits Afrikando Afrikando:

> Sometimes you meet a lot of people from other countries, too. Exchange ideas. I like to ask people every time I see them, "Where are you from?" Yeah, sometimes they're from Togo . . . Ghana, from here. And then, I always wanted—because I didn't have the opportunity when I was in Africa to go to other countries— . . . I like to know different cultures.

On another level of transcultural exchange, Jacques facilitates interaction among non-West Africans and promotes West African culture and businesses. He uses a pan-African identity and informal cross-cultural devices, such as business cards and media under the glass tabletops, and maps and photographs of Africa on the wall, to advertise, network, educate, and attract non-Africans to the restaurant. Jacques seeks a clientele beyond his West African supporters for the survival of his business. Thus, offering a unique and meaningful customer experience also benefits him.

The Gambia International mosque, Masjid Ibadur Rahman, furthers transcultural exchange in this building. Interviewees note that Islam encourages males to regularly attend the mosque closest to their home, so the *masjid* (neighborhood mosque) becomes an important geographical center for Muslim communities. In the culturally diverse Rainier Valley, this results in a very diverse congregation. Abdullai Bayo, the mosque's imam, observes, "Americans come, Somalian come, Gambian—any Muslim." Madou Traore, a Malian regular at Afrikando Afrikando, suggests that the diversity of attendees builds community by offering a place where "you meet everybody different. *L'Arabie Saoudite* [Saudi Arabia], Africa, Guinea, everywhere." Especially after the well-attended Friday prayer, a diverse group of Muslims enjoy fellowship and food together. On weekends, children attend an informal Koranic school at the mosque, adding to the layering of uses and interactions in the building. Additionally, because Muslim West Africans are dispersed throughout the Puget Sound, they also visit mosques in their neighborhoods and connect with those communities.

West African Market and Baraka Information Technology

The second study building (Figure 4.4) houses West African Market, one of a handful of West African groceries in the Puget Sound. Fifi Doumbia and Almamy Sylla (Guinean Malinkes) established West African Market in 2004 in one-half of a 1950s building. Baraka Information Technology opened in 2010 in a small office behind a Filipino grocery in the same building. Dividing a building into multiple and sometimes informal rental spaces is a common theme.

Of the places in this study, the Market attracts the widest range and number of West African immigrants by providing groceries, media, health and beauty supplies, international phone cards, and a wire transfer service. Due to Fifi and Sylla's persist-

Figure 4.4 Uses in the West African Market and Baraka Information Technology building. Photographs and diagram by Rachel Miller.

ence in reaching outside of their ethnic and national connections, customers can find home country products even if they do not share a common religion or local language with the owners. Thus, people from a broader range of backgrounds gather here than at the Afrikando building.

Although there is no clear space for gathering, many chance meetings occur. The store's popularity makes it a great place to advertise and hear about events. As Fifi notes, "Each time [the customers] have something as event . . . we are the first to be notified. And people always pass through here." Customers chat with Fifi or Sylla at the counter before making their purchases, or outside near their cars. Some people meet in the small office behind the counter, but this area is reserved for people who know Fifi and Sylla well. Fifi expresses hesitation about using this space for gathering. She feels that many of her customers treat her business as if it were in their home country, walking into the office without thinking about it being a private business area.

Fifi studies business in the USA and emulates what she sees as best practices. Thus, she is wary of creating an environment that encourages excessive gathering since she feels that is not common in US businesses. Fifi puts some chairs on the sidewalk for necessary waiting, but she does not want to encourage gathering. She wants all customers to feel welcome and recognize the store as a business place, not a social club. She says:

> We like people, we like gathering, but we don't want that to be continuous—like that will bring them a lot, and they will not find anywhere to go. Especially for those people gossiping, we don't want that. . . . Here is a commercial place, and some people, we are being frequented by people, all over. For instance you have some conflict issue, you know, with somebody else that you often meet here . . . you might think that really, that maybe that person is being closer to us. Or we, we Africans, we like talking. And we don't want that here.

She is worried that people will hear gossip at the store, and then say, "Oh, I heard that from Sylla's place," thus making some people feel less welcome or lose trust in Fifi and Sylla. She is also concerned that gathering at her store might suggest that she is close to some people and may alienate people who feel they are not in this "inner circle."

Fifi believes that transcultural interaction has improved her business and that other West African businesses are now learning from her and Sylla. Eager to look outside of her culture for better ways of doing things, she says, "You know, since the African culture is so deep, so why not, I will find out what I'm ignoring, I don't know and so on, find out what I really need again to improve." She attests that she has "learned from the Western world" that building a grocery business can be more than "a small job," as it is believed to be in West Africa. She has found that understanding multiple cultures aids her business:

> We're not seen as, only . . . from West Africa. Actually we like, all over. Because my husband deals with any business—you know—person from all over, he doesn't care. So actually that makes our force. . . . You have to be partners. In

> term of business, it doesn't have to be only people from . . . your area, I mean your community and so on. Because you have to be able to embrace all. . . . So what others have, because actually it's like . . . in French we say, '*la symbiose* [symbiosis].' . . . So where each individual has to take something profitable like they have, each person has to gain something. It's not like getting only on your side. You give, they give, and you know, each person has something to take. Which is really valued.

In addition to working with diverse business contacts, Fifi engages in symbiotic relationships in other ways, as well. She likes to show an authentic African culture in her American life: "I'm obliged to since I symbolize . . . African authenticity . . . so I have to show it, how we live back home." While she demonstrates West African culture in her store, she also exchanges favors with her neighbors. For example, the Filipino woman in the adjacent mini-mart regularly brings *halo-halo*, a Filipino dessert, to her. Fifi also allows her children to wander into the Filipino store, trusting that her building-mates will watch over them. This sharing of cultures and tasks seems commonplace in this neighborhood.

Like Jacques at Afrikando Afrikando, Fifi and Sylla use media to promote West African businesses and events. Business cards for numerous West African enterprises are found on their front counter. Their storefront window also acts as a bulletin board advertising community events.

The most recent business to open is Baraka Information Technology. Frank Magass (Guinean Malinke) rents a small office in the back of the West African Market building to provide computer repairs and rentals, website development, and business technology support. He would have preferred a large storefront where he could hold computer training classes but was financially limited to the small back room. Frank's space was likely intended to be the office for the storefront adjacent to West African Market, but the landlord discovered that it could be rented separately. This overlaying of an informal use with formally recognized businesses is one of the ways hybridity plays out in this community. Frank's office is also an emerging social gathering space. It gets social spillover from West African Market and offers a more private location to chat.

Like Jacques, Frank is reaching outside of the West African community to build his clientele. He advertised his business to nearby apartments and is in contact with a nearby Filipino church to "show them we are neighbors" and offer to share parking. In multiple ways, Frank is presenting a hybrid identity and use of space. Moving to Seattle to study information technology, using these skills to provide a service to a diverse clientele, and combining this with a social gathering space in an informal office demonstrates the essence of the hybridity existing in this neighborhood.

Brighton Beach Autobody and Sale

Edouard Souarez (Spanish–Guinean Malinke) opened Brighton Beach Autobody and Sale in 2006. This business is a major player in building a West African community in the neighborhood. The lobby is a gathering space for West African immigrants, particularly on Saturdays when Sabou Catering, run by a Guinean woman, sells take-out African cuisine out of her van (Figure 4.5). Her food attracts at least ten additional people to the shop. She hopes to open a restaurant near this cluster, but she has not yet found an affordable space. In the meantime, Sabou Catering and Brighton Beach Autobody are mutually supporting each other in a unique way. This informal use of a formally recognized business space is an appropriation of space based on friendship, food, and a desire to see each other's business succeed.

Brighton Beach Autobody, like West African Market, attracts visitors who may not be as interested in the more religious-based gatherings in the Afrikando building. Edouard's assistant, Jean Cisse (French–Guinean Malinke) says, "My grandmother's Muslim and my dad is Catholic. . . . We had to grow up with these two religions. Guinea is very good at walking with both religion; they don't have any major trouble." This openness attracts both Muslims and Christians, and a different ambience can be felt as this is the only place where I have seen alcohol being served (prohibited in Islam). Like Afrikando Afrikando, this is a regularly used gathering space. Even on days without the caterer, Jean's and Edouard's friends often gather in the lobby in the evenings.

Along the lines of Fifi's desire to learn from other cultures and teach the best of her culture, Charlie, another assistant to Edouard, is particularly proud of opening the first West African auto body shop in the Rainier Valley, and believes that other businesses can learn from them:

> It was a very good feeling [to be the first West African business in the area]. In the sense that, when you look at the area, most of the businesses around here are all Asians—Vietnamese, Chinese, you know. So when Edouard bought this place, it was a boost. You know, and then it created competition. It created competition in the sense that the way those people . . . build cars, it's different from the way Edouard does.

West Africans may gather in their lobby, but as a business, they serve more East Africans and other ethnic groups than West Africans. Likewise, they work with neighboring businesses in their industry, such as an Asian-owned auto glass company. Jean feels particularly connected to the Ethiopian owners of a nearby gas

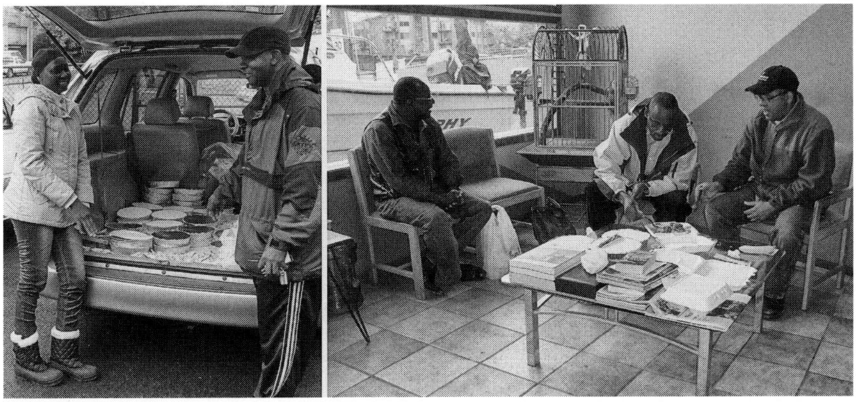

Figure 4.5 Sabou Catering selling West African cuisine outside of Brighton Beach Autobody (left), and social gathering around food in the lobby (right). Photographs by Rachel Miller.

station. The necessary business relationships in a culturally diverse neighborhood seem to contribute to shared meanings and hybrid identities.

Hybrid Identities Beyond the Spaces

As described above, people interact across ethnic and national boundaries in these businesses. However, these are not the only places important to their hybrid identities. Participants also favored workplaces, soccer fields, shopping malls, festival spaces, and dance clubs, which provide additional transcultural opportunities. Ibrahim Bah explains that when he arrived in Seattle fifteen years ago, "there was no West Africans [in Seattle], so most of the people I have as friends is Ethiopians, Eritreans, Somalis, Hondurans, Mexicans, whatever, everybody." Now that the West African community has grown, Ibrahim visits the businesses almost daily while maintaining strong relationships with non-West Africans. Sidikou Oumarou relates a similar story, describing his coworkers, non-West Africans, as his second family, "hang[ing] out there more than half the time, there with friends." Yet he also visits Brighton Beach Autobody nearly every day. All of the interviewees seem to seek a balance between reaching outside of their cultures for social relationships and finding comfort within their cultural group.

However, transcultural exchange can be a challenge, especially in regards to deep-seated feelings about one's tribal identity. One person confides, "Sometimes we don't get along, though, I tell you that, that's true. We don't get along actually because African people talk too much sometimes. They don't mind their business just like back home, so some people don't like that." Fifi notes, "The African culture, it's so deep and confusing. People are from all over, and it's still according to the clans, the tribes." Another discusses his experience with Argas, the Guinean Association:

> 'Cause when they come to Argas, this is just like family punching each other in different tribes, different types. . . . I would love to have Argas work, but it won't gonna work, cause the thing is, it's not about the organization. It's about the tribals that bring their crap up here, 'cause the Fulani community have their organization and that's working among them. But Argas is for everybody, as long you Guineans, or African, whoever, you're welcome there. But the Fulani community, just their community—among them probably, and they don't got no problems.

Conversely, other people spoke about Argas without mentioning such tension and are proud of its welcoming nature. People who spoke of contention are also active members of the community, appreciating and wanting the best for it. Although internal conflicts exist, the hybrid spaces and identities forming in this cluster give hope to places where new understandings and shared meanings can thrive.

Hybrid Places and Transcultural Exchange

The spaces discussed in this chapter support transcultural interactions. West African immigrants engage in regular cultural exchange with diverse populations for a number of reasons: home country products, particularly food, bring people together; tightly shared spaces with informal activities (e.g., children at the workplace, catered food, and social gathering) facilitate small independent businesses and cross-cultural exchange; the *masjid* attracts ethnically diverse Muslims; and the businesses are often dependent on other ethnic groups for their work.

In the three buildings discussed, food draws diverse West Africans and other customers to the area. West African Market's extensive grocery attracts the most people and from the greatest distances. Afrikando Afrikando's food brings West Africans and outsiders into the restaurant, establishing a platform for transcultural exchange. Sabou Catering sells food at the auto body shop, transforming the lobby into a festive gathering space. In each of these cases, West African identities are promoted and sustained, while encouraging further interaction.

A second theme is the multiplicity of uses in small spaces. Informal activities overlap with formally recognized businesses. Frank's micro-space in the back of the Filipino grocery is an informal rental space, but he benefits from the lower cost and spillover from West African Market. Likewise, the three Afrikando building tenants share rent, and the adjacencies of the spaces cause an intermingling of people who enter the building for different reasons. The informal relationship between the auto body shop and the food vendor provides a lucrative venue for Sabou Catering and attracts people to the shop. Informal social gathering and children's play occur in each of the establishments. This tight layering of multiple uses benefits the businesses, encourages social interaction, and efficiently meets the needs of the customers and business owners.

The mosque provides another way to share across cultures. Situated in a diverse neighborhood, it attracts Muslims from many different ethnicities and cultures. The ability to find fellowship there illustrates transcultural processes. In addition, the mosque has seen controversy, but perhaps through these transcultural spaces and Jacques's welcoming nature to "outsiders," the formalized presence of the mosque in the neighborhood can lead to greater cross-cultural understanding.

The diverse neighborhood setting also necessitates business relationships across cultural differences. Dependencies on each other require at least minimal transcultural understanding, and it appears that the shopkeepers are embracing relationships with their neighbors. Likewise, many of them seek non-West African clients, behooving them to reach toward mutual understandings.

Reflections on Varying Hybridities

Clearly, West African immigrants have created places meaningful to them and use their spaces in ways that support transcultural processes. Their spaces also reflect varying levels of transcultural interaction and hybrid identities. For example, the Afrikando Afrikando building is both insular and expansive. Regulars dominate the informal social scenes, and religious ties may limit potential visitors. Gambia International, with its hidden gathering space, is meant for insiders. Like one of the Milwaukee Hmong enclaves (Dearborn, 2008), immigrants are creating a place of retreat and comfort not intended for forging relationships with outside cultures. Although this may seem antithetical to hybridity, it is an important way of ensuring the continuation of the original culture. Immigrants may revisit this space as needed to refresh their connection to the culture. Conversely, these cliques are comprised of vastly diverse West Africans, especially at Afrikando Afrikando. In addition, Jacques seeks non-West African customers and uses his outgoing personality and media to spark conversations that lead to new understandings.

The West African Market and Baraka Information Technology spaces reflect Fifi's and Frank's negotiation of cultures. In contrast to the regular cliques in the Afrikando building, Fifi discourages social gathering at West African Market, emulating Western business practices. She believes in maintaining the ways of her ancestors, while simultaneously living in and understanding Western cultures, saying that "being able to do both is the most important, and the helpful thing." She and Frank use skills gained from US educations, intertwining these influences with West African ways, making businesses accessible and useful to the widest range of people possible.

The Brighton Beach Autobody proprietors were raised in interracial and intercultural families and have been in the USA for many years. Their space caters to general neighborhood residents, while simultaneously supporting West African social gathering. Jean is in an interracial marriage but also appreciates sharing time with people from a similar original culture. Shifting among languages and identities is commonplace in this space.

The multiple layers of uses combined with the transcultural nature of these spaces make them places of hybrid uses and hybrid identities. They serve as models of spaces that support healthy cultural interactions, which can ease immigrants' adaptation process to a new country and help with understanding new ways of doing things (Sonn, 2002; Manzo et al., 2008). As Leonie Sandercock (Sandercock & Attili, 2009, p. 222) argues:

> [T]here can also be a sense of belonging that comes from being associated with other cultures, gaining in strength and compassion from accommodation among and interrelations with others, and it is important to recognize and nurture those spaces of accommodation and intermingling.

This chapter observed what makes this place special to West African immigrants. They represent their original cultural identities in the built environment while reaching across cultures and adding on to their own. As a recently established and fragile community, I hope that this cluster remains a sustained and growing place of comfort and meaningful interaction for West African immigrants and others alike.

Notes

1 The 1996 Illegal Immigration Reform and Immigrant Responsibility Act (IIRIRA) and Anti-Terrorism and Effective Death Penalty Act (AEDPA).

2 The cluster continues to change and grow. Afrikando Afrikando closed in late 2010 (after this study's period of data collection) due to financial difficulties, but reopened within an existing Somali restaurant in 2012. Two Senegalese businesses, a clothing store and a restaurant, also opened recently.

References

AlSayyad, N. (2001). *Hybrid Urbanism: On the identity discourse and the built environment*. Westport, CT: Greenwood Press.

Arthur, J. A. (2000). *Invisible Sojourners: African immigrant diaspora in the United States*. Westport, CT: Greenwood Press.

Bhabha, H. (1994). *The Location of Culture*. London: Routledge.

Byram, M. (2003). Being bicultural. In G. Alred, M. Byram, & M. Fleming (Eds), *Intercultural Experience and Education.* Clevedon: Multilingual Matters.

Community Land Use and Economics (CLUE) Group, LLC. (2009, December). *Retail Development Strategy for Rainier Valley.* Prepared for the City of Seattle's Office of Economic Development.

Davis, T. (1997). The Miracle Mile revisited: Recycling, renovation, and simulation along the commercial strip. *Perspectives in Vernacular Architecture, 7, Exploring Everyday Landscapes,* 93–114.

De Korne, H. D., Byram, M., & Fleming, M. (2007). Familiarising the stranger: Immigrant perceptions of cross-cultural interaction and bicultural identity. *Journal of Multilingual and Multicultural Development,* 28(4), 290–307.

Dearborn, L. (2008). Socio-spatial patterns of acculturation: Examining Hmong habitation in Milwaukee's north-side neighborhoods. *Buildings & Landscapes: Journal of the Vernacular Architecture Forum*, 15, 58–77.

Garde, A. M. (1999). Marginal spaces in the urban landscape: Regulated margins or incidental open spaces? *Journal of Planning Education and Research,* 18, 200–210.

Hadjiyanni, T. (2007). Bounded choices: Somali women constructing difference in Minnesota housing. *Journal of Interior Design*, 32(2), 13–27.

Hall, S. (2003). Cultural identity and diaspora. In J. Evans Braziel & A. Mannur (Eds), *Theorizing Diaspora: A reader* (pp. 222–237). Malden, MA: Blackwell.

Kim, Y. (1988). *Communication and Cross-Cultural Adaptation*. Clevedon: Multilingual Matters.

Manzo, L., Kleit, R. G., & Couch, D. (2008). "Moving three times is like having your house on fire once": The experience of place and impending displacement among public housing residents. *Urban Studies*, 45(9), 1855–1878.

Margai, F. M. (2009). Acculturation and the health of black immigrants in the United States. In Isadore Okpewho and Nkiru Nzegwu (Eds), *The New African Diaspora* (pp. 164–182). Bloomington: Indiana University Press.

Sandercock, L. (2003). *Cosmopolis II: Mongrel cities of the 21st century*. New York: Continuum.

Sandercock, L. & Attili, G. (2009). *Where Strangers Become Neighbors: Integrating immigrants in Vancouver, Canada*. Urban and Landscape Perspectives (Vol. 4). Dordrecht: Springer.

Sonn, C. C. (2002). Immigrant adaptation: Understanding the process through sense of community. In A. T. Fisher, C. C. Sonn, & B. J. Bishop (Eds), *Psychological Sense of Community: Research, applications, and implications* (pp. 205–222). New York: Kluwer Academic/Plenum Publishers.

Sussman, N. (2002). Testing the cultural identity model of the cultural transition cycle: Sojourner's return home. *International Journal of Intercultural Relations,* 26(4), 391–408.

Takyi, B. K. (2009). Africans abroad: Comparative perspectives on America's postcolonial West Africans. In I. Okpewho & N. Nzegwu (Eds), *The New African Diaspora* (pp. 236–254). Bloomington: Indiana University Press.

Winnicott, Donald W. (1971). *Playing and Reality*. New York: Basic Books.

CHAPTER 5

The Sin Oh Dan Street Lion Dance Competition

A Temporary Space for Cross-cultural Understanding

Jayde Lin Roberts

In Myanmar, where the ruling military junta, the State Peace and Development Council (SPDC), oppresses the various ethnic groups living in the country and routinely uses difference as a pretext for repressing community organizations, cross-cultural understanding is a rare occurrence.[1] Decades of political oppression coupled with a history of postcolonial struggle to formulate a national identity has left the populace guarded against unfamiliar people and suspicious of calls to collective action. Before President Thein Sein was elected in 2011, the Unlawful Association Act inherited from British colonialists defined all unauthorized assemblies of five or more people in public outdoor spaces as illegal. Although this law was inconsistently enforced, the threat of jail time in notoriously inhumane prisons discouraged everyday residents from gathering together in any potentially conspicuous manner. Even traditional Burmese festivals such as *Thadingyut* (Festival of Lights) and *Thingyan* (Burmese New Year) were censored, stripped of elements that might be critical of the government. Rather than the customary opportunities to express one's religious beliefs and social/political critique, Burmese holidays had become largely commercialized spectacles that lacked the depth of meaning and community experienced before 1988.

In this political and social context, the "Upper Sin Oh Dan [Street] Family Traditional Chinese Lion and Dragon Dance Exhibition and Competition"[2] (Sin Oh Dan Lion Dance Competition from this point forward) stands out as an unusual cultural performance in a public space that hints at tentative steps towards cross-cultural interaction and understanding. However, these initial forays are best understood as *tactical* maneuvers (Certeau, 1988) to express Chineseness, flexibly and opportunistically performed in a downtown Yangon street with no explicit agenda other than to celebrate Chinese New Year. The organizers of and participants in the competition might harbor desires for a freer society in which different ethnic/cultural identities can be expressed openly, but no one voices those desires publicly. In some cases, individuals were probably not fully conscious of those desires. In others, individuals had learned to self-police, curtailing their speech, especially to a foreign researcher, to avoid drawing unwanted government attention.

And yet, despite the absence of defined goals to promote cross-cultural understanding, the Sin Oh Dan Lion Dance Competition has created a temporary communal space that gathers together people of various ethnicities and encourages young lion dancers to train together, building friendships regardless of cultural heritage. This lion dance competition is an annual event that temporarily transforms a public street into an enlarged family living room, and in doing so enfolds many different peoples (in ethnicity and class) in an updated traditional Chinese celebration that engenders cross-cultural understanding. This analysis is based on two years of spatial ethnographic fieldwork in Yangon (Rangoon), Myanmar (Burma) in which I observed three competitions during the Chinese New Years of 2007, 2008, and 2009.

The larger significance of the lion dance competition that began in 2006 must be understood within the social and political context of Yangon and the nation as a whole. I will first situate the competition in the history of the Chinese in Burma (the Sino-Burmese) and present the political, economic, and social factors that have made the lion dance competition possible. Next, I describe and analyze the lion dance competition and the training that the lion dancers undertake. Lastly, I discuss the efficacy of the competition in promoting cross-cultural understanding and the potential of Yangon becoming a transcultural city.

The Historical Place of the Sino-Burmese in Yangon

Most people of Chinese heritage in Yangon are descendants of migrants who arrived in Burma during the second half of the colonial era (1880s–1948). Their forefathers were mostly destitute peasants who left Fujian and Guangdong (Canton) Provinces because overpopulation, lack of arable land, and internecine warfare drove them to seek a better life outside of China. Rangoon was often the port of last resort after sojourners had tried their luck in other Southeast Asian cities. However, once in Rangoon, Fujianese (Hokkien) and Cantonese migrants found a niche for themselves in the international port and became an established, if still small, segment of the population. Under colonial rule, they lived in a plural society (Furnivall, 1948) where British and Europeans dominated as the ruling elite while Indian merchants, clerks, and coolies composed over 50 percent of the urban population (Pearn, 1939). The Chinese, although the second largest immigrant group, have only constituted 3–4 percent of the total population throughout their history in Burma. Nonetheless, they enjoyed more access to business opportunities than their indigenous fellow residents. Many worked as skilled tradesmen while others became small- and medium-scale merchants. Except for the Kachin and Karen peoples who were favored by the British as martial races suitable for soldiering, other indigenous groups such as the Burman and Mon were numerically significant but excluded from government posts and lucrative commerce (Callahan, 2003).

In their primary pursuit of profit for the empire, the British created a city where commerce became the reigning ethos and the marketplace became a relatively free if unequal space in which people of different ethnicities interacted and came to know each other. Cross-cultural understanding gradually arose by default, not by design, through everyday transactions in open-air markets and through business negotiations. The imperfect foundation laid down during the colonial era served to support further interactions between different ethnic and immigrant populations under independent Burma. However, the mutual understanding established has generally remained at a one-on-one level between close friends. Chinese and Indian settlers have made a home for themselves in Burma but they are not considered to be completely Burmese.

According to contemporary Sino-Burmese, their fortunes in Burma have waxed and waned, at times enjoying great freedom and financial success and at other times suffering extensive oppression. These changes have paralleled the general political, economic, and social shifts in Burmese history. They remember the 1950s as a golden era in which the Sino-Burmese prospered and could live as they pleased, the 1960s and 1970s as a time of collective suffering, and the post-1988 period as one of paradoxical limitations and opportunities. An undercurrent of subtle tension has flowed through the history of the Chinese in Burma and reached an unexpectedly violent climax in 1967. On June 26, 1967, an anti-Chinese riot broke out in Yangon that killed hundreds of people and compelled Sino-Burmese Yangonites to re-evaluate their status in Burmese society (Fan, 2006). As a result, since the late 1960s the Chinese in Yangon have transformed themselves into Sino-Burmese, not simply Chinese, and have actively sought to behave more like their Burmese neighbors.

Waxing and Waning of Chinese New Year Celebrations

In the collective memory of contemporary Sino-Burmese, Chinese New Year has always been celebrated in one form or another throughout their history, but there were periods in which the celebrations were more subdued. Chinese New Year is fundamentally a family affair. The Sino-Burmese, like other Chinese overseas removed from their extended families, organize communal celebrations that in effect gather all Chinese together as one large family. There are no clear records for events before World War II, but since Burmese independence in 1948, Chinese New Year celebrations have waxed and waned in response to the political and economic circumstances.

During the golden era of the 1950s, Chinese New Year was celebrated boldly in public. Chinese Yangonites organized a Culture and Entertainment Carnival (文娱广场) that lasted seven nights and attracted 100,000 revelers including some Burmese and other residents of Yangon. This was during the Parliamentary Period of a newly independent Burma where the inexperienced and overextended government could do little beyond manage the numerous political, military, and economic crises facing the country. Therefore, the actions and institutions of foreign residents were generally unregulated and the Chinese community focused on promoting their own culture and community. This public and large-scale celebration continued for thirteen years from 1951 to 1964 and was only discontinued when successively more restrictive nationalization campaigns terminated all foreign institutions and events. The Socialist government that usurped control in 1962 through a coup d'état was intent on realizing the goal of creating a Burma for the Burmese. It tried to excise all foreign influence within the country and closed its national borders. From 1965 to 1981, Chinese New Year festivities retreated into private homes and surname association halls where those defined as family through bloodline or surname gathered together to welcome the New Year.

Although the Socialist period did not end until 1988, improved relations between China and Burma from 1978 onward emboldened the Sino-Burmese to once again express their heritage and culture more publicly. From 1981 to 1988, the Sino-Burmese Chamber of Commerce sent out a Chinese New Year Greetings Float that was accompanied by lion dance teams (forty to fifty performers in total). They performed throughout Chinatown for nine nights with droves of Sino-Burmese following the traveling show. The Socialist period was one of suppression but people in Burma could still gather in large numbers without political consequences. The organizers had to apply for permission from the township office and clear the streets in order to make room for the passage of the large float, but the government permitted this type of mass celebration on the streets (Yang, 2008).

However, this re-emergence of Chineseness into public space was short-lived. Beginning in March 1988, a series of events and protests culminated in the mass pro-democracy demonstration of August 8, 1988, that was brutally crushed by the military. Soon after, the Socialist government was overthrown and formal military dictatorship was decreed on September 18, 1988. Martial law was imposed until 1992 and the Unlawful Association Act was more strictly enforced, effectively terminating all outdoor festivals. From 1988 to 2005, Chinese New Year parties reverted back to the indoor banquets of 1968 to 1981.

Despite this closing down of the public sphere, the time under martial law was also a period of opening towards international trade. The military dictatorship wanted to exert absolute control over the entire country but it also sought to rebuild the economy by opening Burma's borders to international trade and courted foreign direct investments. In the 1990s, foreign investors were all welcome but Chinese overseas from Singapore, Hong Kong, and Taiwan came in the largest numbers.[3] They brought with them not only capital but an affirmation of Chinese heritage and their own renditions of Chinese tradition. At the same time, foreign influence through mass media streamed into a formerly closed country and exposed all of the residents of Myanmar to popular culture from East Asia, Southeast Asia, and America. The liberalized economic environment revitalized the Sino-Burmese community, gradually enabling Sino-Burmese people to rediscover their cultural heritage and take Chinese New Year celebrations back out into the streets.

The Upper Sin Oh Dan Family Lion and Dragon Dance Exhibition and Competition

In 2006, several men who live on the upper block of Sin Oh Dan Street organized the first-ever lion dance competition. Originally, these men were concerned about the complications they would inevitably encounter at the neighborhood, district, and municipal government offices in order to obtain the proper permits. Myanmar tied with Somalia as the most corrupt state in 2007, and bribery in government dealings is a matter of course.[4] Even for simple matters such as repairing phone lines or paying neighborhood fees, extra money is necessary to facilitate transactions. Therefore, some in the organizing committee were initially pessimistic about their ability to realize the competition and sought out the wealthiest man on the block to serve as their chairman and liaison with the government. As explained by two members of the organizing committee, some things require permission from the local neighborhood government while some require a permit from the city, but it is often unclear which is necessary when. A common solution to this problem is to find someone who has some money and clout to approach the various offices, and then hope for the best. In the case of the lion dance competition, the richest man on the block is a part of a wealthy family that operates multiple businesses. Through their connections, the organizing committee was able to obtain the proper permits as well as support from the local fire department. But even after acquiring the necessary permits, they were still concerned about too much exposure.

The actions of the government can be unpredictable, so the organizers continually tread a fine line between attracting enough participants and audience to create a festive atmosphere and staying below the government's radar. This balancing act is undertaken by all residents of Myanmar, regardless of ethnicity, who wish to do something different or express themselves publicly. The actions of the Sin Oh Dan Lion Dance Competition organizers are notable because they have annually tested the boundaries of what might be deemed acceptable, or, more appropriately, what will be beneath the notice of the government. Each year, through flexible tactics that follow the laws to the extent necessary while maneuvering around them to achieve their goals, they transform a major street in Chinatown into an open-air stage and living room.

In the hierarchy of streets in downtown Yangon, Sin Oh Dan is a second-level street, half the width of the primary north–south and east–west thoroughfares. It is an important street in Chinatown known for the Hokkien Guanyin Temple that marks the southernmost end of the street, and the numerous surname and martial arts associations along its length. Through years of daily interaction, the men of Sin Oh Dan conceive of themselves as a family that shares a street, and together they wanted to revitalize Chinese New Year. They lamented the fact that Chinatown had grown quiet during Chinese New Year and wished that their fellow Sino-Burmese would come out onto the streets again to celebrate. They named their event "The Upper Sin Oh Dan Family Lion and Dragon Dance Exhibition and Competition" to literally reflect their sense of community and set the goal of re-enlivening Chinese New Year by showcasing a cultural practice they all enjoyed in their youth. Most of these men practiced martial arts and performed lion dances when they were young. They were also inspired by the lion dance competitions held in Singapore and Malaysia and adopted that overseas Chinese practice to revitalize Chinese New Year in Yangon.

Each year since 2006, the southern entrance of the upper block of Sin Oh Dan is transformed into a festive gateway with a red banner and lanterns that announce the competition many days before the event (Figure 5.1). It is also marked by a large

Figure 5.1 Gateway for the Sin Oh Dan Lion Dance Competition. Photograph by Jayde Lin Roberts

screen onto which the dances are projected for those who cannot squeeze in close enough to watch the actual performances. Around six o'clock in the evening, to begin the nightly celebrations, one or two dragon dance teams perform for the audience and receive red envelopes in return. After these opening acts, the competing lion dance teams arrive one by one, each with a customized stage composed of a series of three- to six-foot tall posts (Figures 5.2 and 5.3). Each team must set up their stage before they perform and the competition usually continues until midnight.

In 2007, 2008, and 2009, approximately 700 to 1,000 people gathered on Sin Oh Dan to watch the competition. They filled the entire width of the 50-foot street for two-thirds of the length of the upper block. The audience was packed so tightly into the street that moving in or out of the performance area required great determination and the willingness to be skin-on-skin against other people. The organizers became concerned that the audience might become thirsty and hungry during this long period of time and took it upon themselves to provide refreshments. Treating the audience as their guests, they bought water, juice boxes, and sweet rolls to distribute to the spectators on every night of the competition. This concern for everyone present transformed the street into the organizers' living room and transfigured the dragon and lion dances from impersonal spectacles into personal invitations to jointly celebrate Chinese New Year. They could have used the competition as an opportunity to make money or raise funds for the competition itself but chose to treat all of the spectators as their guests. When asked about the food and drink provided, most of the organizers said that they envisioned the competition

Figure 5.2 Lion dance team performing on their elevated stage as a large audience watches. Photograph by Jayde Lin Roberts.

Figure 5.3 Lion dance acrobatics. Photograph by Jayde Lin Roberts.

as a service to other Sino-Burmese and as an opportunity to share their culture. Furthermore, as the Sin Oh Dan Upper Block family, they were in essence welcoming all of these people into their family space.

Since 2006, the Sin Oh Dan Lion Dance Competition has taken place each year and grown in scale. In 2006, four teams competed over two nights. In 2009, nine teams competed over four nights and part of the competition was featured on national television. The organizers were initially concerned about such broad exposure but ultimately allowed Myawaddy Television to film during one night only of the competition. Such visibility in Myanmar was potentially problematic but the organizers decided that this sharing of traditional Chinese culture would not be perceived as a threat.

Lion Dancing—A Cohesive Force Among the Marginalized

In contemporary Yangon, lion dancing has become an acrobatic sport pursued by marginalized young men of various ethnicities (Figure 5.4). In 2007, 2008, and 2009, all of the team members who competed were unemployed or worked sporadically and viewed Chinese New Year as an opportunity to earn some much-needed, if insufficient, income. The prize money for the first, second, and third place teams in the Sin Oh Dan competition in 2009 was 800,000, 600,000, and 300,000 kyat (approximately US$800, $600, and $300) respectively, which served as a major incentive for the formation of new teams. In addition, each business or family that

Figure 5.4 Ease and camaraderie between the lion dancers. Photograph by Jayde Lin Roberts.

invites a team to perform provides a red envelope averaging 10,000 to a few hundred thousand kyat (approximately ten to a few hundred US dollars) depending on the reputation and performance of the team. The total income for the traditional performance period of eight to nine days can be significant, sometimes thousands of dollars. However the official sponsor[5] of the team usually takes a large cut, leaving little to the lion dancers. The two main dancers and two understudies of a well-known team can usually earn a few hundred dollars during the New Year season but often have to pay their own medical expenses. Falls from the three- to six-foot-tall posts that compose the stage are common and can lead to serious injuries. One of the best lion dancers in 2008 and 2009 severely injured his knees and lower back and had to pay for all but his first visit to the doctor and for all his medication. He was so disappointed after that experience that between the Chinese New Year of 2008 and 2009, he found an alternative sponsor for himself, his lion dance partner, and their understudies.

These lion dancers do not fit the bill of a stereotypical Sino-Burmese person, perceived to be money-driven and often rich. In fact, they seem to be at a loss as to how to improve their finances and devote their time to perfecting an athletic art form that only brings benefits around Chinese New Year. Among the various teams, there is a mixture of Sino-Burmese (including those whose forefathers are all Chinese and those who are of mixed Chinese and Burmese descent), Indo-Burmese, and

Burmese members who are all struggling financially. Although the lion dancers interviewed did not explain their situations quite in these terms, their individual poverty combined with the depressed national economy gives them very few options. This makes the risky sport of lion dancing an attractive if impractical long-term pursuit as they have little to lose.

Each year, a few months before Chinese New Year, lion dancers begin to organize themselves and reinitiate the teams they performed with previously or establish new ones. Lion dance teams are temporary entities with a life span of about two to three months and there is obvious alteration from year to year. Lion dancers pull together the best teams they can, finding new teammates, new coaches, and/or new sponsors. Although each dancer is loyal to his teammates, there appears to be a sense of camaraderie that reaches across team boundaries, tying all dancers together as one large lion dancing community. Ko Myo, captain of the 2009 Red Ruby Kung Fu Club, explained that once they started practicing the new form of lion dance from Malaysia and Singapore, everyone, regardless of membership in specific teams, became close friends. "We all know each other and help each other. I have taught many people in many teams how to do the lion dance. And we have all belonged to different teams. For example, I used to be in Dragon Head but now I'm in Red Ruby."[6] These young men of various ethnicities do whatever it takes within severe economic and social limitations, using flexible tactics to practice an athletic art form that they love, and thereby to briefly exercise some control in their lives.

This new form of lion dancing on posts arrived in Yangon via videodisks along with other cultural influences from Southeast Asia. Prior to its introduction around 2002, lion dancing was performed on the ground and was the exclusive purview of martial arts schools and secret brotherhoods. The older form was used to showcase the martial skills of the men in these organizations, and should two opposing teams meet on the street, bloody fights ensued. In contemporary Yangon, lion dancing has become an art form that brings people together. The lion dancers must rely on their ingenuity to create new routines and have little guidance other than the videos from Malaysia and Singapore. Their common struggle and long hours of training have fostered strong bonds between the young men. Ko Japan, a former lion dancer who now serves as a judge in lion dance competitions, commented, "The lion dancers are all close friends. They hang out together even when they are not training and flow back and forth between the teams. No one fights anymore. Instead of forming gangs, these youth are forming lion dance teams."[7] This transformation of lion dancing from a practice to assert dominance to a celebratory cultural sport is opening up a space for athletic young men to learn about Chinese culture through their training and for audiences in Yangon to become more familiar with Chinese tradition.

International Trade as a Facilitator for Interstitial Tactics

As noted above, the opening of the Myanmar economy ushered in foreign influences that significantly altered the cultural landscape in Yangon and Burma as a whole. Under the insular policies of the socialist government (1962–1988), Burma was effectively closed off from the world. After 1988, as foreign media flowed in and Burmese people traveled abroad, residents of Myanmar realized the extent of their former isolation and, in their words, lack of modernity. All of the Burmese people encountered said they are "*hkit ma mi bu* (not caught up with the times)" and yearned to catch up with the world. They look to places such as Singapore as modern cities to emulate and this perception makes the lion dancing from Singapore attractive, not only as spectacular acrobatics but also as something more advanced. In addition, television series from Hong Kong and Taiwan that featured dazzling martial arts were broadcast weekly in the mid 1990s; this portrayal of Chinese culture has created a positive link in the minds of the Burmese between Chinese people and Kung Fu. These two factors—the appreciation of Chinese Kung Fu and the admiration of Singaporean culture—have made lion dancing on posts an alluring sport for Sino-Burmese and other Burmese.

Since the creation of the Sin Oh Dan Lion Dance Competition in 2006, lion dancing has grown in popularity and new teams have been formed each year. According to the competition organizers, none of this would have been possible if the SPDC had not opened up the national economy, setting in motion a series of developments that have enabled the men of Sin Oh Dan to sponsor an annual competition. All of the men in the organizing committee are living comfortably and some have become wealthy in the post-1988 economy, providing them with the financial ease and leisure to pursue non-essential activities such as the lion dance competition. Each year, members of the organizing committee donate money to finance the competition. They must pay for numerous items including: two new lion dance costumes (approximately US$500 each), banners and decorations, wiring to light the competition, refreshments for the audience, red envelopes for all of the dragon and lion dance teams, red envelopes for various local officials, trophies, and prize money for the three winning teams. None of the organizers were willing to divulge the exact amount necessary to realize the event but one organizer indicated the cost was about US$10,000 per year.[8] This is a significant amount of money in a country where the GDP per capita is about US$300.[9]

Although pro-sanction activists define all trade with Myanmar as an unethical practice that only serves to perpetuate the rule of the military junta, international commerce in Myanmar operates at multiple levels and many of the small- and medium-scale businesses are untethered from the generals. These smaller enterprises have been operated with much flexibility, and their successes through tactical maneuvers have opened up unexpected spaces for self-expression. The opening of the Myanmar economy has created some wiggle room for the residents of Yangon and emboldened the Sino-Burmese to express themselves more openly. While the merits of mini skirts and rebellion against Burmese tradition can be debated, the possibility of gathering in an outdoor public space to celebrate the culture of a minority population must be seen as a step in the right direction. This could be interpreted as a breaking down of heteronormativity into more diverse social arrangements as temporarily enabled by capital trade (Deleuze et al., 2009). This decoding of Burmese social norms is facilitating new interactions that could serve to foster inter-ethnic understanding. On each of the four evenings of the lion dance competition, hundreds of Sino-Burmese, Indo-Burmese, and Burmese Yangonites crowded into Sin Oh Dan Street and remained for four to five hours to watch the dragon and lion dancing. They probably did not understand the various gestures of the lions and dragons or their cultural significance but they all enjoyed the performance and contributed to the celebration of Chinese New Year through their joyful appreciation. At a minimum, the 700 to 1,000 audience members on each of the four nights have helped to make Chinese New Year more festive, as desired by the competition organizers.

Conclusion

Viewed within the historical and political context of Myanmar, the Sin Oh Dan Lion Dance Competition is a space of appearances where people can be seen and heard by others (Arendt, 1968). The risk that the organizers and lion dancers take in exposing themselves on a public street enables the repeated formation of an imperfect public realm where Yangonites can gather as fellow residents but where politics is rarely discussed. Since 2006, their daring has created temporary gatherings that bring many different Yangonites together in one place. Whether this brief four-day congregation around February of each year will engender deeper understanding between the different ethnicities in Yangon remains to be seen. Cultural events such as the celebration of Chinese New Year are fairly innocuous and demand minimal participation or self-exposure from the audience. However, as stated at the beginning of this chapter, until the end of 2011, public assembly in Myanmar was potentially dangerous and everyone I encountered considered their public actions carefully. At present, this temporary four-day space of appearances has not accumulated enough strength or safety to foster political action. However, it is slowly nurturing an "acting in concert" through which a public shared world is annually recreated and enfolds increasing numbers of participants. In the six years since 2006, the pool of active participants has grown through the number of organizers and volunteers and the number of lion dance teams. How this gathering of people supports the current developments towards political action and democracy also remains to be seen.

In many ways, this street turned stage and living room is like the marketplace in colonial Burma where everyday residents interacted and came to understand each other incidentally, not intentionally. It is a temporary space of personal interactions that can build trust between individuals and perhaps foster transcultural fellowship but does so without explicitly challenging the government. In a politically oppressed society where the populace is generally distrustful, cross-cultural understanding can only progress incrementally through small initiatives. As such, even under these unfavorable circumstances, the risks taken by individuals or small groups of people to express themselves publicly can catalyze change. The annual competition has encouraged many youth to try lion dancing and inspired a group of Sino-Burmese to produce and sell a set of videos about traditional Chinese lion and dragon dancing. These videos are intended for a Burmese audience. They explain the art and history of lion and dragon dancing in Burmese, not Chinese, and are sold in many shops throughout Yangon, not just in Chinatown. This is a very small step towards cross-cultural understanding but it is nonetheless a step forward. As Aung San Suu Kyi, now a member of parliament, continues to address ethnic minorities in her pursuit of a democratic Burma, cross-cultural understanding between the numerous ethnic groups living within Burma will be absolutely essential.

Notes

1 The names of Burma and Myanmar are highly contested (Houtman, 1999). See Mary Callahan's explanation in her book *Making Enemies: War and State Building in Burma* (2003, p. xvi) for an excellent explanation. I will employ these names in their chronological context—Burma for events before 1990, or in both periods; Myanmar for events during and after 1990. After forcefully subduing the August 8, 1988, popular uprising, military leaders established a new government, the State Law and Order Restoration Council (SLORC). This government was renamed as SPDC in 1997.

2 This long name is a direct translation of the Burmese title given by the organizers. I have included the name in full because the explicit inclusion of the word "family" is significant and will be discussed later. It should also be noted that the competition only included the lion dances.

3 Interviews with Sino-Burmese merchants and officers in the Myanmar Chinese Chamber of Commerce (2007–2009). The Myanmar government publishes some data regarding foreign investments in the country but the reliability of these figures is under question. Furthermore, it appears that overseas Chinese businessmen often found ways to sidestep official regulations by establishing businesses under the names of local Sino-Burmese. Those ventures would not be captured in official data.

4 Economist Intelligence Unit (EIU) Country Report—Myanmar (Sept. 2011), http://www.eiu.com.offcampus.lib.washington.edu/report_dl.asp?issue_id=1848397569&mode=pdf (accessed August 15, 2011).

5 All of the teams that competed had a sponsor. As poor and unemployed men who want to do the lion dance, the dancers must find someone who has the income and connections to (1) purchase the necessary lion dance costumes, (2) pay for the manufacturing of the steel performance stage, (3) set up a practice space, (4) pay for the printing and distribution of announcements to advertise their availability to perform, (5) hire the one or two trucks necessary to transport the stage and the dancers.

6 Interview on January 18, 2009.

7 Interview on January 24, 2009.

8 Conversation with Dai (his surname in Chinese but few people know him by his Chinese name) on February 16, 2008.

9 Economist Intelligence Unit, Country Report—Myanmar, http://www.eiu.com.offcampus.lib.washington.edu/index.asp?layout=displayIssueArticle&issue_id=378324022&article_id=798324064 (retrieved August 1, 2011).

References

Arendt, H. (1968). *Between past and future; eight exercises in political thought* (Enl. ed.). New York: Viking Press.

Callahan, M. P. (2003). *Making enemies: war and state building in Burma*. Ithaca, NY: Cornell University Press.

Certeau, M. (1988). *The practice of everyday life* (S. Rendall, Trans.). Berkeley: University of California Press.

Deleuze, G., Guattari, F., & Foucault, M. (2009). *Anti-Oedipus: capitalism and schizophrenia*. New York: Penguin.

Fan, H.-W. (2006). 1967 年緬甸 (6, 26) 排華事件與緬華社會研究 (Analysis of the June 26, 1967 Anti-Chinese Riots and Myanmar Chinese Society) *台灣東 南亞學刊 (Taiwan Southeast Asia Studies Journal),* 3(2), 47–72.
Furnivall, J. S. (1948). *Colonial policy and practice: A comparative study of Burma and Netherlands India* [microform]. Cambridge: Cambridge University.
Houtman, G. (1999). *Mental culture in Burmese crisis politics: Aung San Suu Kyi and the National League for Democracy*. Tokyo: Institute for the Study of Languages and Cultures of Asia and Africa, Tokyo University of Foreign Studies.
Pearn, B. R. (1939). *History of Rangoon*. Rangoon: American Baptist Mission Press.
Yang, M. Q. (2008). Sino-Burmese New Year's Greetings Group. Retrieved from: http://mhhzhh.blog.163.com/blog/static/521600962008111184050129/

PART II

PLACEMAKING IN THE SPACE OF FLOWS

CHAPTER 6

The Korean Diaspora in Philippine Cities

Amalgamation or Invasion?

José Edgardo Abaya Gomez, Jr.

In major Philippine urban centers like metropolitan Manila, Baguio City, Angeles, and Davao, it is not uncommon to hear the locals exclaim how "the Koreans are taking over!" with grim-faced guardedness or cheerful resignation. For over a decade now, a steady stream of students and small entrepreneurs has been flowing in from the Republic of Korea (South Korea, hereafter), and has introduced visible and not always well-integrated changes in the city fabric. If only because of the proliferation of commercial signs written in Hangul popping up in the most unlikely street corners and lower-end neighborhoods, signs which are of course unintelligible to the vast majority of Filipinos, the Korean population is making its presence abundantly palpable. This is a significant transnational phenomenon, as a more disciplined culture of Northeast Asians attempts to plant itself in a historically more relaxed Southeast Asian culture influenced strongly by Spanish/Latino-American and North American attitudes. Urban changes are noticeably manifested in the form of corner groceries, evangelical churches, travel agencies, and English language schools catering to Koreans. Even odder to the host culture is the fact that Koreans are actually staying for years at a time—despite numerous perceived inefficiencies and travails presented by the developing cities.

For instance, due to the recent hostage fiasco that involved the fatal, botched rescue of eight Hong Kong tourists in August 2010, Metro Manila again became known for the risks it poses to foreigners. Yet Koreans are to be found setting up shop far from the luxurious gated communities used by other expatriates, brushing aside the inconveniences of smog, potholed streets, muggers, and occasional extortion by wayward cops. Moreover, it should be noted that transcultural communication is inadvertently at the heart of this process, as Koreans often come to learn English, a language that Filipinos themselves absorbed (albeit imperfectly) under colonial rule by the United States. The locals wonder why the Koreans, who are perceived to be able to afford migration directly to America, choose the Philippines. Yet the urban society remains open to such waves of foreigners, especially because many bring in hard cash, and long-term investments—even though the visitors have not always behaved like guests. This chapter seeks to examine this phenomenon, through a snapshot of both the apparent physical changes in urban areas, as well as the possible attitudes and policies that lie beneath this intermingling of cultures.

Multiple Transitioning Cultures and the Use of Space

The term *transcultural* connotes change or movement across or beyond a prevailing state of culture, possibly driven by the arrival of external influences. Such influences, as brought about by migrants or transients, may be absorbed into a largely homogenous recipient culture, or result in a more multicultural setting—which may or may not be harmonious, and which is becoming more commonplace in a globalizing world. Qadeer (1997) argues that policy-makers and urban planners addressing such realities need to hold standards up to the light of different social values and recognize the legitimacy of ethnic neighborhoods that form a mosaic (e.g., Canada since the 1971 Trudeau doctrine), rather than a melting pot (e.g., the United States, historically). Qadeer (1997) adds, "old residents come calling on planning institutions to protect their interests. The new arrivals seek fair treatment and accommodation of their needs and preferences," such that this may result in multilingual signage and more access to a range of housing choices. This attempt to institutionalize multiculturalism is inherent in the way cities assemble identity and generate difference, which may be recognized, encouraged, or otherwise tolerated, depending on the stance of government, and which may conceal underlying tensions (Uitermark et al., 2005; Rex & Singh, 2003). Historically, the maintenance of immigrants' cultural identity is often accompanied by formation of what Funkhouser (2000) calls a primary enclave, with dense geographic concentrations of immigrants, where an enclave economy is formed out of the synergistic relationship between a given ethnic group and a sector-specific economic structure; for example, successful Korean businesses in Los Angeles, which employed some 80 percent of their laborers from within the community in the 1980s (Kaplan, 1997). In the literature, the ethnic enclave is a voluntary grouping as opposed to a ghetto, which is often an imposed segregation in a certain part of the city (Varady, 2005). As this "stepping-stone" economy matures, immigrants or younger generations disperse into secondary enclaves farther away from the core (Funkhouser, 2000; Kaplan, 1997), especially when they are no longer limited by language barriers and other constraints on leisure (Scott et al., 2006).

The phenomenon of cultural intermingling speaks to issues about the appropriation and definition of space. Rahder and Milgrom (2004) assert that global mobility makes cities more socially and culturally diverse—a process with no end in sight, that results in the *uncertain city*, where all solutions are partial and temporary, often the result of planners working with others. This unfinished, even disorderly quality, seemingly antithetical to conventional planning that regulates urban space, can be seen as a manifestation of democratic processes that allow varying forms of public address, to which people react differently, depending on their understanding of local norms (Staeheli, 2010).

In the Asian context, this could be interpreted as a reaction to the state's monopoly of cultural space, and a movement towards decoupling of culture, nation, and government. It is precisely this contestation of space by different groups that challenges national identities, and brings into play various groups at the center and periphery of a given society (Tiryakian, 2003). Yet at the same time, a globalizing state that pays lip service to the ideas of equity and social justice must work towards ensuring what Parnell and Pieterse (2010) call "the right to the city," especially for the poor and marginalized—which takes a strange twist in this case, as Korean visitors are rarely penniless, and do not remain at the margins of Philippine communities for too long, while their local neighbors may stay mired in abject poverty. Urban spaces, in summary, become constitutive elements of the reproduction of the social—they inform actions of inhabitants, but are also outcomes of actions, and no other institution but the state is constructed and equipped to shape both the space and the activities therein in such a definitive manner (Zierhofer, 2005).

Globalization and International Migration in and around Asia

Although international migration has a long history in Asia, it has acquired an unprecedented scale, diversity, and significance, as voluntary or forced movement of both skilled and unskilled labor has significant impacts on both the sending and receiving states (Hugo, 2005); these latter having been caught in the wave of economic liberalization and global change that enriches some and socially excludes others (Beall, 2002). For example, expatriate employment has grown for developed countries like South Korea, but has caused a massive brain drain from developing countries like the Philippines, which exports, and relies on remittances of, the nearly 10 percent of its population that serves overseas.

Migration in Asia today is characterized by (1) the dominant role of contract labor migration; (2) recruitment and placement left largely in the hands of commercially motivated agencies; and (3) active promotion by the governments of emigration countries, even if there are far more people seeking to emigrate than there are destination states to accept them (Abella, 2005). Such movements of people can also be internal to a single state across a wide national area, such as in Indonesia, where mobile upper classes form layers of inter-island migrants (Vickers, 2004), or between close neighbors, like the states of the Korean peninsula and Japan. Japan distinguishes between the *kitachosen* from North Korea, and the *kankoku* from South Korea, who, like other migrants to Japan, have few practical opportunities for full integration (Haines et al., 2007). Acceptance by host states is controlled in the form of policies, such as in Singapore, which is highly selective of its emigrants, in contrast to the elaborate cultural mediation that it provides for tourists (Hugo, 2005; Ooi, 2001).

Ultimately, according to Friedman and Lehrer (1997), the impacts of such foreign migration are experienced at the local level, to which local and national governments must eventually respond. In contrast to the predominant pattern of expatriation in which workers from developing countries work as low-wage labor in developed or newly industrialized nations, the phenomenon of Koreans in the Philippines presents an anomaly. Nevertheless, the impact of migration is still felt and has to be negotiated locally.

Push and Pull Factors: Korean Aspects

South Korean society is changing with the world around it. The ever-present threat of invasion from the North, plus a flow of refugees and defectors, has burdened an otherwise robust economy. Unemployment is among the lowest in East Asia, but the economy structurally makes it difficult to find a job for workers over 50, fresh university graduates, women in general, and the handicapped. Employment is often monopolized by networks of Koreans with shared regional origin or university affiliation (Bidet, 2009). At the same time, in the last decade it has become increasingly acceptable for Korean men, particularly never-married rural bachelors and divorcees, to marry foreign women (Lee et al., 2006). Taking advantage of the proliferation of cheap airlines in the region, Koreans range far and wide to find work and spouses. Simultaneously, the United States has been declining as a destination-of-choice for Koreans, with Canada representing a multiculturally friendly alternative (Han & Ibbott, 2005). This is possibly because Korean immigrants to the USA have cited higher levels of job discrimination, owing to language difficulties (Gee & Ponce, 2010), and have been documented to come into conflict with Blacks and Hispanics (Pyong Gap Min, 2007). Meanwhile, countries of Southeast Asia offer accessible options, with the Philippine islands regarded as a viable alternative for affordable quality education and acquisition of functional Anglophone skills.

The seminal work by Miralao (2007) explains that despite the increasing number of Koreans that has arrived in the Philippines since the 1990s, the phenomenon is largely temporary. On the other hand, Miralao explains that unlike Korean diasporas to other countries in the past (i.e., soon after World War II), the waves of Koreans to the Philippines are the result of the last decade's prosperity at home that has driven business and networks to expand outwards, especially to where cheap labor markets for Korean companies and the opportunity to learn English coincide.

Historical and Urban Profile: The Philippines

The archipelago that makes up the Philippine Republic has historically been skirted by the great civilizations and religions of Asia, though records from China indicate that their mariners did visit the islands, en route to markets in the Moluccas. By 1565, when Spanish *conquistadors* returned to claim the islands that Ferdinand Magellan had first set foot upon in 1521, Islam and Hinduism had established only footholds in central and northern settlements, allowing soldiers and missionaries to colonize many of the islanders with Hispanic culture and Roman Catholicism. By 1898, Spain had ceded the Philippines to the United States, a new colonizer who began a process of "benevolent assimilation" that led to government reforms, territorial consolidation, public works, and, most significantly, public education with English as the medium of instruction. Given this European–American legacy, the Philippines emerged as an independent state in 1946, and began forming its own ties with other Asian states in the diplomatic, sociopolitical, and economic realms. As a nation, the Philippines arguably lies at a border between "oriental" and "occidental" influences, which has allowed it to exploit ties with the USA but has also left it still struggling to forge a unique identity in Southeast Asia, where the Philippines' more open culture contrasts with inward-looking states like Myanmar or Laos.

Not all Philippine cities are like Metro Manila, the prime national capital region of some 11.5 million inhabitants (2007 Census), which commands the lion's share of wealth, education, and entertainment, but remains fraught with urban ills.

Quite to the contrary, other areas offer diversity (Baguio), idyllic coastal settings (Dumaguete), and crime-free streets (Davao), making those potent alternatives for immigrants wishing to settle in the laid-back tropics. Over 50 percent of settlements are coastal, often sited away from the battering Pacific Ocean. Many town layouts possess a plaza of Hispanic provenance. Around this core, modern buildings, roads, and real estate projects sprawl, often bankrolled by Chinese magnates and Filipino industrial giants. The result, in recent years, has been the proliferation of glass-&-steel commercial centers and residences that are not affordable to most citizens, but which offer excellent living conditions for wealthy foreigners.

The Korean Wave

Is the Korean migration to the Philippines a result of directed national policies or laws pushing or pulling Koreans to Philippine shores? Apparently not, as explained by the Director for the Northeast Asia Division of the Philippines' foreign ministry. Policies on granting tax holidays to investors prevail for all states having diplomatic relations with the Philippines. However, what did surface often enough to merit attention in terms of high-level bilateral diplomacy were two elements: first, there is the importance that older Koreans accorded to the Philippine Expeditionary Forces to Korea (PEFTOK), which fought as part of the United Nations when the Korean War broke out in 1950. This sentiment has added a distinct strength to South Korean–Philippine relations. Second, there is the possibility of a North Korean attack; Seoul lies within striking distance of missiles from Pyongyang and has hinted at the Philippines as a remote alternative refuge in case of attack, after Japan and Taiwan. At present, in response to the increased economic opportunities for Koreans, there are talks between ministries to provide Koreans with pre-departure briefings and more police protection upon their arrival, given the rise in organized crime targeting wealthy Koreans. It should be noted, however, that interviewees consistently emphasized that "not all Koreans are rich." In fact, the Philippines has become a fallback for middle-class parents who, despite limited budgets, aspire to keep their children competitive by having them learn English. This last note hints at a larger transcultural phenomenon that integrates the Philippines as a "stepping stone" for Korean go-getters, whose object is actually often a more remote destination, like the United States.

Moreover, the Department of Tourism has a program for enticing foreigners to retire in the Philippines. According to the Department's records, Korean tourist arrivals surpassed Americans in 2006 to become number one, with approximately 936,000 in 2010, up from half a million at mid-decade. As for residents, there are already 115,000 Koreans around, according to figures from the Bureau of Immigration—this is up from a 2001 estimate of 24,618 (Kutsumi, 2004). It should be noted, however, that some who enter the country with a tourist visa do not actually intend to depart soon, but find their way into the network of dormitories and enclaves that will be discussed in the next section.

Realities on the Ground

Typical Korean Blocks

"Koreanization" has been observed in at least three distinct urban spaces: residential neighborhoods, university districts, and non-central business district (CBD) commercial areas. In most cases, Korean establishments are identifiable by the distinctive signage and dangling ornamentation that portray the *Taeguk* or the *Sam-Taeguk*,[1] and by modern boxlike buildings embellished with colorful signs. In the first instance, Koreans become visible in neighborhoods when they begin to cluster along a street, and little signs in Hangul begin appearing on mailboxes or fences. The definitive proof comes when large residences (lot sizes upward of 300 square meters) in gated subdivisions are rented and converted into informal dormitories, and when shoe-racks at the doorway can be glimpsed by Filipinos (who remain shod indoors). A similar phenomenon takes place near universities, where several apartment high-rises within a 2 to 3 kilometer radius are occupied by Korean students, who are sometimes distinguished by the way they dress, and by their lengthy study hours. Commercial spaces on the ground floors of such buildings soon turn into laundromats, convenience stores, or Internet cafés for this clientele.

The settlement process seems to follow a pattern. After pioneer establishments take root in sleepy corners, succeeding waves of Koreans tend to cluster there, instead of—and this makes them noticeable—in larger formal districts where stores of wealthy locals and Filipino-Chinese tycoons already dominate. In spatial terms, the density of Korean establishments that line a community road varies from 1 per 200 meters (in Barangay[2] Holy Spirit, Quezon City) to 1 per 30 meters (along Aguirre Avenue in BF Homes Subdivision, Parañaque City). Below these densities, they would

Figure 6.1 Typical mixing of Korean and English signs along a residential road in the Philippines. Photograph by José Edgardo Abaya Gomez, Jr.

Figure 6.2 In this renamed neighborhood, Koreans rule where American soldiers used to hang out for R&R. Photograph by José Edgardo Abaya Gomez, Jr.

probably not be noticeable. Above these densities, they would make up a "Korean Town," such as the one in Angeles City, where a kilometer-long street has been thoroughly *Koreanized* on both sides. Oddly, this latter phenomenon has taken place in a honky-tonk district beside the former Clark Air Base, where US military personnel once loitered.

Koreans Share their Experiences

Stories of both visitors and hosts help to illuminate human motivations behind urban manifestations. One respondent, who has lived in Metro Manila since 2003, explained that her children had wanted to learn English, and that she had heard life in the Philippines was much less stressful than in Korea, where pupils often only got five hours of sleep a night because of heavy study loads. So she came with her husband, rented a house, and soon set up an informal dormitory-cum-school where children of other Koreans could learn English. It took her only three months to adjust, though she confessed that she has no business with Filipinos at all, after hearing stories of how locals swindle hapless Koreans. She ventured to say that many pioneers follow the same pattern: they set up shop where operating costs are cheaper, but serve only their countrymen, with whom compatriotic bonds are extremely strong, so that even families back in Korea entrust their grade-schoolers to dormitories run by their friends. Besides, those who set up thriving businesses with Filipino interactions are usually Korean men whose Filipino wives help them get through the bureaucracy and establish rapport with the locals.

One of the respondent's most memorable cultural shocks took place when the couple hired laborers to extend some house renovations till dusk, not an unusual practice in South Korea. An irate neighbor stormed their gate, holding his shotgun and demanding that they stop the din. "My husband had never seen a shotgun before!" she related (and had certainly not been threatened with one). From this, they learned that the noise of work was *verboten* after sunset. However, they find it strange that their neighbors carouse all night, singing pop-songs on the karaoke, unconcerned about dormant co-villagers. On a more positive note, the couple's daughter said that she got along well with Filipino friends in a private high school.

Another "survivor" does missionary work together with her pastor husband in Bulacan province, north of Metro Manila. Following the established routine, they settled with the help of her husband's contacts. Over the last half-decade, they constructed ten churches, and penetrated the poorest neighborhoods where they soon won a large following, as their feeding programs have provided hot meals for up to 900 children twice a week—all subsidized by Korean money. In contrast to other Koreans, this respondent and her husband assert "we have many local friends," and foresee themselves staying to consolidate the gains of Protestant proselytization, though their two teenage children will eventually be sent to the United States for university.

An older student, about to finish his Master's degree in Special Education, arrived about seven years ago, originally to learn English. He was from rural Koeje-si in the south of the peninsula, and although he had Korean friends who helped him adjust, he was not well advised. He wound up enrolled in a school for less-privileged classes, where the students spoke English poorly and spent their after-school hours "drinking and hanging around." He stayed in a Korean-run dormitory until he had sufficiently adapted, then moved on to a more reputable university.

Yet another entrepreneur married a local, and tried to run a *Norebang,* or karaoke business, in Pasay (in central Metro Manila). He closed up shop after three years, as he claimed that the authorities were harassing him. In 2009, he moved on to try his luck in Subic Bay, the former US Naval Station northwest of the capital, which has been transformed into an economic zone with incentives for foreign investors.

Whether their interest is entrepreneurial or religious, all migrant Koreans need to learn to communicate locally, and English becomes the idiom of choice as well as the overarching reason for settling down. This is one level where the transcultural process begins: a local English instructor is sought out, and a relationship develops, depending on how personably the parties behave towards one another.

Filipinos Ride the Korean Wave

Understandably, many Filipinos began their dealings with Koreans after frequent requests for English tutors. Koreans were perceived as paying "modest" to "low" sums of money, ranging from 200 to 400 pesos (US$4 to US$8) per hour, but the income was easy to obtain and informal (hence untaxed). Plus, after developing a good working relationship, Koreans would add in some perks like food and gifts on the side, as in the case of one housewife-respondent who has become "like family"

Figure 6.3 This once rundown street has been thoroughly Koreanized. Photograph by José Edgardo Abaya Gomez, Jr.

to her tutees after seven years of interaction. She remarked that she was the one who learned their ways, such as how to eat and spice their noodles, how to get value for money out of work hours, and how to keep the anteroom spotless.

On the other hand, there are Filipinos who did not form any long-term close relationship with their employers. Another respondent found a "sideline" by working for a Korean businesswoman, who operated an on-line tutorial center for elementary children. The Filipina respondent prepared booklets and manuals similar to TOEFL[3] reviewers, for a good wage of ₱100 ($3) per hour of writing. Most of the operations were conducted at night, in a largely vacant building in the Ortigas business district of Metro Manila. The author visited the nondescript premises, which were covered with dust and loose wiring on the outside but contained state-of-the-art workstations inside a cramped air-conditioned room. In this room Filipino tutors sat in front of screens, coaxing bored-looking Korean schoolchildren to pronounce English words. The setup was lucrative but exhausting for the Filipinos, and eventually the interviewee left to take on another job, even though she had by then won the trust of her employer, who occasionally asked for advice on such matters as setting up a "summer camp" for Korean children.

Blue-collar workers, however, have complained that Korean masters are harsh, and have been known to yell at their maidservants, or deduct from employee wages for the tithes of the Koreans' religious congregations. The author has witnessed this in commercial areas, where Korean overseers keep a sharp eye on carpenters, plumbers, and other day-laborers, fretting until the tasks are finished exactly as envisioned.

Micro and Macro Processes of Transculturation

One cannot resist a bio-ecological analogy here: the Korean arrivals can be likened to a hardy plant that sinks roots into soft rock (the Philippines) already overlain by moss and lichens (i.e., Chinese, Japanese, Spanish, and American cultures), and persists in spreading its layer over earlier sheaths, forming commensal, rather than symbiotic, relationships with the Filipino host. Adults are averse to mixing with the locals—the exception invariably being Korean men wedded to Filipinas. Without exception, all the Koreans interviewed, and even those whom the author has met outside of the research framework, have listed "learning English" as their first and most enduring reason for sojourning. Whether they do this at the most personal level (e.g., hiring a tutor), or at the institutional level (entering a university), depends on personal circumstances. Those in a hurry will take a crash course, then return home or move onward to developed countries—unless they succeed at business or work for a Korean multinational in the Philippines. Others, like missionaries of Korean Christian churches, may stay to preach, while sending their kids to study through secondary school and university. That is, school enrollment appears incidental to, or comes subsequent to, language acquisition. After learning English, Korean youth may continue education in an Ivy League institution, or find jobs back in South Korea. Even missionary work is an offshoot of economic forces, as churches would come initially to minister to Korean congregants, and then venture to convert the locals.

There are at least two ways to look at these experiences: the first is to conclude that there is actually very little in the way of transcultural exchange taking place on the ground, as the sample of Koreans interviewed universally admitted that beyond the necessities of learning English and obtaining household services and government permits, they and their countrymen eschew mixing on a personal level with the locals. Direct evidence of Koreans learning Tagalog and mingling with Filipinos has been observed by the author only among shopkeepers and students. Some Filipinos, on the other hand, have developed a taste for *japchae* glass noodles, and follow dubbed *koreanovelas* (soap operas) like *Oh Feel Young, Queen Seon Dok* and *My Name is Kim Sam Soon*. These limited exchanges of ideas, tastes, or artifacts show how the races learn from one another, inadvertently absorbing attitudes and ideas that transform them.

The second, simultaneous, way to interpret the phenomenon is to realize that this Korean wave is a manifestation of larger transcultural processes, in which the Philippine state plays its historical role as a bridge between east and west. The targeted social milieu is embodied by countries like the United States or Canada, which promise a degree of prestige. It makes sense to obtain a foretaste of the west in the Philippines, especially since the cost of living is lower and enables middle-class Koreans to thrive there luxuriously. The urban manifestation of Korean immigration is an enclave, consisting of dormitories where successive batches of Korean youths spend up to a decade of their lives learning a foreign language once-removed. It will take another study to trace whether or not this strategy actually helps those youths integrate into western contexts after they leave the Philippines for good.

At the most impersonal level, there is economic gain for Filipinos, specifically in lower-income neighborhoods where the surge in direct cash transactions alleviates

Figure 6.4 A bit shabby at first glance, but business is booming, especially because many Koreans buy only from their fellow Koreans. Photograph by José Edgardo Abaya Gomez, Jr.

residents' poverty immediately. Where Koreans are living—staying an average of four years if they are students—they contribute to local markets by spending for daily needs. This is in addition to bilateral development aid and high-level investments such as the location of an overseas headquarters of Hanjin Shipping in Subic Bay.

From an urban planning and development perspective, the Korean "invasion" is a positive force that drives renewal and new construction, which would otherwise be unaffordable to many locals. This renewal and construction is, however, unplanned or loosely regulated by local authorities. Neighborhoods, after all, turn into Korean quarters incrementally, and officials at city hall rarely find the time to make the rounds to check on urban design and inter-functionality with pre-existing structures and activities.

What remains questionable from a longer-term social planning and governance perspective is the de facto creation of an "alien" space that is literally unintelligible to locals. Municipal ordinances still fail to address this, with politicians blithely ignoring the phenomenon as a non-issue, as long as the Koreans, like the Chinese with their *Hanzi* script in Chinatown, pay up. It is also unusual, but appreciated by local governments, when Koreans take over impoverished locales, rebuild, and transform the streets into little slices of Seoul. The most practical explanation is that some Koreans are opportunists of a fairly daring nature. They tend to grab a good deal when they see one, especially in the many centrally located idle strips scattered around. These strips are essentially frozen either because the government has no funds to evacuate and resettle squatters, or because the cautious private sector, consisting of Filipino businessmen, is holding back and waiting for grayfield prices to appreciate. That more risk-taking Koreans arrive in bands to scoop up the real estate, paying their way with cold cash, is often enough to satisfy long-time occupants and send them on their way. This is not random land speculation—there is usually another magnet of activity nearby, such as a university or a strategic mercantile location. These actions therefore constitute a method, deliberate and thoroughly rational from an economic point of view, to explain the "Koreanization" of urban areas.

Consequently, one must relate to this the issue of social intermingling. Many Filipinos and Koreans will admit that integration of the latter into Philippine society is slow, at best, as the Koreans generally do not mix with the locals except on a functional basis. Accounts from Philippine history narrate a similar pattern among Chinese and Japanese immigrants, who started out in ghettos and did not meld into local society until the second and third generations, which did not stand out as much as their parent stock as foreigners. Thus, while present relations still manifest in the form of ethnic "turf," interracial understanding and mixing will probably improve over time.

While it cannot be disputed that South Koreans in general appear to be more driven, hierarchical, and stubborn (Rhie, 2002) than their Filipino counterparts, the ability of the former to thrive under Philippine conditions suggests that they may at least share some traits or tolerances that enable them to find niches in the host society. Indeed, there is neither hard-line segregation nor hurtful exclusion, as has been practiced elsewhere by European colonizers of the past. Aubrey (2009) posits a list of Korean traits that resonate partially with local Filipino dispositions: a preference for full-flavored food; value given to speed, function, and getting things done with little heed to detail; tolerance for low cleanliness in public places; and adherence to Christianity. It is perhaps also through the combination of Filipinos' hospitality, or need for income, and the South Koreans' determination to learn and succeed that partnerships are engendered, despite the inconveniences of floods, power outages, traffic, and other urban ills that affect local and foreigner alike.

Looking Ahead

Korean presence in the Philippines is still growing, though it remains to be seen how and when the energy of population migration will burn out. As international economic fortunes rise and fall, local interaction continues at a modest pace, and Koreans disperse rapidly across the archipelago, unavoidably entering into professional and personal transactions with their hosts and helpers. This is ironic, because some Koreans, despite their aloof haste, catalyze the means by which transcultural exchanges can take place, simply by finding homes and building shops in more urban centers—without government prodding. For this author, such accidents of circumstance promise to provide more transcultural exchange rather than frustrate it, since Philippine history has shown that eventually a good number of the newcomers will put down roots, intermarry, and graft their culture onto the local sociopolitical stem. This points the reader in the direction of future studies that look at the progress of the Korean wave, and at the spread of its effects occurring in other parts of the country, as well as in the rest of Southeast Asia.

Notes

1 *Taeguk* refers to the red and blue "Yin-Yang" symbol, found in the center of the Korean flag. *Sam-taeguk* is a similar symbol with a yellow field, and colors arranged in a semi-spiral.
2 The *barangay* < *ba-rang-gai* > is the smallest political unit in the Philippines, equivalent to a village.
3 Test of English as a Foreign Language.

References

Abella, M. (2005) Social issues in the management of labour migration in Asia and the Pacific. *Asia-Pacific Population Journal,* 20(3), 61–86.

Aubrey, S. (2009) A cross-cultural discussion of Japan and South Korea and how differences are manifested in the ESL/EFL classroom. *Asian Social Science,* 5(5), 34–39.

Beall, J. (2002) Globalization and social exclusion in cities: framing the debate with lessons from Africa and Asia. *Environment and Urbanization,* 14(11), 41–51.

Bidet, E. (2009) Social capital and work integration of migrants: the case of North Korea and defectors in South Korea. *Asian Perspective,* 33(2), 151–179.

Friedman, J. & Lehrer, U. (1997) Urban policy responses to foreign in-migration: the case of Frankfurt-am-Main. *Journal of the American Planning Association,* 63(1), 61–78.

Funkhouser, E. (2000) Changes in the geographic concentration and location of residences of migrants. *International Migration Review,* 34(2), 489–510.

Gee, G. & Ponce, N. (2010) Associations between racial discrimination, limited English proficiency, and health-related quality of life among 6 Asian ethnic groups in California. *American Journal of Public Health,* 100(5), 888–895.

Haines, D., Makito, M., & Shinji, Y. (2007) Transnational migration in East Asia: Japan in comparative focus. *International Migration Review,* 41(4), 963–967.

Han, J. D. & Ibbot, P. (2005) Korean migration to North America: some prices that matter. *Canadian Studies in Population,* 32(2), 155–176.

Hugo, G. (2005) Migration in the Asia-Pacific region. Paper presented to the Policy Analysis and Research Programme of the Global Commission on International Migration, September.

Kaplan, D. (1997) The creation of an ethnic economy: Indochinese business expansion in Saint Paul. *Economic Geography,* 73(2), 214–233.

Kutsumi, K. (2004) Koreans in the Philippines: a study of their social organization. Paper delivered at the PMRN Regional Conference on Transnational Communities in the Philippines, Richmonde Hotel, 28 May.

Lee, Y., Seol, D., & Cho, S. (2006) International marriages in South Korea: the significance of nationality and ethnicity. *Journal of Population Research,* 23(2), 165–182.

Miralao, V. A. (2007) Understanding the Korean diaspora to the Philippines. In Miralao, V. A. & Makil, L. P. (Eds), *Exploring transnational communities in the Philippines*. Quezon City: Philippine Migration Research Network and Philippine Social Science Council.

Ooi, Can-Seng. (2001) Decoding and manufacturing cultures: interpreting Singapore for tourists. *CAUCE: Revista de Filología y su Didáctica,* 24, 617–635.

Parnell, S. & Pieterse, E. (2010) The "right to the city": institutional imperatives of a developmental state. *International Journal of Urban and Regional Research,* 34(1), 146–162.

Pyong Gap Min. (2007) Korean–Latino relations in Los Angeles and New York. *Du Bois Review,* 4(2), 395–411.
Qadeer, M. (1997) Pluralistic planning for multicultural cities. *Journal of the American Planning Association,* 63(4), 481–494.
Rahder, B. & Milgrom, R. (2004) The uncertain city: making space(s) for difference. *Canadian Journal of Urban Research,* 13(1), Supplement 27–45.
Rex, J. & Singh, G. (2003) Multiculturalism and political integration in modern nation-states. *International Journal on Multicultural Societies,* 5(1), 3–19.
Rhie, Won-Bok. (2002) *Korea unmasked.* Seoul: Gimm-Young International.
Scott, D., Lee, S., Lee, J., & Kim, C. (2006) Leisure constraints and acculturation among Korean immigrants. *Journal of Park and Recreation Administration,* 24(2), 63–86.
Staeheli, L. (2010) Political geography: democracy and the disorderly public. *Progress in Human Geography,* 34(1), 66–78.
Tiryakian, E. (2003) Assessing multiculturalism theoretically: e pluribus unum, sic et non. *International Journal on Multicultural Societies,* 5(1), 3–19.
Uitermark, J., Rossi, U., & Van Houtom, H. (2005) Reinventing multiculturalism: urban citizenship and the negotiation of ethnic diversity in Amsterdam. *International Journal of Urban and Regional Research,* 29(3), 622–640.
Varady, D. (2005) *Desegregating the city: ghettos, enclaves, and inequality*. Albany: State University of New York Press.
Vickers, A. (2004) The country and the cities 1. *Journal of Contemporary Asia*, 34(3), 304–317.
Zierhofer, W. (2005) State, power, and space. *Social Geography,* 1, 29–36.

Interviews

M. S. Kim and S. H. Song—15 October 2010, Quezon City.
D. H. Park—21 October 2010, Pasay City.
K. S. Ryung—26 October 2010, Quezon City.
L. D. Choul—mid-2009 (informal).
Mrs. M. D. Oliva, English Tutor—intermittent, October–December 2010.
Mrs. K. Suñega-Cruz—intermittent, January 2011.
Director J. Ignacio, Director for Northeast Asian Division, Department of Foreign Affairs—27 October 2010.

CHAPTER 7

The Transcultural Production of Space

Making "Little Shanghai" in Sydney

Duanfang Lu and Hongguang He

The inflow of migrants has played an important role in shaping the social and physical landscapes of Australian cities (Jordans, 1997; Saeed & Akbarzadeh, 2001; Burnley, 2001; Jupp, 2007). Yet despite the palpable and widespread spatial effects brought by immigration, little is understood about how urban spaces are made and remade by migrant communities in Australia. This chapter aims to fill the gap by investigating Chinese migrants' roles in the transformation of public spaces in Ashfield, a multicultural suburb in the inner-west of Sydney, in the past two decades. In this time, Ashfield's streetscape has been through radical changes. While its main commercial streets had a mix of Anglo, Italian, and Greek shopping before the mid-1990s, about 85 percent of the shops are now Chinese small businesses, including restaurants, supermarkets, barbershops, and bookstores. As most of these shops are run by migrants from Shanghai, the area is widely known as "Little Shanghai" (Figure 7.1).

The re-territorialization of migrants has involved complex socio-spatial processes and taken various forms (Lu, 2000; Castles, 2000; Tewdr-Jones & Allmendinger, 2006; Hillier, 2007). With the continuing interest in the inclusion of migrants into the host society, many migration studies are dominated by the notions of assimilation and integration, which assume that the process of dropping one culture in favor of another is natural (Qadeer, 1997; Reeves, 2005). Binary categories such as sending society/host society and tradition/adaptation serve to describe migration as a linear and unidirectional movement, with little room left for more complicated orientations (Sandercock, 2003; Watson, 2006). Drawing upon recent developments in the understanding of the plurality of knowledge, this study aims to go beyond assimilationist or integrationist models of migration by examining the coexistence of distinct and equally legitimate bodies of knowledge in the transcultural production of public spaces.[1] It is our hope that the findings will contribute to the development of a new framework which highlights epistemological diversity in urban transformation and contribute to more sensitive models of urban governance that promote social sustainability by going beyond the "deficit model" of immigrants.

This chapter is divided into four sections. It will first offer a theoretical discussion on travelling knowledge, then discuss the streetscape of Ashfield as a spatial product resulting from a series of transcultural practices. The third section will focus on how Chinese migrants coped with the cultural and economic disjunctions they experienced in Ashfield. It will conclude by providing a summary of the main findings and sketching the beginnings of a knowledge approach to migrant spaces.

Figure 7.1 Shanghai restaurants in Ashfield, Sydney. Photograph by Duanfang Lu (2010).

Travelling Knowledge

All social practices involve knowledge. Each society has its own criteria regarding what counts as knowledge and what it means to know. There are different knowledge systems and different modes of knowing. For Plato, knowledge is justified true belief. In contrast, Confucius stated that "When you know a thing, to hold that you know it; and when you do not know a thing, to allow that you do not know it; this is knowledge" (Waley, 1938, 2:17). There is no definitive way of ordering, classifying, and explaining processes, entities, and relations between things (Santos et al., 2007).

Knowledge travels. In his book *The World, the Text, and the Critic*, Edward Said (1983) introduces the concept of travelling theory to address the migration of ideas in an international sphere. According to Said, the travel of theories and ideas occurs in four major stages. First, there is "the point of origin": the location at which the idea first entered discourse. Second, there is "a passage through the pressure of various contexts" as the idea moves to a new time and place. Third, there is a set of conditions of acceptance which allows the introduction or toleration of the transplanted idea. Fourth, the now fully or partly accommodated idea is somewhat transformed by its new uses and new position (Said, 1983, pp. 226–227). Said's concept of travelling theory establishes the spatial movements of ideas as an important aspect of the production of knowledge, although the concept fails to account for the means through which knowledge travels. It should be noted that ideas always travel to new places via vehicles such as travelling individuals and literature.

This chapter proposes seeing migrants as one of the principal vehicles through which knowledge is circulated across cultural or territorial boundaries. What should be emphasized here are the mutual dynamics between the different bodies of knowledge involved. While the image of "knowledge transfer" implies a one-way process, "travelling knowledge" entails a more fluid mode of exchanges and cross-overs. It emphasizes momentary cross-cultural transgression in which ideas intersect, combine, and ramify through myriad daily practices. In that process various layers of boundaries are dismantled, and new ideas and practices arise.

The notion of "travelling knowledge" recognizes the legitimacy of different bodies of knowledge and the positive effects that may be produced through border-crossings between them. It therefore provides a way of looking at migration that is radically different from the "deficit model" of migrants, which implies that the modes of knowledge that migrants have brought from their home country are "inferior" and must be replaced by the knowledge of the host country. Such assumptions are indeed the very barriers for the development of cross-cultural understanding. Only if we recognize the values of other knowledges can we foster mutual respect and create more opportunities for transcultural interaction and connectivity.

It is widely recognized that having been socialized in their home countries, migrants arrive with a collection of know-how, social and cultural understandings, and epistemological approaches. What is often neglected in much of the discussion is that these attributes are often internalized in individuals as a set of inexplicit dispositions that incline them to act and perceive in certain ways. It is here that Pierre Bourdieu's notion of *habitus* can help us. For Bourdieu, one does not freely choose to be what one is, but receives through numerous mundane processes of socialization a *habitus*—a set of internalized dispositions that unavoidably reflect the social conditions within which they were acquired (Bourdieu, 1977). These dispositions are ingrained in such a way that they are not readily amenable to conscious reflection and modification (Thompson, 1991, pp. 12–13). *Habitus* is thus a structured structure: an embodiment of the social system in which one is inculcated from birth through the family, the educational system, and other individuals encountered in one's life (Bourdieu, 1984, p. 170). As Bourdieu puts it:

> The habitus, a product of history, produces individual and collective practices—more history—in accordance with the schemes generated by history. It ensures the active presence of past experiences, which, deposited in each organism in the form of schemes of perception, thought, and action, tend to guarantee the "correctness" of practices and their constancy over time, more reliably than all formal rules and explicit norms.
>
> (Bourdieu, 1990, p. 54)

With the *habitus* continuously structuring new experiences in accordance with the structures produced by past experiences at every moment, there is an "ontological complicity between the *habitus* and the social field" (Bourdieu, 1990, p. 194), which brings about an integration of the experiences common to members of the same class.

Bourdieu's conceptualization is useful in that it illustrates a cultural field in which actors improvise their actions with the feel of the game provided by their *habiti*. Yet his model privileges highly localized cultural inscription and thus creates some uncertainties when applied to the situations of cross-cultural interactions. How can fluid dialogue take place across the multiple fields without being trapped in the cracks between them? How can a new order be established amid differences, which will allow meanings and actions to become mutually constitutive? And how does the *habitus* continuously maintain its coherence while operating in a highly disjunctive world? This chapter will address these questions by examining the production of migrant spaces in Ashfield.

The Transformation of Ashfield

Located nine kilometers southwest of Sydney's central business district (CBD), Ashfield is an old federation suburb with a population of 39,667 (Ashfield Council, 2012). It was home to the Wangal people prior to the arrival of Europeans. Early British settlement started in the late eighteenth century. Ashfield gradually developed into a highly desirable residential location; many grand Victorian houses were built during the late nineteenth century (Coupe & Coupe, 1988). For example, Mei Quong Tart (1850–1903), a prominent businessman and community leader who migrated from China in 1859 and owned a network of tearooms in Sydney, lived with his family in the Ashfield mansion known as "Gallop House" (Travers, 1981).

With most rich families moving to the North Shore in the 1910s, many grand homes in Ashfield were demolished and replaced with smaller houses or blocks of flats for working-class families in the 1920s and 1930s. While early residents were predominantly Anglo-Celtic, sizable Italian, Greek, and Polish populations settled in Ashfield during the post-World War II migration boom, followed by migrants from the Philippines, Turkey, and Lebanon in the 1970s and 1980s. The 1990s saw a large influx of migrants from China and India. According to the 2006 Australian Bureau of Statistics Census, 60 percent of Ashfield residents were born outside Australia, with 9.9 percent born in China and 3.2 percent born in India (Table 7.1). Currently, approximately 14 percent of the population spoke a Chinese language at home, with 8.8 percent speaking Mandarin and 4.9 percent speaking Cantonese (Australian Bureau of Statistics, 2006).

The prevalence of Chinese migration to Ashfield is related to a number of factors. Australia opened its education market to China in the late 1980s. As a result, a large number of Chinese from Shanghai, Fujian, and Guangdong arrived in Australia on student visas. Many of them discovered that Ashfield was a connection node to both the city and the western Sydney industry employment opportunities. It has a convenient train station and is in the first zone on the train line. With an affordable "red weekly train ticket," passengers can travel cheaply between Ashfield and the city, while the fare to go to the next train station, Croydon, is higher. Ashfield is also

Table 7.1 Population and housing in Ashfield (Local Government Area) and Sydney (Major Statistical Region), 2006

	Ashfield	**Sydney**
Population	39,667	4,119,190
Median weekly income	AU$478	AU$518
Born in Australia	48.4% (19,183)	60.4% (2,486,710)
Born in China	9.9% (3,912)	2.6% (109,142)
Born in India	3.2% (1,280)	1.3% (52,795)
Total private dwellings	17,074	1,643,676
Living in separate houses	38.9% (6,194)	61.7% (939,075)
Living in semi-detached, row, or terrace houses, and townhouses	13.1% (2,085)	11.8% (180,162)
Living in flats, units, or apartments	47.1% (7,509)	25.7% (390,679)

Source: Australian Bureau of Statistics (2006); Ashfield Council (2012).

a hub in a network of well-known and developed suburbs with Rockdale and Hurstville to its south, Chatswood and North Sydney to its north, and Parramatta to its west. Right after the 1989 Tiananmen Square political rupture, the Australian government granted permanent residency to some 42,000 Chinese students who were studying in the country. These visa holders were then able to bring family from China under Australia's then family reunion migration scheme. Some bought businesses and made Ashfield their home. The availability of Chinese shopping and cheap flats continues to attract students and migrants from China to live in the area.

The Chinese presence has radically transformed the streetscape of Ashfield. Up until the mid-1990s, the area had a mix of shops: "Italian delicatessens, Greek fish shops, small fruit shops run by Lebanese and Greeks, real estate agents that were maybe Anglo-Saxon or of European origin" (Dasey, 2003). Since 1990 more and more Chinese migrants have taken over the street-front, small-scale retail spaces by offering higher rents to their owners. With the preponderance of restaurants and shops run by Shanghainese, "Little Shanghai" has won its fame. The new Chinese business establishments have helped to revive the once-obsolete shopping strip on Liverpool Road. Shoppers and visitors of diverse cultural backgrounds come to Ashfield from all of Sydney. For many, Ashfield is known as a highly functional place for grocery shopping. It only takes 20 minutes by express or local trains to travel from Sydney's CBD to Ashfield. Unlike the downtown Chinatown, where parking space is expensive and difficult to find, free parking is available at the Ashfield Shopping Centre. Few local supermarkets can compete with the range, freshness and price of fruit and vegetables available at stores in Ashfield. Some cosmopolitan gourmets name Ashfield the best place to taste authentic Shanghai cuisine. Its restaurants not only offer meals with flavors different from the Cantonese dim sum variety in the old Chinatown but also boast much lower prices for similar-quality food. Visitors are also attracted to Chinese-related activities during festivals such as lion-dancing in the Spring Festival. Recent Chinese migrants who are still struggling with culture and language find it convenient to shop in Ashfield as most shop-keepers can communicate with them in Mandarin; apart from the familiar food, the books, magazines, and DVDs from China available at bookstores and supermarkets in Ashfield also make them feel very much at home (Figure 7.2).

The current shape of "Little Shanghai" is not only the result of individual business decisions, but also the product of collective spatial struggles at some critical moments during its formation. The parking protest in 2003 illustrates one such moment. Entering into the corner of Brown Street and Bland Street, near the Ashfield railway underpass, there was a free open parking area that could accommodate 101 cars. It was also a convenient loading zone for local shops. In 2003, under the pressure of their yearly budget, Ashfield Council auctioned this area to a developer for the price of AU$2.6 million. The decision encountered unyielding resistance from the Chinese community, as local business would be heavily disturbed by the loss of free parking space. The resistance, however, was initially ineffective due to lack of organization. According to one of our interviewees:

> At the beginning, we were completely ignored. Mayor and other Council members paid little attention to us. In retrospect, I think this was because we didn't organize ourselves properly.

The situation was changed after JM, a Chinese businessman who had worked in Ashfield since 1990, established the Sydney Inner West Chinese Merchants Association (SIWCMA) among Chinese shop owners in Ashfield and led the negotiation.[2] A compromise was eventually reached; the developer agreed to reserve a three-hour free parking area for 99 cars (now at the basement of the built residential tower). This victory was decisive for the subsequent healthy development of Ashfield's shopping strip on Liverpool Road. With membership covering two-thirds of Ashfield Chinese businesses, the SIWCMA has since played an important role in

Figure 7.2 "New China Bookshop" in Ashfield, Sydney. Photograph by Duanfang Lu (2010).

bringing Chinese shop owners together and organizing a number of significant initiatives and negotiations. For example, it managed to protect shopkeepers from the interference of local racketeers, which was a common problem in the early 2000s.

Our interviews with JM reveal how his *habitus* has influenced his life trajectory and continuously evolved through a series of cross-cultural practices. A graduate of the Huadong Science and Technology University, JM was a project manager for the redevelopment of a chemical plant in Shanghai before migrating to Australia in 1990. He worked as a storage keeper for an electronics wholesaler during his first five years in Sydney. As a well-trained chemical engineer, he displayed superior memory and efficiency in keeping things in order. This won him trust and respect from the manager, which allowed him to introduce other Chinese migrants to work for the company. By the mid-1990s, JM was ready to start his own business. The leadership skills he developed while working in Shanghai and the networking he established after moving to Ashfield not only helped him develop and continuously expand the business, it also drove him to become a community leader from the early 2000s onward.

Our conversations show that his perceptions of the local political landscape were shaped by both the knowledge that he acquired as a project manager in Shanghai and the knowledge he acquired by participating in local politics. While describing his experience of working with Ashfield Council, for instance, JM drew some parallels between the Council and the Street Office (*jiedao banshichu*), the sub-district urban government in China. His understanding of the latter necessarily led to some unique views about the former. In comparison to the Street Office in Shanghai, for example, JM found the policy making process of Ashfield Council rather slow, as it had to seek balance among different interest groups. He learned to cope with a very different working pace and obtained a feel of the local political game over time, which in turn helped him play an active role in various transnational networks such as the Australia–China Economic Development Foundation and the Australian Shanghai Townsmen Association. His past experiences in Shanghai and new opportunities created by migrant networks together shaped the direction of his career and social ambitions. He managed to draw official connections in Ashfield and Shanghai to develop new international trading and exchange activities between Australia and China. He also played a leadership role in a variety of philanthropic activities in recent years. It was through the efforts of fluid subjects like JM that passages between distinct compartments multiplied, and the cultural and physical landscape of Ashfield gradually evolved.

Habitus, Time, and the Transcultural Production of Space

Much discussion on Ashfield has focused on its cultural diversity, while the class dimension is largely ignored. Amanda Wise's study of elderly Anglo-Celtic long-term residents' reactions to the rapid changes brought by Chinese migrants in Ashfield provides one such example (Wise, 2004). Based on interviews with English-speaking residents, her research finds a "direct correlation between age, declining physical mobility and feelings of alienation from the 'new' Ashfield." In particular, older residents' resentment was directed towards the lack of street signs in English, lack of design quality, and the predominance of Chinese language menus displayed in the windows of restaurants along the Ashfield main street. Her findings became the basis for the "Building Neighbourhood Community Harmony" program in 2006, which aimed to achieve the objectives of helping Chinese shopkeepers make their premises more amenable to non-Chinese shoppers, increasing the presence of English signage, and promoting intercultural interactions. It is reported that the program has cleared some misunderstandings and helped build greater cross-cultural trust (Singerman, 2006).

Wise's study highlights cultural differences as the cause of contestation in regard to Ashfield's environmental quality. Yet her inattention to the class status and the dynamic nature of the *habiti* of Chinese migrants in Ashfield has missed some important aspects of the picture. Our research instead finds that the unfashionable character of shopfronts in Ashfield has more to do with the fact that most shopkeepers possess little capital when they start their businesses; such a condition is not static but continuously changing as shopkeepers' *habiti* and resources available to them shift over time. Bourdieu (1986; quoted in Lu, 2006, p. 159) categorizes the main components of social resources which define power and the capacities of the actors in the field into economic, cultural, and social capital. Economic capital not only consists of capital in the Marxian sense but also of other economic possessions. Cultural capital has three forms of existence: first, it exists in the *habitus*; second, as manifested in cultural articles such as dress and bodily decoration; third, as institutionalized via educational institutions. Social capital is a resource that is connected with group membership and social networks and based on mutual recognition. Although many shopkeepers were well-educated professionals who possessed reasonable amounts of economic, cultural, and social capital before they came to Australia, migration radically changed their class status. With poor language skills, it was difficult for them to find a white-collar job that matched their original expertise. Many of them became blue-collar workers with low salaries. Some managed to establish small businesses with limited capital. In either case, the various types of capital they accumulated before migration did not help them secure their previous class status. There was an acute sense of socially marginal status among Chinese migrants.

Ashfield has been a working-class suburb since the 1930s. It has lower median household income than Sydney at AU$478 per week. Of the 8,664 occupied private dwellings counted in 2006, 59 percent were flats, much higher than the Sydney figure of 26 percent and the national figure of just 14 percent. The high number of flats contributed to a higher than average number of people renting (45 percent); many tenants were international students and recent migrants with relatively low income and high mobility. Like locals, Chinese migrants were unsatisfied with the environmental quality of Ashfield and disappointed at being trapped in a lower-class suburb. As one shopkeeper commented, in comparison to the city of Shanghai, "Ashfield is the countryside." In the past, there was a time lag between the sending society and the host society; people working abroad possessed higher levels of economic and cultural capital. A return trip home, as Paul Siu (1952, p. 39) describes, would show that [s/he] was "a person to be admired, to be appreciated, to be proud of, and to be envied." With rapid economic development in China in recent years, many Chinese migrants in Ashfield experienced a reverse time lag.

Some of their friends and relatives at home were now much better off in terms of income level and living standards. The sending society appeared more developed and fashionable than that of the host society in terms of environmental quality. A number of interviewees stated that Ashfield was like Shanghai ten or twenty years ago. They considered the latter to be "more modern" than the former. Some also noticed that Ashfield failed to keep up with other suburbs. As one of our informants pointed out:

> I have lived in Ashfield for the ten years since I arrived in Australia. There have been very few changes in Ashfield. I don't feel that I am attached to this place. Even some marginal suburbs such as Blacktown now have a Westfield [shopping mall], but nothing has happened here yet.

People have coped with the time lag differently. Some accepted the fact that Ashfield was a functional place for low price shopping and dining rather than anything else. The pocket-friendly environment of Ashfield is acknowledged by both local residents and visitors. A cosmopolitan gourmet writes in his online column (Howard, 2008):

> Seriously, I refuse to pay a premium price for my beloved dumplings. While Din Tai Fung has grace, class and attentive customer service, Ashfield's small and dodgy looking establishments make it close enough to being the dumpling capital of Sydney, if not Australia. Keep in mind, it's the great tasting dumplings at a fraction of the price of Din Tai Fung which makes me coming back for more at either Shanghai Night, or New Shanghai.

Quite a few Chinese shopkeepers told us that their modest-looking shopfront was a strategic choice; they did not wish to push away shoppers who believe that an expensive-looking shop tends to be expensive. Others sought to improve the environmental quality of their shops once their economic status improved. TN was one of them. A long-term resident of Ashfield, TN established a dumpling restaurant in 2005. While it was small, the restaurant was clean and elegant, which attracted both Chinese and non-Chinese diners. Unlike some other Chinese shopkeepers who tended to deal with customers briskly, TN loved to engage in conversation with clients and made a lot of local friends. The small restaurant came to be a nodal point of the inter-ethnic information network over time. As TN was interested in interior design, she sought advice from customers regularly apart from exposing herself to related books and magazines. With the good rapport that had been established between TN and her clients, the latter offered many ideas and suggestions, which TN used to gradually upgrade the aesthetic quality of the shop's interior design. A German–Australian, for example, noticed that the restaurant featured a unique-looking lamp, but it was too close to the ceiling. He suggested lowering it a bit to reveal its full beauty and to add another layer of depth. TN adopted his advice in the redevelopment of the restaurant, which worked well. Little by little, such inter-ethnic knowledge exchange has not only allowed TN to adapt her *habitus* to the new environment, but has also helped her improve the design quality of the restaurant. TN's active engagement in transcultural learning on a daily basis paid off. Her restaurant has won the "Best Shopfront Prize" several times since Ashfield Council established the prize in 2007.

Instead of accepting the reverse time lag as the status quo, some struggled to use knowledge and skills from home to fill in the disjunction. More often than not, such efforts generated positive results and helped create new transcultural spaces. MI, a hairdresser who recently migrated from Shanghai, found that clients in Ashfield were not as picky as those in Shanghai and did not usually ask for the most fashionable hairstyles. She used to have many chances to observe the newest hairstyles and exchange ideas with other hairdressers while practicing in Shanghai. There were fewer opportunities for her to do the same after settling in Ashfield. She worried that her hairdressing expertise would degrade if she were unable to work on new hairstyles or update her knowledge regularly. Whenever possible, MI would convince her clients to try new hairstyles, although designer styles often required more time and attention. A combination of her superior hairdressing expertise, quality service, and reasonable price won MI many repeat customers from different ethnic backgrounds. The latter in turn offered helpful feedback for MI to further develop her hairdressing knowledge and hone her skills for the local context. MI is not alone. In fact, in addition to restaurants, hair salons have mushroomed in Ashfield in recent years. Their chic shop design and hairdressing expertise from Shanghai have received popular recognition, attracting both local residents and visitors from other suburbs. These hair salons have arisen as a new type of transcultural space where inter-ethnic negotiation and mutual learning takes place.

Knowledge transmission between parents and children also plays an essential role in facilitating the transcultural production of space in Ashfield. Although most first-generation migrants actively absorb knowledge from the host society through everyday practice, the dispositions they acquired in their sending society often prevent them from achieving an intimate understanding of the host society. With limited language skills, many continue to read newspapers and books in Chinese and only hang out with Chinese friends. With marginal economic and social status, their spatial experiences also remain limited. Working hard at their blue-collar jobs or labor-intensive small businesses to provide financial support to their families, many of them have few opportunities to explore the environment beyond their homes and workplaces. As a result, first-generation migrants experience the reduction of "touch sensitivity" to the host society, as if they were wearing gloves while they are fumbling through the new place. Under these circumstances, their children become an important "cultural sensor" for them to receive and respond to newer developments in the host society. While parents still play a dominant role in family decision-making, it is their children who know Australia more intimately, having been schooled there since early childhood, and who therefore play a crucial role in providing constructive suggestions that help bridge the gap between first-generation migrants' *habiti* and the host society.

One of our interviewees, for example, established a restaurant close to the municipality of Ashfield Council. The shop design idea came from one of his sons, ED. Born and bred in Sydney, ED has a Bachelor of Arts from the University of Sydney. He insisted that an attractive shop space should convey the sensitivity of traditional

Chinese culture. His father took his advice and decorated the walls with paintings of traditional Chinese cuisine. The discussion between father and son also led to another innovative idea: the romantic stories behind selected dishes were explained in English to help incite the curiosity of cosmopolitan customers. These design strategies are quite effective; the restaurant has been very popular among both Chinese and non-Chinese customers since its establishment. The mutual learning process between first-generation migrants and their children within domestic space, while largely ignored in existing literature on placemaking, turns out to have considerable implications for shaping transcultural spaces in Ashfield.

Conclusion: Beyond Assimilation and Integration

The discourse of ethnic community life in Australia has a tendency to use "culture" as the sole trope through which all aspects of migration are understood (Jayasuriya, 1997, p. 2). This study provides new insights into how the intersection of cultural differences and economic opportunities has shaped the development of its streetscape. It reveals that the built environment is a dynamic field where elements such as identity, knowledge, and capital are played out through the everyday practices of social actors. It demonstrates that the re-territorialization of migrants not only involves complicated social and political processes but also entails significant epistemological shifts. The latter cannot be easily categorized as a unidirectional movement towards the dominant knowledge system, as is often imagined in the model of assimilation. Instead, examples from Ashfield show that the knowledge brought by migrants from the sending society and the knowledge of the host society may be reformulated not as an evolutionary series but as coexisting bodies of knowledge. It is through the constant exchange, negotiation, and contestation between them that the social and physical environments of a multicultural society are continuously made and remade.

The findings of this research have important implications for rethinking the issues of assimilation and integration that have dominated current studies of migrant spaces. The assimilation model is a powerful one as it matches the social and political expectations of dominant societies and the nation-based perspectives of social sciences (Wood & Landry, 2008). The model is also deeply embedded in the modern assumptions of knowledge. The rise of modern society has been associated with the hierarchization and homogenization of knowledge. During the processes of colonization, in particular, the knowledge of dominant societies was constructed as the only valid knowledge while the knowledge of other people was dwarfed as irrational narratives that should be exorcized for lack of epistemological validity (Lu, 2010, p. 20; Lu, 2012). The repression of other knowledges has enduring consequences up until today. Under the rubric of multiculturalism, the knowledge of the host society is constructed as "modern," "positive," and "liberating," while the knowledge of the sending society is constructed as "traditional," "negative," and "restricting." Based on a refusal of the knowledge brought by migrants from their sending society, the assimilation model tends to portray migrants as people restricted by the bonds of their culture, waiting to be set free through the process of integration.

This chapter, instead, proposes an approach to migrant spaces based on the recognition of other cultures at the level of epistemology. On the one hand, its formulation of travelling knowledge attempts to go beyond the understanding of culture as bounded, essentialized, and totalizing. By looking into the daily practices of knowledge production, such an approach provides an alternative interpretation of migration not as a process of assimilation in which a static set of characteristics and values are meant to be replaced by those of the host society, but as an ongoing process of generating new knowledge and meanings through inter-ethnic communication and learning. On the other hand, this approach highlights how differently subjects and their embodied knowledges are played out in tangible economic, political, and social contexts. Going beyond the "deficit model" of immigrants, this chapter considers the knowledge brought by migrants from their sending society simultaneously with knowledge offered by the host society. It is our hope that a more open and sensitive model based on the plurality of knowledge will help to build a dynamic local community that encourages reciprocity amongst different cultural understandings.

Notes

1 This chapter is based on field research conducted in Ashfield in 2010–2012, which involved semi-structured interviews, participant observation, historical research, and visual analysis. We would like to thank all the participants of this research for sharing their experiences with us. We are also very grateful to Ashfield Council for the kind assistance we have received. The research of this chapter is supported by Worldwide Universities Network (WUN) research funding for the project "Transcultural Cities: Urban Space as Vehicle for Cross-cultural Understanding" (2010–2011) and seed funding for the research cluster "Making Migrant Space" from the Faculty of Architecture, Design and Planning, University of Sydney (2009–2012).

2 For reasons of anonymity, the names of the interviewees have been changed by using a coding system in this chapter.

References

Ashfield Council (2012) Key Statistics, available at: http://www.ashfield.nsw.gov.au/ (accessed 9 February 2012).

Australian Bureau of Statistics (2006) National Regional Profile, available at: www.abs.gov.au (accessed 9 February 2012).

Bourdieu, P. (1977) *Outline of a theory of practice*. Cambridge: Cambridge University Press.

Bourdieu, P. (1984) *Distinction: A social critique of the judgment of taste*. Cambridge, MA: Harvard University Press.

Bourdieu, P. (1986) The forms of capital, in J. Richardson (ed.), *Handbook of theory and research for the sociology of education*. New York: Greenwood Press, pp. 241–258.

Bourdieu, P. (1990) *The logic of practice*. Cambridge: Polity.

Burnley, Ian H. (2001) *The impact of immigration on Australia: A demographic approach.* South Melbourne: Oxford University Press.

Castles, S. (2000) *Ethnicity and globalization: From migrant worker to transnational citizen*. London: Sage.

Coupe, S. & Coupe, R. (1988) *Speed the plough: Ashfield, 1788–1988*. Ashfield, NSW: Ashfield Council.

Dasey, D. (2003) A home that's streets ahead, http://www.smh.com.au/articles/2003/12/25/1072308628877.html?from=storyrhs (accessed on 12 December 2010).

Hillier, J. (2007) Multiethnicity and negotiation of place, in W. Neilljind & H.-U. Schwedler (eds), *Migration and cultural inclusion in the European city*. Basingstoke: Palgrave, pp. 74–87.

Howard (2008) Who needs Din Tai Fung? http://www.eatshowandtell.com/2008/12/31/shanghai-night-ashfield/ (accessed on 12 December 2010).

Jayasuriya, L. (1997) *Immigration and multiculturalism in Australia*. Nedlands, WA: University of Western Australia.

Jordans, A. M. (1997) *Alien to citizen: Settling migrants in Australia: 1945–75*. St. Leonards: Allen and Unwin in association with the Australian Archives.

Jupp, J. (2007) *From White Australia to Woomera: The story of Australian immigration.* Cambridge: Cambridge University Press.

Lu, D. (2000) The changing landscape of hybridity: A reading of ethnic identity and urban form in late-twentieth-century Vancouver. *Traditional Dwellings and Settlements Review,* 11(2), 19–28.

Lu, D. (2006) *Remaking Chinese urban form: Modernity, scarcity and space, 1949–2005.* London: Routledge.

Lu, D. (2010) Introduction: Architecture, modernity and identity in the Third World, in Duanfang Lu (ed.), *Third World Modernism: Architecture, development and identity*, London: Routledge, pp. 1–28.

Lu, D. (2012) Entangled modernities in architecture, in G. Crysler, S. Cairns & H. Heynen (eds), *Architectural theory handbook*, London: Sage, pp. 233–248.

Qadeer, M. (1997) Pluralistic planning for multicultural cities: The Canadian practice. *Journal of the American Planning Association,* 63(4), 481–585.

Reeves, D. (2005) *Planning for diversity: Policy and planning in a world of difference*. New York: Routledge.

Saeed, A. & Akbarzadeh, S. (eds) (2001) *Muslim communities in Australia*. Sydney: University of New South Wales Press.

Said, E. W. (1983) *The world, the text, and the critic*. Cambridge, MA: Harvard University Press.

Sandercock, L. (2003) Planning in the ethno-culturally diverse city: A comment, *Planning Theory and Practice,* 4(3), 319–323.

Santos, B. S., Nunes, J. A., & Meneses, M. P. (2007) Introduction, in B. S. Santos (ed.), *Another knowledge is possible: Beyond northern epistemologies.* London: Verso, pp. xvix–lxii.

Singerman, D. (2006) Easing tensions in Sydney's Little Shanghai, http://www.eurekastreet.com.au/article.aspx?aeid=851 (accessed on 28 November 2010).

Siu, P. (1952) The sojourner, *American Journal of Sociology*, 58, 34–44.

Tewdr-Jones, M. & Allmendinger, P. (eds) (2006) *Territory, identity and spatial planning: Spatial governance in a fragmented nation*. New York: Routledge.

Thompson, B. (1991) Editor's introduction, in P. Bourdieu, *Language and symbolic power.* Cambridge: Polity.

Travers, R. (1981) *Australian mandarin: The life and times of Quong Tart*. Kenthurst, NSW: Kangaroo Press.

Waley, A. (1938) *The analects of Confucius*. New York: Vintage Books.

Watson, V. (2006) Deep difference, diversity, planning and ethics, *Planning Theory,* 5(1), 31–50.

Wise, A. (2004) Contact zones: Experiences of cultural diversity and rapid neighbourhood change among Anglo-Celtic and long term elderly residents in Ashfield. Report of Centre for Research on Social Inclusion. Sydney: Macquire University.

Wood, P. & Landry, C. (2008) *The intercultural city*. London: Earthscan.

CHAPTER 8

Listening to Transcultural Voices, Watching out for Trans-Asian Places

Kampung Kanthan in Transition

Shenglin Elijah Chang and Yenchew Foo

Chinese Malaysian contestant Jess Lee was singing Beyoncé's *Listen* at the finals for *The Avenue of Super Stars*, the Taiwanese version of *Britain's Got Talent,* on the evening of January 30, 2011. Malaysian TV stations were broadcasting the final round to domestic audiences in Malaysia. That night, at the moment when Jess Lee was announced as the new *The Avenue of Super Stars* champion on stage in Taiwan, Malaysian audiences cheered in front of their televisions in their Malaysian homes. Jess, a Chinese Malaysian exchange student who studies bio-science at the National Taiwan University, was born and raised in Seremban, West Malaysia, where 45 percent of the city's population is ethnically Chinese. The Seremban area has been a historically Chinese Malaysian dominated region since the 1970s, perhaps even earlier. Many New Villages, the Malaysian version of Chinatowns, are located in this region.

Beyoncé Knowles is a daughter of African American and Creole (a blend of African, Native American, and French ethnicities) parents. Her track, *Listen*—especially conveyed through Jess Lee's Asian voice, face, and body—conjures migration memories and experiences for Chinese Malaysian audiences. When Jess sang, "I am alone at a crossroads. I'm not at home in my own home . . .," her voice explicitly spoke to the deepest feelings of many Chinese residents in Malaysia's New Villages. These Chinese New Villagers' lives have been marginalized by deliberate political and economic policies. The young generation of Chinese New Villagers would rather jump overseas and relocate to other countries, e.g., Singapore, Taiwan, Japan, Canada, the United States, etc., than stay in Malaysia, where they do not see any hope for future prosperity. Jess Lee is among the young generation of Chinese Malaysian New Villagers who seek opportunities outside Malaysia. She made her success on the stage of *The Avenue of Super Stars* in Taiwan. However, not everyone has been as lucky as Jess. Some young people had to be smuggled into other countries and were forced to explore their dreams underground.

As young Chinese New Villagers seek opportunities abroad, other Asian immigrant workers from Nepal, Indonesia, Thailand, and India, etc., have relocated to areas neighboring Chinese New Villages to work in the industrial parks that were developed after the 1980s. These multiple directions of transnational migrations and relocations have opened up different opportunities and provided unexpected circumstances for cultures to blend into each other. More importantly, some cultures have entered into a transcultural phase in which people's lifestyles, landscapes, and identities may not be dominated by a single distinguishable cultural practice, but rather by sets of hybrid cultures that intermingle.

The terms *transcultural landscape*, *lifestyle,* and *identity* are derived from the evolving concept of transculturalism. In his essay "Transculturality: The Puzzling Form of Cultures Today," Wolfgang Welsch (1999, p. 195) points out that traditional concepts of cultures—including interculturalism, multiculturalism, and the classic single culture—assume that "every culture can be distinguished and remain separated from other folks' cultures." However, transculturalism "sketches a different picture of the relation between cultures, not one of isolation and conflict, but one of entanglement, intermixing, and commonness" (1999, p. 205). Welsch argues that we already live in a transcultural context. "Lifestyles no longer end at the borders of national cultures, but go beyond these . . . There is no longer anything absolutely foreign . . . Today, in a culture's internal relations—among its different ways of life—there exists as much foreignness as in its external relations with other cultures" (1999, pp. 197–198). Welsch's conception of transculturalism as an entangled, intermixed relation between cultures is a useful starting point for analyzing the relationship between placemaking and identity transformations.

Investigating Malaysian society through the theoretical lens of transculturalism, we recognize that Malaysian history is one of combining various cultures, different colonial rulers, and mixed migrations. Malaysia was colonized by Portugal, the British Empire, and Japan before becoming independent.[1] However, transculturality is not a default Malaysian cultural norm. When scholars in Southeast Asian studies try to analyze multiple layers of ethnic and cultural interactions within a society like Malaysia, they refer to it as an environment with plural cultural practices. For example, in the early twentieth century, Furnivall (1956) argued that this is a country containing a plural-ethnic population. Even in this century, Embong (2002) stated that Malaysia is a society of multi-ethnicities, not multi-civilizations. Diverse types of ethnic peoples coexist and practice their own cultures, but do not necessarily blend into each other's customs or beliefs. Among its 28 million citizens, Malay, the largest group, comprise a little over half of the population (54.9 percent). The majority of the remaining population includes Chinese, who make up almost a quarter of Malaysia's population (24.5 percent, 6.4 million), and Indians (7.4 percent). In addition, there are more than 13 percent aboriginal and "others." This mixed population constitutes the base for Malaysia's multi-racial, multicultural, and multi-religious environment. Malay, English, Mandarin, and Hindi are languages one hears daily, while Islamic-style buildings, Chinese architecture, mosaics with modern-style high-rise towers, and other styles of built form are one's visual surroundings. Residents worship the Islamic Muhammad, Christian Jesus, or Chinese gods, according to individual choice rooted in ethnic values. It is reminiscent of what Sandercock (2003) points out—that we are living in global cities where cultural and economic integrations are not controlled by any single country or ethnic group. She argues that this global integration is "drastically restructuring cities and regions, [and] is now a familiar refrain" (Sandercock, 2003, p. 3). In the case of Chinese Malaysian New Villages, we wonder how Chinese and other Asian cultures have been mixed or recreated when residents are busily relocating from one home to another and from one country to the next.

Unlike the Chinatown models of immigration in northern American or European societies (Yip, 1985; Anderson, 1991; Zhou, 1992), Chinese Malaysian New Villages started as concentration camps during the Emergency period (1948–1960). In this period, under the "Briggs Plan," the British relocated about 1.2 million rural dwellers and squatters, including Chinese, Malay, Indian, and *Orang Asli* (aboriginal), into about 600 new settlements (Pan, 2004). The establishment of these resettlement schemes, commonly known as "New Villages," involved forced movement of populations from scattered villages in rural areas, rubber plantations, and tin mining regions. The New Villages were planned for supposed national security reasons and carried out hurriedly in order to curb communist insurgents from approaching scattered villages for recruits or logistic support (Ngah, 2010). The "Briggs Plan" resettled squatters into New Villages surrounded by fences and police posts that cut the communists off from their sources of food, supplies, and manpower.[2] This massive resettlement program mainly affected Chinese immigrants and redefined the residential patterns of Malaya (Tan et al., 2005).[3] With Malaysia's independence, the country has grown up economically. During the 1980s, the Malaysian government imported Asian immigrant workers from Indonesia, Thailand, and Nepal to fill the labor needs of the country's rapid industrialization. The government's part in the resettlement of British and Asian immigrant workers will be discussed later in the chapter.

Using qualitative methods, we investigate how trans-Asian migrations stimulate transcultural placemaking around the New Villages in the Kampung area.[4] The New Village of Kampung Kanthan is located in West Malaysia not far from Jess Lee's home region, the Seremban area that we mentioned at the beginning of the chapter. In the following section, we first introduce the context of Chinese immigrant history and the formation of New Villages, as well as the socio-economic transformation of Chinese Malaysians within those villages. Second, we analyze the trans-Asian migrations within the younger generation of Chinese Malaysian New Villagers, while we study the migration of Asian workers outside Malaysia into the New Villages in the Kampung area. We focus on how the younger generations of Chinese Malaysians blend their cultural practice in their newly settled countries, and how other new Asian workers have been negotiating new boundaries and transcultural spaces within these Chinese New Villages. In conclusion, we address how trans-Asian newcomers redefine the landscape, and examine opportunities to reshape transcultural places in ways that allow hybrid identities to emerge.

The Center of Chinese Immigrants on the Margin of Malaysian Society

A wave of Chinese immigration in Malaya (now named Malaysia) began in the early eighteenth century. Due to crises related to famine, exploitation, and overpopulation, Chinese from the Fujian and Guangdong provinces migrated to Malaysia seeking new opportunities. Three dominant settlement forces structured these Chinese immigrants' residential patterns: urban port, mining, and rural agriculture settlements (Lee & Tan, 2000, pp. 1–5). The British encouraged rural agricultural development. They needed Chinese labor to open up the jungle and farm, especially for cultivating economic plants—e.g., rubber. When British rule brought stability and order to the Malay states, the Chinese came in greater numbers, as laborers, traders, merchants, shopkeepers, and farmers. Some made their fortunes and then returned to China, but many more remained (Tan et al., 2005, pp. 9–16).

During World War II, Japan invaded Malaya and defeated the British government. The Japanese occupation (1942–1945) brought harsh political and economic changes. To avoid the Japanese, thousands of urban unemployed Chinese laborers, miners, and estate laborers took to farming the unoccupied lands on the jungle's fringe. Some of them took up arms, hid in the hills and jungles, and created trouble for the Japanese. Many of them became members of the Malayan Communist Party (MCP) (Chor-Swang, 1985, p. 35).[5] After World War II, the British returned to re-colonialize Malaya. The MCP demanded and fought for Malayan independence. The British declared the MCP an unlawful "terrorist" organization, and this led to the state of emergency, also called "The Emergency." The British officially proclaimed "The Emergency" in Malaya on June 16, 1948, following the outbreak of an armed Communist revolt (Nyce, 1973, p. xxix). Chinese New Villages were established during the Emergency period (1948–1960).[6]

Malaysia had 450 New Villages on record as of 2002 (Lin & Fang, 2005). Chinese make up more than 82 percent of the New Villages' population in all states.[7] New Villages have been marginalized in political, economic, and social development in Malaysia.[8] Particularly regarding the economic conditions, Chinese New Villagers were left behind in the multicultural national economic picture of the 1970s. Malaysia's New Economic Policy (NEP) was first announced in 1970 as the principal policy response to the post-election race riots of May 1969,[8] which also resulted in a significant regime change. The NEP was supposed to create conditions for national unity by reducing inter-ethnic resentment due to socio-economic disparities. In practice, the NEP policies were seen as pro-Malay, the majority ethnic community (Jomo, 2004). The economic opportunities for the Chinese urban and rural working classes actually became worse under the NEP. The problem of socio-economic under-development in the New Villages was largely ignored by the government. The allocation of funds from the government to New Villages was only of a small proportion. According to one newspaper, the Malaysia Plan only allocated a total of 33 million ringgit (around US$10.5 million) to New Villages from 1971 to 2000 (Lin, 2009). In reality, however, the Department of Housing and the Local Government granted 206 million ringgit to the 450 New Villages. Each village had an average of 590,000 ringgit, totaling 3.1 percent of all allocated funds under the Eighth Malaysia Plan (Chinese Associations' Recommendations for the Ninth Malaysian Plan, 2005).

In and Out of Kanthan New Village

Planned during the Emergency period, Kampung Kanthan is one of the Chinese *Hepo Hakka* (dialect group) New Villages located in Perak State, West Malaysia. After mining declined, the area became famous for its cassava plantations between 1960 and 1970. However, farming experienced a drastic decline, and most tapioca products derived from cassava were imported from Thailand. The agricultural economy could not support second-generation villagers, many of whom migrated overseas to countries such as the United States, the United Kingdom, Japan, or Taiwan for career opportunities.

Asian newcomers emigrated to Kampung Kanthan in the Perak region in the 1980s when the area was industrialized.[10] Due to emerging urbanization in the Perak region, many areas of Kampung Kanthan were rezoned into industrial parks and resident housing from the 1980s through the 2000s. In general, we define the Kampung Kanthan New Village and the new industrial parks and surrounding communities as the Kanthan area. The Perak Hi-Tech Park program from the Ninth Malaysia Plan was among the newly rezoned parks.[11] In the early 2000s, large numbers of Asian immigrant workers, from Indonesia, Thailand, Nepal, Myanmar, India, Cambodia, and Bangladesh, relocated to the Perak region to seek better labor-intensive jobs.

Indeed, in the Kanthan area in the twenty-first century, many young Chinese Malaysians are heading overseas while young Asian immigrant workers are taking their place (Figure 8.1). As explained earlier, Chinese New Village settlements have been outside the mainstream of economic and social development. Young New Villagers are put at a disadvantage, e.g. by not having ownership of their houses in the New Villages or prospects for prosperous careers in Malaysia. Like the Malaysian contestant Jess Lee, who we introduced at the beginning of our chapter, waves of young New Villagers have—legally or illegally—migrated to other countries in search of their dreams. Popular destinations include Japan, the United Kingdom, the United States, Canada, Australia, Taiwan, Singapore, and China. Compared to Thai workers who relocate to Denmark, several of these countries are not too far away from home.[12] However, the Officer of the Public Services and Complaints Department,

Figure 8.1
The multicultural streetscape in the Kanthan area where commercial signs are written in different languages including Malay, Chinese, Tamil, and English. Photograph by Yenchew Foo.

Datuk Michael Chong Ten Soo, revealed that up to 42,000 Chinese Malaysians (most are New Village residents) had become illegal immigrant workers in Japan in 1993 (Oriental Daily, 2010/02/16).

The young New Villagers' transnational migration created transnational families within the New Villages. In the case of the transnational family phenomenon, many Chinese Villagers' families have children, siblings, or relatives scattered in different countries. Much more complicated than Levitt's (2001) Transnational Villagers, whose intertwined lives connected Boston, USA, and the corner of Mayor Carlos Pena's feed store in the Dominican Republic, Chinese Malaysian New Villagers' family members are scattered in multiple countries. Maybe the term "multinational families" better depicts the phenomenon. For example, a typical scenario could be as follows: a family has three siblings; the elder sister works in Singapore, or maybe Japan; the younger studies in Taiwan; the little brother stays with grandparents in the Kanthan New Village. During the Chinese Lunar New Year, they all return to their home in the New Village and go to their local Malaysian karaoke bar to sing pop music from Taiwan, Korea, Japan, and China. Therefore, we will identify karaoke singing and other activities as *movable transcultural performances,* and we will analyze movable performances after we explain how Asian laborers join the transnational movements.

Meanwhile, in terms of Perak's regional labor market, large numbers of Malaysian Chinese ventured out to developed countries. However, this led to a domestic industrial need to bring in fresh young labor to fill the gap; the shortage in the workforce created a pull for Asian newcomers. The Malaysian government imported Asian immigrant workers to fulfill the needs of the rapid industrialization of the 1980s. Newcomers were from countries including: Indonesia (66.5 percent), Nepal (9.2 percent), Bangladesh (8.0 percent), India (4.5 percent), and Myanmar (4.2 percent). Immigrant workers were employed in all major sectors of the economy, with manufacturing accounting the largest share at 30.5 percent, closely followed by services (25 percent), agriculture (24.7 percent), and construction (19.8 percent). Immigrant workers in the service sector were mainly employed in restaurants, hotels, and as domestic maids (2004 report, Ministry of Finance, 2006). The 2007 statistics report from the Immigration Department of Malaysia shows that the number of immigrant workers was more than two million (Department of Statistics, 2009). Compared to Chinese New Villagers, the newly arrived Asian immigrant workers have been placed in even more marginal socio-political ranks than their Chinese predecessors.

Both Chinese Malaysian New Villagers working overseas and Asian newcomers working in the Kanthan area struggle with their foreign environments. In particular, the Chinese Kanthan New Villagers have faced strong discrimination abroad. Chia Heng is a Chinese Malaysian from the Kanthan area. He became an illegal construction laborer in Japan from 1993 to 1997. He said when he was first working on a construction site in Japan; he was mistreated due to the language barriers. His boss hit his head, and cohorts called him "an idiot" or "bastard." Malaysian workers were often discriminated against in Japan. Japanese made fun of Chinese Malaysian workers and their food.[13] In the meantime, Chinese villagers in Kanthan Village also excluded other newly arrived Asian workers. Many newcomers lived in crummy

Figure 8.2 Burmese workers dressed in their home clothes at their dormitory room. Photograph by Yenchew Foo.

apartments adjacent to the Kanthan Village (Figure 8.2); few of them lived in the New Village with Chinese families around. When these Asian newcomers hung out in Chinese New Villages, for example playing basketball, shopping, and gathering, Chinese villagers felt threatened. Chee Wei was afraid of immigrant workers because he thought they intruded into *his* village. A young mother Lan Mei said, "a long time ago, the village was peaceful. Now I don't let my daughter go cycling in front of the house, as I worry about her getting kidnapped by an immigrant worker."

Indeed, Chinese villagers' feelings about their new neighbors are a mix of fear, worry, concern, and sympathy. When stories like a Nepalese worker being robbed and injured by an unknown Indian worker came up, the locals showed their humanity. Lan Mei thought some immigrant workers are "good" guys; others are "bad." Her family members are also immigrant workers in other countries, and may also face abuse by their employers. She realized all of them are in the same boat, whether they are inside or outside the Kanthan New Village, even though passengers on the boat may be of different nationalities.

Movable Cross-cultural Performances for Emerging Transcultural Spaces

The "same boat" that we pointed out in the last paragraph serves as a perfect metaphor to describe the situation for both Chinese villagers and new workers who live in the Kanthan area and in other countries. Within this situation, physical and material spaces may not be the mechanism that glues people from different cultures together. Instead, cultural events, actions, activities, and performances serve to transform daily places into transcultural stages. Similar to the trans-Pacific lifestyles defining Taiwanese Americans' dual homes between the USA and Taiwan (Chang, 2006), this cross-cultural Kanthan "boat" is not constructed by solid materials, but by actions; that is, a series of *movable cross-cultural performances*, bodily based movements that break boundaries of distinctive cultures. Movements can include, but are not limited to, music, sports, festivals, or ritual practices in ordinary circumstances or special occasions. Indeed, these spaces change completely when several performing activities take place, whether in public areas or private places. For example, music in KTV and music in public spaces can both be transculturalized, even though the former is closed and private and consumes space, and the latter is an open, public space for festivals and parties. When music is present and songs from different countries and cultures are sung, different cultural characteristics are engendered. For the cross-cultural Kanthan boat, three layers or locations of movable performances should be addressed: sport fields, public spaces or private karaoke centers, and traditional Chinese religious temples.

Kanthan Area's Sport Fields as Transcultural Space

Sports and the landscape of sport fields facilitate encounters between people from diverse cultures. Newly arrived workers cross the Kanthan area on their way to the leisure park at Taman Sri Kanthan, where they have created a comfortable space for themselves to play soccer, volleyball, and to chat. Occasionally, workers from different countries play together, and the field is transformed into a cross-cultural space. At sport fields in the Kanthan area, Chinese Malaysians, Thais, Nepalese, Indonesians, or Bangladeshis play their favorite sports. In general, Chinese Malaysians love basketball, but other Asians enjoy soccer. Soccer games are popular for Asian newcomers. The soccer field of the Kanthan area often turns into an Asian United Nations when games start. For example, Kuwat, an Indonesian worker, joins games with workers from other Asian countries after work, and Ahmad goes to Taman Sri Kanthan to play soccer and chat with friends. The experienced soccer players can easily identify the movement styles of players from different cultural backgrounds. Usually, Malaysian players enjoy team play, and they value good players only if the individual can support the group. Thais and Indonesians, deeply influenced by their traditional game "Sepak Takraw," tend to shoot the ball high in the air, but are less capable in handling ground-level ball movements. Workers from Myanmar (or Burma) historically struggled with and suffered from wars; they tend to have a very tough attitude and make formidable opponents. When a player is a ball carrier, the defenders follow him as closely as his shadow.

Each player's body movements also allow him to reconnect to certain defining moments in their memories—memories of playing in their own home countries, e.g., the triumphant moment of shooting the goal, the moment of being a goal keeper who successfully blocks a shot, the post-game experience of lying on the soft grass watching clouds in the sky, and so on. Memories of these bodily engaged feelings bring the newly arrived worker to mentally revisit their homelands.

Transnational Music Cultivating Transcultural Bodies

The cross-cultural lyrics in music—whether played in public spaces or private karaoke centers— nurture transnational personal attachments for both the newly arrived workers in the Kanthan area and transnational Chinese Malaysian young people in different countries and societies. When Jess Lee, the Chinese Malaysian singer, sang Beyoncé's *Listen*, the lyrics of the song go far beyond Beyoncé's individual life experiences and cultural interpretations. The broadcast of *The Avenue of Super Stars* in Taiwan captivated Chinese Malaysians' attention around the globe. At the moment when Jess sang, memories, emotions, life experiences, and personal meanings emerged from each listener's heart. Transcultural spaces surfaced in people's homes and in commercial clubs as the channel played and each audience member engaged in Jess's *Listen*.

Music, particularly karaoke, is an important link in the transcultural process for Chinese Malaysian young people working overseas. Jane, who worked in Singapore as a project coordinator, liked to shop and sing on weekdays. Chinese Malaysian Yoke Qi also said she always gathered with friends and sang at KTV or cooked tasty Malaysian food when she worked in Japan. Chia Heng worked in Chiba Prefecture in Japan and enjoyed singing at KTV. Many Chinese Malaysians sing popular songs at KTV in Japan or Singapore as a way to blend different cultures in a closed space. In our own experiences of living in Malaysia and Taipei and traveling in China, our friends from different countries all enjoy KTV for the access to songs from Taiwan, Hong Kong, Korea, China, and the United States. The pop songs serve as our common language, and each of us has personal memories and meanings associated with particular songs.

Asian newcomers love to sing, too. Indonesian immigrant workers living in the Kanthan area often play guitar outside their dorms (Figure 8.3). Indonesian music has many similarities to Malay language and rhythm. Music is a bridge for immigrant workers to adapt and enjoy their new lives overseas. However, the language of lyrics embodies socio-cultural status and, therefore, sets boundaries between different ethnic groups. For example, Chinese Malaysians seldom sing Malay songs. They prefer to sing popular Hong Kong, Taiwan, or Western songs. Pei Pei, a Chinese villager, told me she didn't sing Malay songs because she thought it was difficult for Chinese speakers to pronounce the Malay language. More importantly, Chinese media networks often broadcast news and pop music from Hong Kong, Taiwan, and China. Therefore, local Chinese Malaysians associate themselves with Pan-Chinese pop culture much more than with Malaysian youth culture.

The Transcultural Stage in Front of the Kwan Tee Meow Temple

Religious spaces and festivals provide the third opportunity to transform plural cultures into intertwined forms. At the traditional Chinese religious temple, the multilingual music and dance creates a barrier-free zone for different religious believers. Horton (1995) observed an amazingly transcultural July 4th Cinco De Mayo in Monterey Park, California, where Asian and Latino Americans celebrated the US Independence Day with Chinese dragon dances and Latino music and traditional dress. The multicultural festival at Monterey Park is a clear example of diverse

Figure 8.3 This Indonesian worker loves to play guitar and sing with his friends after work. Photograph by Yenchew Foo

cultures mixing in a significant event that was originally only meaningful for one particular cultural group. In the Kanthan area, the traditional Chinese Kwan Tee Meow Temple, religious center for villagers for decades, invites newly arrived workers with different religious beliefs to join the Chinese audience through pop music and hip-hop performances by bikinied dancers. Every year, the Temple committee coordinates the Mid-Autumn Festival on August 15th of the Lunar Calendar, dedicated to the Guan Tee god (*Guan Ti* or *Guan Gong* idol found in the hall of most Buddhist temples). The festival's events include *Teo-Chew* and Cantonese Opera, as well as dancing and other performances.

During the 2010 Moon Festival, Chinese villagers and migrant workers were not very interested in the traditional *Teo-Chew* performance. However, these two groups became extremely engaged in the popular dancing and singing portion of the program, even though it was rainy (Figure 8.4). That day, a bikini-dressed performer was praying for *Kwan Tee* to bless all attendees with peace and fortune. After that, six young women danced and sang to praise the *Kwan Tee* god. The rhythm of the song was adapted from a very popular song that everyone was familiar with. Hip-hop dancers worshipped the god with energetic movements that young audiences loved, Chinese villagers and migrant workers alike.

Ahmad, an Indonesian worker, told us that although he had already lived in the New Village for two years, he still felt *seronok* (happy or excited) when he went to watch the performance. Geni and Ali, also Indonesian, said they had watched the performance in Kwan Tee Meow. Ali expressed that while he did not go inside the hall or eat the food served there because he is a Muslim who follows halal dietary

Figure 8.4 Local Chinese New Villagers and other Asian newcomers watch girls hip-hop dancing at the Moon Festival on a rainy evening. Photograph by Yenchew Foo.

laws, he still enjoyed the show from outside. Even Muslim workers standing outside the hall with umbrellas remained nearby the hall railing or sat behind villagers. Some immigrant workers also joined the crowd and sat at the edge of the hall. Although the villagers and immigrant workers have different religious beliefs, popular music culture blurred the boundaries between them. As such, Kwan Tee Meow Temple momentarily switched into a transcultural space for villagers and immigrant workers.

Conclusion

Throughout this chapter, we argued that a transcultural place cannot be materialized unless certain culturally meaningful actions take place. We observed a series of movable cross-cultural performances that helped construct transcultural places. In our case studies, we identified three major performances: (1) culture-based body manners in soccer games, (2) cross-cultural lyrics and personal attachments in public spaces or private karaoke centers, and (3) music and dances at the traditional Chinese temple.

Our observations revealed that transcultural processes can occur anywhere, at any time. However, in an economically marginalized and resource-hungry community, like the Chinese Malaysian New Village, the daily mood of Chinese villagers may *guard*, not *share*, their public spaces with outsiders. Only when important festivals are taking place and people are in a celebratory mood do the locals become generous and willing to share their spaces with others.

Before closing, we suggest that everyone listen not only to their own hearts but to others' as well. As transnational migrations and transcultural exchanges take place more frequently, as we noted at the beginning of our chapter, we may all feel what Jess Lee sang: "I am alone at a crossroads. I'm not at home in my own home . . ." Indeed, we will feel lonely and insecure if we continue to embrace the old idea that one distinct culture or practice is *the* way to do things. This "not at home in one's own home" feeling in our neighborhoods, communities, and cities could be relieved were we to cultivate our understanding in multiple ways and use a transcultural lens to view and learn from our surrounding world.

Notes

1 Malaysia is a federal constitutional monarchy with thirteen states and three federal territories (Kuala Lumpur, Labuan, and Putrajaya).

2 "The Malayan Emergency (1947–1960)." Retrieved 31 October 2010.

3 Due to differences in geographical layout, historical background, and data information the states of Sarawak and Sabah in eastern Malaysia are excluded from the analysis. Our analyses focus mainly on New Villages in western Malaysia.

4 Based on qualitative methods, we applied participant observation and in-depth interviews in the field during a twelve-month period. The majority of the in-depth interviews were conducted between July and October 2010. Combining the "snowball" method with random selections, we interviewed more than thirty interviewees, including Chinese villagers and immigrant workers.

5 With the British defeated, the MCP became the spearhead of resistance to the Japanese. With the assistance of the British, the MCP retreated into the jungle to organize an anti-Japanese movement, and slowly built up a powerful fighting force, known as the Malayan People's Anti-Japanese Army (MPAJA) (Nyce, 1973, p. xxxiii). According to Tan and other researchers (Tan et al., 2005, pp. 44–46), after World War II when the Communists launched their armed struggle against the British, the previously non-political squatters called Min Yuen (mass movement) became an important source of political and material assistance to the guerrillas.

6 In the portfolio of the Ministry of Housing and Local Government (KPKT), New Villages were largely made up under the Emergency period, but they include a few Chinese villages that pre-existed the Emergency and a few others that were constructed afterwards.

7 "The Over View." Retrieved 30 October 2010.

8 From 1948 to 1969, the planning and development of New Villages was under military command. Many villagers were forcefully uprooted into this new environment. During the period after 1963 independence, as Francis Loh Kok Wah (2000) highlights in an overview of the distribution and population growth of the New Villages, five major problems faced the majority of the New Villages in the 1970s: lack of access to land, low income levels, inadequate educational opportunities, housing and overcrowding, and the lack of social facilities and physical amenities (Loh, 2000, p. 262).

9 Inter-ethnic political crisis of May 13, 1969, in Malaysia.

10 At the time the Malaysian economy was progressing from a heavy dependence on commodities towards industrialization, Perak was developing its own economy in the same direction, particularly since the heydays of tin mining were long gone (ASLI, 2003).

11 These rezoned areas include Taman Sri Kanthan, Taman Naga Kanthan, and Taman Perindustrian Kanthan DTI and more. The government acquired the land of Chinese New Villagers, because the villagers neither have ownership of the land they live on nor permits to develop industrial areas.

12 When I visited Copenhagen, Demark, in the winter of 2007, I learned to my surprise that the Thailand government has established an official immigrant workers'

program with the Danish government. Many hotel service providers are Thai workers, while Thai restaurants are popular in tourist districts. A Thai waitress informed me that there were nonstop flights between Copenhagen and Thailand at that time.

13 Yoke Qi went to Japan when she was 21 years old. She became an illegal housemaid and factory worker. She said that Japanese mock Malaysian rice as bird food because they believe that Japanese rice is more expensive and tasty.

References

"The Malayan Emergency (1947–1960)." Retrieved 31 October 2010, from: http://www.historyofwar.org/articles/wars_malaya.html.

"The Over View." Retrieved 30 October 2010, from: http://ekgbaru.kpkt.gov.my/overview_history.php.

Anderson, K. J. (1991). *Vancouver's Chinatown: Racial Discourse in Canada, 1895–1980*. London: McGill-Queen's University Press.

ASLI (Asian Strategy and Leadership Institute) (2003). The Perak Economic Summit: Repositioning for Competitiveness and Growth. Ipoh, Perak, ASLI and Perak State Government, 8–9 September.

Chang, S. (2006). *The Global Silicon Valley Home: Lives and Landscapes Within the Trans-Pacific Taiwanese Culture*. Palo Alto,CA: Stanford University Press.

Chinese Associations' Recommendations for the Ninth Malaysian Plan (2005). Online, available: http://web.jiaozong.org.my/doc/2009/rnr/wenxian/2005-9plan.pdf (accessed, November 23, 2010).

Chor-Swang, N. (1985). Reproductive Decisions and Contraceptive Use in a Chinese New Village in Malaysia (Culture, Women, Children, Pregnancy, Breastfeeding). PhD dissertation, Department of Anthropology, University of California, Davis.

Department of Statistics, Malaysia (2009). Labour and Human Resource Statistics—Series 10, February 2009. Online, available: http://www.statistics.gov.my/portal/download_Labour/files/BPTMS/PST-Siri10.pdf (accessed, November 23, 2010).

Embong, A. R. (2002). Malaysia as a Multicivilizational Society. *Macalester International*, 12: 36–58.

Furnivall, J. S. (1956). *Colonial Policy and Practice*. New York: New York University Press.

Horton, J. (1995). *The Politics of Diversity: Immigration, Resistance, and Change in Monterey Park, California*. Philadelphia, PA: Temple University Press.

Jomo, K. S. (1 September 2004). The New Economic Policy and Interethnic Relations in Malaysia. Geneva: United Nations, Research Institute for Social Development.

Lee, K. H., & Tan, C.-B. (2000). *The Chinese in Malaysia*. New York: Oxford University Press.

Levitt, P. (2001). *The Transnational Villagers*. Berkeley: University of California Press.

Loh, F. K. W. (2000). Chinese New Villages: Ethnic Identity and Politics. In K. H. Lee & C.-B. Tan (Eds), *The Chinese in Malaysia*. New York: Oxford University Press: 255–275.

Ministry of Finance. (2006) "Foreign Workers and the Malaysian Economy," extracted from the Economic Report 2004/2005. Kuala Lumpur: Ministry of Finance, Malaysia.

Ngah, Ibrahim, D. P. A. N. A. (2010). Current Planning Priorities in Rural Villages in Malaysia: Learning from the new Malaysian Village Action Plans. Paper presented at the Innovation and Sustainable Development in Agriculture and Food (ISDA) symposium, 28 June–1 July. Montpellier, France.

Nyce, R. (Ed.) (1973). *Chinese New Villages in Malaya: A Community Study*. Singapore: Malaysian Sociological Research Institute.

Sandercock, L. (2003). *Cosmopolis II: Mongrel Cities of the 21st Century*. New York: Continuum.

Tan, T. J., Ho, W. F., & Tan, J. L. (2005). *The Chinese Malaysian Contribution*. Kuala Lumpur: Centre for Malaysian Chinese Studies.

Welsch, W. (1999). Transculturality: The Puzzling Form of Cultures Today. In M. Featherstone & S. Lash (Eds), *Spaces of Culture*. London: Sage: 291.

Yip, C. L. (1985). San Francisco's Chinatown: An Architectural and Urban History. PhD dissertation, Department of Architecture, University of California, Berkeley.

Zhou, M. (1992). *Chinatown: The Socioeconomic Potential of an Urban Enclave*. Philadelphia, PA: Temple University Press.

林友順，Lin (2009–07–05) 大馬華人新村最新夢想。 亞洲周刊 **二十三卷二十 六期**。

林廷輝，方天養，Lin & Fang (2005) 馬來西亞新村——邁向新旅程。 吉隆坡， 策略分析與政策研究所 (INSAP)

潘婉明， Pan (2004) 一個新村，一種華人？ 重建馬來(西)亞華人新村的集體回憶。 馬 來西亞，大將。

Oriental Daily 謝仲洋 (2010/02/16). 濫用跳飛機 遊學簽證喊停. 東方日報. 馬來西 亞. . A12.

CHAPTER 9

Everyday Places that Connect Disparate Homelands

Remembering through the City

Clare Rishbeth

> Oh, this reminds me I must buy some dates. Everything you want you can find here. Do you have some money here? Oh no (laughs) I am always on credit here. They want to find some . . . big ones, they are good but the problem is the good ones. The big ones somewhere as well. I'll go for these ones it's £6.90. Like this you can pick this up in Arabia for £1.
>
> (Mahmud from The Yemen)

Understanding the diverse experiences of first generation migrants can provide a vital insight into connections between personal histories and place. How do the dynamics of displacement, arrival, and acculturalization shape landscape memory, support affiliations, and highlight difference? That migrants adapt their private homes to reflect their cultural identities has been well documented (Boym, 2001), but less well understood is the shared negotiation of practice and preference in public space. By being outside, issues of visibility and performance take on primary importance, and can reflect gender, social status, religious and ethnic identities.

The potential for flash points of disaffection (in the UK, for example, the 2001 disturbances in northern cities)[1], and an increasing fear of home-based terrorism, have intensified the political and policy debate about the challenges of multi-ethnic communities (Amin, 2002; Commission on Integration and Cohesion, 2007) and underlines the need for social science inquiry to inform practice. Sandercock's *Cosmopolis II* poses the question of how genuine integration and engagement can be recognized in urban neighborhoods (Sandercock, 2003a; also Binnie et al., 2006). Amin (2002) argues that the outdoor public realm is limited in its potential for meaningful sustained interaction, though Dines et al. (2006), in their report on public space use in East London, found evidence of value given to serendipitous cross-cultural encounters and points for lingering.

The idea of a *sense of place* being marked and defined by long-term residents' knowledge can be a pervasive one, and raises the question of how new arrivals find their own way of belonging, or "know their place" (Altman & Low, 1992; Armstrong, 2004). Equally, a political debate around citizenship and national affiliations needs to be informed by a deeper sociological understanding of how places that are initially unfamiliar can become identified as home, and how outsiders become insiders (Hudson et al., 2007). How might engaging in the public landscape increase personal and community confidence, and the development of customary knowledge?

Figure 9.1
Shami and her daughter. Photograph by Clare Rishbeth.

Within professional fields of planning and design, there is need for a greater critical understanding of the diversity of ways in which residents from different cultural backgrounds conceptualize, value, and use their locality. Landscape Architecture, both as an academic discipline and as a profession, has much to add to these debates; in particular contributing an understanding of the interactions between the materiality of place and social practices.

The Walking Voices Project

This chapter outlines the methods, findings, and implications of a research project called Walking Voices located in Sheffield, UK. The overarching aim of the project was to investigate perceptions and values of local neighborhoods by first generation migrants. We developed an audio and walking research method, which would allow us to more deeply understand relationships between memories of participants' homelands and development or absence of place attachment to the adopted city. Through working with key participants, from different cultural backgrounds but all living in the same neighborhood, we explored diversity in patterns of use and affordances of the public realm. Ultimately we hope to inform design and planning practice in urban transcultural locations.[2]

The research process generated on-site accounts of participants walking and thinking. In focusing on the mundane practices of urban living, the Walking Voices project relates to an emergent body of research that employs innovative methodologies to address temporal, multi-sensory, and embodied qualities of landscape (Butler, n.d.; Ingold and Vergunst, 2008). It is shaped by a belief that the detail of "social exchanges, memories, images and daily use of the material setting" is useful in understanding people–place relationships (Low, 2000, p. 128). It also responds to Sandercock's discussion of the "multiplicity of stories" that recognizes the importance of connecting immediate experiences with personal meaningful resonances, and developing these as a means of sharpening theory and informing policy (Sandercock, 2003b; Sarkissian, 2005; Armstrong, 2004; Low et al., 2002).

Focusing on One Urban Neighborhood

The project was located in Burngreave, Sheffield, a large city (pop. 500,000) in the north of England. The neighborhood is characteristic of many British inner city urban areas in reflecting patterns of British migration over the last sixty years. In the years after the Second World War, Britain needed overseas workers to work in industry and in the health sector, and actively recruited from countries within the former British Empire, especially Africa, the Caribbean, and the Indian Subcontinent (Mason, 2000). In Burngreave, a Yemini community of steel workers was established in the 1950s, and it became a key housing area for Pakistanis and Indians from the 1970s onward. More recently, the UKs immigration pattern has changed, reflecting the expansion of the European Community, and the human impact of wars in the Middle East and Afghanistan. Many Eastern Europeans, most notably Polish, have taken advantage of reciprocal rights of work to move to the UK. Burngreave is also one of the main areas in Sheffield where asylum seekers are housed. Though the official record of ethnicities other than "White British" in the UK is 9 percent of the national population (Office of National Statistics, 2001), in urban areas, especially those with low-grade housing, there is a significantly more diverse ethnic profile. The 2001 UK Census categorizes the predominant population groups in the Burngreave ward as White (56 percent), Pakistani (19 percent), Caribbean (6 percent), and African (5 percent) Almost one-quarter of Burngreave residents registered in the census were born overseas.

The housing is a mix of late nineteenth-century terraces and housing estates built over the last four decades, with 48 percent of homes defined by two or more aspects of disadvantage (education, health, employment and housing quality (ONS, ibid.). There are a few designated parks, but most of the green space is municipal mown grass around the housing blocks in the newer estates, and areas of shrub woodland that accommodate steeper slopes. At the time of the fieldwork a ten-year national funding initiative was running to improve social, economic, and environmental aspects of the neighborhood.[3]

Representing Diverse Experiences

The methods used aimed to address some of the limitations in traditional qualitative research techniques, specifically when working with people from diverse cultural backgrounds. With the Walking Voices method, we chose to work with a few people over an extended period of time, so it was particularly important that the selection process of our key participants was rigorous, and that we were mindful of the demands and power relationship of the research process.

People living in areas of multiple deprivation are often "over-researched" through a combination of academic and community planning initiatives. This can particularly affect people from ethnic community backgrounds who are regular users of cultural community centers. Recruiting solely through community centers can also introduce a bias towards the voices of people not in work and with strong links to their cultural heritage. Instead, during the first months of the project we immersed ourselves as broadly as possible in the neighborhood, discussing the project themes with more than sixty contacts (individuals and organizations) in a wide range of locations: healthy walks, library users, allotments, places of worship, park users, shops, community centers, and adult education.

From these contacts we recruited eleven participants who lived in or near Burngreave. The five women and six men came from five different countries of birth: The Yemen, Jamaica, Pakistan, Iraq, and Somalia. They had a diverse range of life contexts: employed/not in work, large family/single person, strong community networks/relatively isolated, intention of returning to home country/long-term settled in the UK.[4]

Figure 9.2
Mahmud showing us his street. Photograph by Clare Rishbeth.

Recording of On-site Commentaries

The participants were loaned mini-disk recorders for up to three months spanning late summer to early winter. Their brief was to regularly record on-site commentaries during their everyday walks or journeys, talking about observations, thoughts, and choices prompted by the changing scene. We worked in collaboration with BBC Radio Sheffield, with the aim that selected recordings would be broadcast, and participants received training from BBC staff.

The process of making the recordings varied greatly between participants. Some were largely independent and returned recordings regularly. A few required more support; we found that a researcher walking alongside them ("active listening" rather than interviewing) gave them more confidence. Most participants made recordings infrequently, but with longer commentaries than expected, and in total we achieved an extended body of qualitative data (58 recordings, average length 30 minutes).

The ethnographic approach, where we were in contact with participants on a weekly or fortnightly basis over a period of months, allowed for trust to build between participants and researchers. Research topics could be gradually introduced, discussed, and understood within the wider context of the participants' lives. Walking practices varied; in particular, some of the male participants preferred to make recordings while in their cars as they did not often walk places. However, regular contact ensured a reflective loop for researchers to discuss emerging findings with participants. The majority of the recordings were produced while the researcher was absent, allowing the direction and priority of content to be decided by the participant. As such, we ceded some of the social scientist's traditional powers: the right to ask questions and to impose consistency of method.

Informing the Research Themes

> There used to be a river flowing out, but when I came you could hear the water. The river behind all these bushes and it sounded so lovely. I'm assuming it would be cold as well . . . I thought I heard a trickle of water. It doesn't look as lush. I mean there's a lot of bush and green and everything but it's not as lush. It's dried and portions of it are dried . . . ahhh there's some people coming in front of us and it's very narrow. We'll make it somehow.
>
> (Thelma from Jamaica, walking in the countryside near Sheffield)

The content of the self-recorded commentaries were shaped by participants' own interests and daily practices, and provided accounts of place that often reflected sensory and experiential details. Many recordings took the form of extended monologues, and a number drifted from a reaction to the immediate situation (feeling cold, not very safe) to thoughts on their own migrant identity, childhood experiences, and stories of their arrival in Britain. The descriptive and reflective qualities of the transcripts exceeded expectations, and provided an "insider ear" to the emotional dimensions of migration and place. Aspects of cultural identity were mentioned, sometimes obliquely, and at other times related to the broader range of recordings. The life history interview was not part of the project, so our understanding of participants' backgrounds was dependent on whether participants volunteered this information. By using on-site moving recordings we supported multi-sensory descriptions that responded instinctively to the unpredictable changing characteristics of specific places (visual, social, weather-related, physicality of walking).

The task of talking to oneself maintained an internal dialogue of perceptions and value judgments, and triggered memories and ideas, and seemed to be significantly different in character to descriptions of landscape in indoor interview settings (Scott et al., 2009; Finney and Rishbeth, 2006). "Walking . . . inevitably leads to other subjects," suggests Solnit (2001, p. 8). We indeed found that this alternative to one-off interview formats supported gradual revelations that enabled participants to express ambiguities and contradictions; both common discrepancies between intentions and actions, and those specifically reflecting the complexity of cultural hybridity for first generation migrants. Listening to the actual recordings (as opposed to reading the transcripts), the quality of evoking a sense of place is striking, and prompted a focus on audio content within the dissemination strategy which was largely achieved by means of the website (www.walkingvoices.group.shef.ac.uk).

Reflective Analysis and Dissemination

From working intensively with a small group of participants we recognized the need to contextualize and confirm the findings. We scheduled interviews with a selection of participants to discuss an overview of their use of public spaces and travelling, and to probe deeper into some of the issues arising from their recordings. We circulated "draft implications" to a range of local and national environmental and community organizations (many of which we contacted in our initial networking) and undertook interviews to incorporate their perspectives on the issues raised. Through transcription, coding, and ongoing discussion within the research team it was possible to build up profiles of preferences, concerns, and motivating factors for each participant.

The collaboration with BBC Radio Sheffield was integral to participant recruitment and community dissemination, with the ultimate intention of creating a radio program. Participants were attracted to receiving training and insider insight into a high-profile partner, and took pride in learning new skills with professional quality equipment. The recordings provided a media-friendly focus for disseminating the research within a local context. Though ultimately a stand-alone program was not achieved, a few participants were interviewed on existing mainstream programs, and most had their clips broadcast on air. Participants were consulted both prior to and after the fieldwork with regard to consent to use their recordings, with care taken to discuss the implications for differing media: a one-off radio broadcast, a publicly accessible website with images, and academically focused papers. Some chose to remain anonymous, while others wanted to be credited for their creative output.[5]

Critique of Methodological Approach

The Walking Voices method was developed to attempt to replicate as nearly as possible the everyday experience of being outdoors, in trying to reach beyond the distancing effect of an indoor interview or the staged nature of researcher-led trips to specific leisure destinations. It aimed to give space for participants to develop their own ideas and issues to communicate, with an absolute minimum of direction from researchers. The gradual gaining of an overview of the participants' lifeworlds was vital for the careful analysis of emerging themes of the recorded material, and to demonstrate a genuine interest and respect for the people and communities involved. However, the intensity and demands on both researchers and participants restricted the practical scope of numbers involved.

Within the context of social science research the method offers a precise detail and tactile quality to previous work on place attachment. The unusual and engaging research activity attracted a varied group of participants; certainly a different sample than might have been recruited to take part in interviews or questionnaires. The ethnic range is unusual in academic studies, which have a tendency to focus on specific ethnic groups rather than interrogating diverse views of a common place (Glick Schiller & Çağlar, 2009). An approach that mostly absented the researcher from the production of the primary data allowed participants freedom to express values and interests, and was key to our ideas about ethical relationships in combining creative processes and research, but did limit some of the means by which the data could be evaluated.

Findings: Experiencing Transcultural Places

The Walking Voices project generated narratives which contribute new understandings of landscape and migration in two important respects. In focusing on the urban landscape experience of first generation migrants it presents an experiential focus on the role, qualities, and opportunities of public space. Second, the use of immersive participant-led recordings allows a vital insight into person–place interactions, providing a landscape-specific framework for analysis. The findings highlight the potential and problems of ethnically diverse neighborhoods as places where migrants can establish a sense of belonging (Binnie et al., 2006; Amin, 2002; Alexander et al., 2007).

Comparisons and Connections to the Country of Origin

> This morning was quite shiny, hot, sunny, and some people came out without jacket—the thing is now it's raining. When it gets to about 12-something the rain starts, and that's why I said earlier on you don't trust the weather. Great Britain can be anything, sometimes it can be hot, sometimes it can be ok, one time it can be winter. Unreliable, the weather in Great Britain is unreliable. And that's why you have to put your jacket on, that's what I do always.
>
> (Osman from Somalia)

Changing country of residence can lead to feelings of culture shock and reduce a person's ability to undertake everyday tasks, and this study found the context of the local environment is a strong contributing factor. Disorientation, difficulty in understanding social situations, and lack of familiar leisure activities can be exacerbated in the public realm and contribute to initial negative feelings about residence. However, going outside to undertake daily tasks can support practical re-skilling of new migrants. By observing others, reading information, visiting different shops, and noting landmarks, newcomers engage in experiential learning and pick up customs and competencies related to their local neighborhood and beyond. Amassing insider knowledge supports a sense of belonging and rekindles an ability to feel relaxed and "at home," one of the dimensions of place attachment.

Memories of migrants' home countries are sometimes prompted by outdoor places in the city. New landscapes and experiences are compared both positively and negatively to those in the home country. Views, social practices, and activities can all signify a break with the past ("This is very different from . . .") or a means of continuing an identity ("This reminds me so much of . . .").

Finding aspects of commonality with previous places of residence is generally a restorative experience, mostly prompting feelings of pleasure, sometimes with aspects of loss. Activities within public space are important. Continuing habits of socializing—men sitting outdoors having a chat, or greeting people in the street—appeared to be particularly significant as they reflect deeper personal values of the importance of community. "Socialize is like you socialize back home. Place where everyone meet and everybody sits down, have a talk" (Abdullah, from The Yemen, talking about the small green in the center of Burngreave). This connection of memory is an important contribution to theoretical inquiries into the nature of

Figure 9.3
Ellesmere Green, a small central green space near the mosque, library, post office, and bus stops. Photograph by Mark Powell.

place attachment, and broadens the understanding of familiarity with regard to landscape.

The research found that migrants were creative in the means by which they reinterpreted activities that they enjoyed in the past, adapting locations of walking, driving, or socializing to a new context. This implies that the participants were engaging at times in "reflective nostalgia" (Boym, 2001), a fluid interpretation of past and present experiences in a way that provides new insights into the nature of their located identity.

Performing Individual and Community Identities in Public

> I hated that road . . . I am very wary of going there because there are a lot of Arab men and women always looking at you and judging you, seeing what you got on, seeing how you look, seeing if your scarf is on properly.
>
> (Shireen from The Yemen)

Many first generation migrants live near to others of their own ethnic community, and this proximity means there can be a high level of awareness that by being outdoors they are responding to the public gaze of that community. The research underlined the importance of performing social roles in response to the expectation of one's own community, often framed as a need to reflect well on your family, and this dimension of performance can directly impact on personal experiences of public space (Powell and Rishbeth, 2012).

Figure 9.4 Qualities of walking with a pushchair. Photograph by Clare Rishbeth.

Cultural codes of how you are seen, and who you are seen to be, are complex, and relate to one's own perception of visibility in a public setting. While being outside in their local neighborhood provides a social context for migrants, this can become burdensome at times, either by restricting opportunities or by having too many obligations. Making occasional or regular trips out of the neighborhood was a strategy used by participants in our research for temporarily escaping a known public gaze and finding some personal privacy while outdoors.

For most first generation migrants, the visibility of being different from the white British population increases when in less ethnically diverse neighborhoods, such as rural localities or more affluent parts of the city. Visiting these areas can trigger feelings of vulnerability to racism, especially if this has been previously experienced. Most of the participants considered themselves more socially comfortable living in areas of ethnic diversity, and identified positively with this aspect of their home neighborhood.

Life Choices Reflected in Place Attachment and Sense of Belonging

> I know this is an area predominantly that is talked about being full of crime and violence. I lived here for a year, and I never really witnessed any incidents really and I know there are things that happen but I never seen anything. And today is another Sunday evening like everywhere else really. I can see the sunset across the horizon and it's quite beautiful, bits of gold glowing, soft background of purple clouds.
>
> (Lola from Jamaica)

Living in a city offers migrants a diverse range of experiences beyond that found in their immediate neighborhood or own ethnic community. Exploration of different places and activities can inform aspirations for the future, sometimes coupled to their motivation for initially moving to the UK. Their life choices are often shaped by a tension between keeping contact with their own ethnic group, respecting and enjoying its cultural traditions, and a desire to embrace new opportunities. Negotiating aspects of individual and community identities can take a spatial form, concerning both daily movements and long-term residence. Movement and exploration can represent a confident development of hybrid identities, mixing elements of affiliation and association that both reflect and inform a complex sense of self (Rishbeth and Powell, 2012).

The recordings clearly represent the complexity and ambiguity with which migrants engage with place. The participants identified a range of positive and negative feelings about their locality, some of which were informed by their experience of difference, and comparisons with their previous home. Through becoming familiar with a place, and finding means of connection with their personal values, they develop a sense of themselves as locals. The characteristics of super-diversity of the neighborhood give emphasis to a shared sense of belonging that is less dependent on country of birth. Attachments to migrants' current homes exist alongside transnational links and options of future moves (within the UK or internationally).

In understanding issues of integration and belonging in multi-ethnic neighborhoods there is a need to recognize the importance of non-verbal interactions. Movement through and spending time outdoors were connected with emotional understandings, social interpretation, and the physicality of the bodily experience (Ingold, 2000; Massey, 2005). We suggest that political discourses that highlight separatism at the local scale often fail to recognize the nuances of social networks, place attachment, and engagement in the public realm experienced by migrants of different ethnic backgrounds and life situations.

Guidance for Practitioners: Placemaking in the Space of Flow

The implications arising from this research were formulated by careful discussion of the fieldwork, in particular contextualizing the recordings of the key participants with information provided by local and national community organizations, and our own understandings of use of the open spaces in Burngreave from our site visits. I also drew on my own professional background in landscape architecture, responses to presenting the findings in professional contexts, and awareness of current practice in these areas. As such, the guidelines are less a direct outcome of the recordings, and more a dialogue stemming from them, a critical engagement that seeks to give practitioners the opportunities to deepen their understanding of migrant experiences. The implications for practitioners are also summarized in a leaflet, distributed in hard copy and available in digital form from the Walking Voices website.

The key points are outlined below. There is no apology that many of these points are also relevant to improving the experience of public open space for all users, that which is common between people being as important to that which differs. But I suggest that because of the common past experience of many (not all) migrants—disorientation, unfamiliarity, feeling unskilled or isolated, experience of racism—attention to these issues might be more important for this user group than the non-migrant population.

Facilitate overlapping and flexible use of space. Frequent use of parks, streets, public transport, and urban centers can support individual relationships and provide opportunities over time for acquaintances to strengthen into friendships. For this to happen, public spaces need to be safe and attractive to use. Visual permeability—multiple ways of viewing in and out of a place—can be particularly important, both in terms of a personal sense of safety, and of increasing the visible presence of different ethnic communities. Priorities for improvement should be given to locations where people gather due to common interests that cross over ethnic groups: playgrounds, bus stops, sports facilities, and entrances to schools and libraries, waterfronts and large parks. Design and plan for spaces to be flexibly used by different user groups, with the possibility of informal appropriation regarding seating, types of activities, and seasonal uses, as well as formal temporary activities such as markets and festivals.

Figure 9.5
Car washing in the street. Photograph by Mark Powell.

Be aware of culturally defined behavior and legitimize where possible. Acceptable behavior within the public realm is culturally defined, and as such can present misunderstandings, and occasionally conflict, between different ethnic communities. For example, men from some cultural backgrounds are used to meeting in groups outside of the home, and often appropriate street spaces for socializing. This can sometimes be perceived as threatening by others, especially if the places used to meet are degraded or associated with crime (for example derelict plots). Increasing potential locations for social gatherings that are explicitly designed and well maintained, and providing comfortable seating which accommodates groups rather than couples, can help reconfigure these associations and tacitly legitimize such gatherings. Cultural values can vary with regard to litter, noise, and acceptable use of street space, potentially leading to petty irritations which exacerbate or reinforce ethnic stereotypes. It may not be possible or appropriate to design out these differences, but supporting community political and social structures that enhance communication could ease the sometimes awkward quality of urban life, that of "living among strangers" (Sandercock, 2003a).

Support diversity without resorting to ethnic stereotypes. The "critical mass" of one's own ethnic community was seen as crucial in providing support, resources, and social contact. The mix of nationalities was also often cited as a positive characteristic, and the diversity of residents was something that, at best, provided a rich social and entrepreneurial resource for the area. A quality of super-diverse neighborhoods is a high level of transitory residents, with an inflow of newcomers hoping to make a new life. Planning policies that can support informality of land uses, short-term enterprises, and micro-businesses may be appropriate to ensuring the vitality and strengths of this residential mix. Promoting an area as a "cultural" or multicultural destination can give a neighborhood particular protection and increase financial resources. However, caution needs to be exercised. Inter-community relations can be damaged if it is perceived that one ethnic group is being given preferential treatment. More subtly but intrinsically problematic is the process of urban designers or planners naïvely confirming stereotypes of bounded ethnic identities, sometimes by use of particular icons from the country of origin, which fails to recognize the flux of transnational links that characterize individuals and communities.

Tackle inequality. Though living in an ethnically diverse community was highly valued, the low socio-economic status of the area meant that residents were, often reluctantly, planning to move to another neighborhood at some future date. As well as flashpoint issues of crime, education, and housing, there were also frustrations at a smaller scale: lack of provision of cashpoints and supermarkets, and poor maintenance of the streetscape; residents were aware that in other parts of the city there was a much better quality of urban environment. A permanent move to another part of the city, often motivated by a desire for better schooling and a safer environment, was characterized by participants as a significant loss of contact with their religious and ethnic communities.

> Everything in this area is . . . low level . . . so why would I be living in an area with low level? Sometimes . . . because my community live here, my people, you cannot associate with other people, and if you want to go to [place name: a predominantly white community], you get yourself isolated and lonely . . . lot of people don't think of this area as bad. They enjoy the area, they want to be here.
>
> (Osman from Somalia)

For first generation migrants the conflict of push and pull factors can be particularly sharp: sometimes a sense of shame that they have not accomplished a better standard of living, against the relief of growing familiarity and attachment to this new location that they now call home. Environmental inequality is well documented in academic and policy arenas, but is still an issue that is far from resolved. Decoupling the links between areas of multiple disadvantage and communities of high ethnic diversity is an urgent challenge for urban professionals across design, planning, and politics.

Summary

This chapter has looked at some of the ways in which place is experienced and envisioned specifically within a neighborhood which is ethnically mixed. The qualities of this transcultural place are not shaped by one population shift, but by a complex web of incomings and outgoings, and an ongoing history of short-term residents becoming long-term residents. The dynamics of how these cultures are represented in the urban environment is reflected through physical adaptations, flexible appropriation, patterns of socialization, and both shared and divergent values of what makes the area "home." Working on planning and design projects in neighborhoods such as these require professionals to be culturally agile, good listeners, observant to nuances of use, and advocates for those with less political and economic power. Above all, these communities need skilled optimists who can recognize and improve the potential of urban spaces for communication, empowerment, and diverse ways of belonging.

Notes

1 These were disturbances in some northern English towns in areas characterized by racial residential and educational segregation and a context of high unemployment. Neighborhoods affected had majority populations of Pakistani and Bangladeshi residents. Violence was, by and large, provoked by white racist gangs targeting Asian areas, with subsequent retaliation. The police force suffered high casualties (Kundnani, 2001).

2 The fieldwork was undertaken by the author and a Research Associate, Dr. Mark Powell, during the period 2006–2007. The research was funded by the UK Economic and Social Research Council.

3 Burngreave New Deal for Communities (BNDfC) was a government-funded organization set up to regenerate the Burngreave area. The Department for Communities and Local Governments' Neighbourhood Renewal Strategy awarded Burngreave £52 million, to be spent on the regeneration of the area over a ten-year period from 2001 to 2011 (http://www.bndfc.co.uk/).

4 We were not able to involve first generation migrants who did not speak reasonable English, or those who were not confident in learning simple new technologies. Within the scope of the project we addressed this by conducting one individual and two group interviews with older members of the Pakistani community.

5 All participants' names are anonymized within this chapter.

References

Alexander, C., Edwards, R., & Temple, B. (2007). Contesting cultural communities: Language, ethnicity and citizenship in Britain. *Journal of Ethnic and Migration Studies, 33*, 783–800.

Altman, I. & Low, S.M. (Eds). (1992). *Place Attachment.* New York: Plenum.

Amin, A. (2002). Ethnicity and the multicultural city: Living with diversity. *Environment and Planning A, 34*, 959–980.

Armstrong, H. (2004). Making the unfamiliar familiar: Research journeys towards understanding migration and place. *Landscape Research, 29*, 237–260.

Binnie, J., Young, C., Millington, S., & Holloway, J. (2006). *Cosmopolitan Urbanism.* London: Routledge.

Boym, S. (2001). *The Future of Nostalgia.* New York: Basic Books.

Butler, T. (n.d.). *Memoryscape audio walks.* Retrieved December 7, 2010, from: http://www.memoryscape.org.uk/

Commission on Integration and Cohesion. (2007). *Commission on Integration and Cohesion—Our shared future.* Retrieved December 7, 2010, from: http://collections.europarchive.org/tna/20080726153624/http:/www.integrationandcohesion.org.uk/Our_final_report.aspx
Dines, N., Cattell, V., Gesler, W., & Curtis, S. (2006). *Public Spaces, Social Relations and Well-being in East London.* Bristol: Policy Press.
Finney, N. & Rishbeth, C. (2006). Engaging with marginalised groups in public open space research: The potential of collaboration and combined methods. *Planning Theory & Practice, 7*, 27–46.
Glick Schiller, N. & Çağlar, A. (2009). Towards a comparative theory of locality in migration studies: Migrant incorporation and city scale. *Journal of Ethnic and Migration Studies, 35*(2), 177–202.
Hudson, M., Phillips J., Ray, K., & Barnes, H. (2007). *Social Cohesion in Diverse Communities.* York: Joseph Rowntree Foundation.
Ingold, T. (2000). *The Perception of the Environment: Essays on livelihood, dwelling and skill.* London: Routledge.
Ingold, T. & Vergunst, J. (Eds). (2008). *Ways of Walking. Ethnography and practice on foot.* Aldershot: Ashgate.
Kundnani, A. (2001) From Oldham to Bradford: The violence of the violated. *Institute of Race Relations.* Retrieved August 10, 2011, from: http://www.irr.org.uk/2001/october/ak000003.html
Low, S. (2000). *On the Plaza: The politics of public space and culture.* Austin: University of Texas Press.
Low, S., Taplin, D., Scheld, S., & Fisher, T. (2002). Recapturing erased histories: Ethnicity, design and cultural representation—a case study of Independence National Historical Park. *Journal of Architecture and Planning Research, 19*, 282–299. Retrieved from: http://www.lockescience.com/
Mason, D. (2000). *Race and Ethnicity in Modern Britain*. Oxford: Oxford University Press.
Massey, D. (2005). *For Space.* London: Sage.
Office of National Statistics (ONS). (2001). *Burngreave Ward.* Retrieved December 7, 2010, from: http://www.neighbourhood.statistics.gov.uk/dissemination/LeadDatasetList.do?a=7&b=6174550&c=burngreave&d=14&g=365020&i=1001x1003&m=0&r=1&s=1291727935184&enc=1&domainId=16
Powell, M. & Rishbeth, C. (2012). Flexibility in place and meanings of place by first generation migrants. Tijdschrift voor economische en sociale geografie, 103(1), 69-84. DOI 10.1111/j.1467-9663.2011.00675.0078
Rishbeth, C. & Powell, J. (2012) Place Attachment and Memory: Landscapes of Belonging as Experienced Post-migration. *Landscape Research*. DOI:10.1080/01426397.2011.642344
Sandercock, L. (2003a). *Cosmopolis II. Mongrel cities of the 21st century*. London: Continuum.
Sandercock, L. (2003b). Out of the closet: The importance of stories and storytelling in planning practice. *Planning Theory and Practice, 4*, 11–28.
Sarkissian, W. (2005). Stories in a park: Giving voice to the voiceless in Eagleby, Australia. *Planning Theory and Practice, 6*(1), 103–117.
Scott, A., Carter, C., Brown, K., & White, V. (2009). Seeing is not everything: Exploring the landscape experiences of different publics. *Landscape Research, 34*(4), 397–424.
Solnit, R. (2001). *Wanderlust: A history of walking*. London: Verso.

PART III

BRIDGING SPACES OF DIFFERENCE

CHAPTER 10

"We Are the Fruit Bowl"

Place, Cultural Identity, and Social Ties among Immigrant Residents in Public Housing

Lynne C. Manzo

> They [HUD] only gave a small percent [of the HOPE VI grant] for translating languages. I think it was one percent of the budget, and I said "That's impossible for us" [and the HUD staff person said]: "Oh you just need to translate it into Spanish" and I go "no!" and he goes "No? What do you mean, no?" [I said] "There's eighteen different languages!" and it just blew the whole room. They could not comprehend it. We are one community and we are a mixture of people. We are the fruit bowl.
>
> (Resident and community leader)

Toward Transcultural Cities

To consider a city transcultural implies that it is more than multicultural; it is a place that provides opportunities for the spatial expression of multiple cultural identities and the chance to traverse different worldviews. A transcultural city enables new cultural encounters and spatial practices to emerge and offers opportunities to engage in cosmopolitanism as Hannerz (1996) defines it—as an orientation and willingness to engage with the Other (p. 103). It calls for a greater appreciation of the fluidity and hybridity of urban space, enabling critical encounters with difference within the same local urban space. The goal of achieving the transcultural city calls for us to be greater than our individually situated selves, to bridge spaces of difference, not to create isolated ethnic enclaves, and to move our cities toward greater spatial justice. That is, transcultural cities do not merely exist because diverse populations coexist in urban spaces; transcultural cities are something toward which we can actively strive in an effort to achieve social and spatial justice. In this view, then, the transcultural city is an aspirational project and will be treated as such in this chapter.

In the current context of neoliberalism, however, the allocation of spatial, political, and economic resources favors economic growth rather than wider social benefits (Fainstein, 2010) and diversity is less important, irrelevant, or worse, contrary to that agenda. Nonetheless, increased cultural diversity is redefining immigrant-receiving cities throughout the world, making multicultural planning imperative for a "socially cohesive and sustainable urban environment" (Pestieau & Wallace, 2003, p. 257). There is evidence that urban transcultural "moments" and locations are achieved in cities and such moments are described in this and other chapters in this volume. While some work (see, for example, Cahill, Chapter 14 in this volume) shows that such transcultural moments can be locations of contestation, in other instances, such as the case I discuss here, transcultural moments can be models for traversing worldviews and redefining common ground in urban space. While it is no small task to embrace diversity in a way that would move our cities to be truly transcultural, doing so is necessary to move our urban centers toward being more than just cities (Fainstein, 2010).

Geographers, urban theorists, and planners have been considering how to achieve the transcultural city for at least two decades (Al Sayyad, 2001; Fainstein, 2005; Hannerz, 1996; Rishbeth, 2004; Sandercock, 1998; Sandercock, 2003; Sandercock & Attili, 2009). For example, Sandercock (2003) offers an image of *cosmopolis* as "a construction site of the mind and heart, a city in which there is genuine acceptance of, connection with, and respect and space for 'the stranger' . . . in which there exists the possibility of working together on matters of common destiny, of forging new hybrid cultures and communities" (p. xiv). This is similar to what Al Sayyad (2001) describes as "hybrid places"—places where cultural identities meet and transform each other. Sandercock and Attili (2009) argue that such cross-cultural transactions can offer not only a sense of belonging that derives from being associated with other cultures, but a "strength and compassion from accommodation among and interrelations with others," noting that it is important to recognize and nurture those spaces in which accommodation and intermingling occur (p. 222). This chapter tells the story of one such hybrid space, a space of accommodation and intermingling—a public housing community of ethnically diverse residents on the brink of redevelopment.

One of the earlier writers on cultural diversity, Ulf Hannerz (1996) put forward seven arguments for its value: (1) the moral right to one's culture; (2) the ecological advantage of different orientations; (3) a form of cultural resistance; (4) pleasurable experience of different worldviews; (5) possibility of confrontation that can generate new cultural processes; (6) a source of creativity; and (7) a fund of tested knowledge (pp. 56–57). Similarly, Glaeser (2000) argues that close proximity of people of diverse backgrounds creates "intellectual spillovers" outside of the marketplace that may be especially important for marginalized communities. Here, the sharing of information outside of the marketplace sparks creativity, innovation, and mutual support among members who are alienated from the marketplace (Miller, 2011). More recently, Emily Talen (2006) observed that "diversity is a positive force in a global society, constituting a mode of existence that enhances human experience" (p. 236). She identified four benefits of diversity to the city—place vitality, economic health, sustainability, and social equity. This chapter illustrates how these benefits unfolded in a community that was labeled by housing providers as distressed, yet was the site of remarkable transcultural mutual support networks.

Before turning to that case, however, there is one more theory to consider to help us understand the dynamics within that community, and this is Lefebvre's (1991, and as summarized in McCann, 1999, p. 172) three moments in the production of space. First are *representations of place*—this is the space of planners and bureaucrats. Constructed through discourse, it remains purely conceived rather than directly lived, yet it shapes how we conceptualize ordered space. Second is *representational space*—the space of imagination through which life is directly experienced and lived through its inhabitants. This includes the construction of counter-discourses. Finally, there are *spatial practices*— the everyday routines and place experiences that produce their own social spaces. These practices are mediated by representations of space and representational space. As I will demonstrate, formal bureaucratic representations of space are often at odds with lived representational space, and notably at odds with spatial practices, particularly those among poor communities of color. It is in examining the tensions between these different aspects of the production of urban space that we can better understand how to move closer to achieving a truly transcultural city.

This three-pronged framework is especially useful for understanding the kind of urban restructuring program taking place in US cities today in mixed-income developments and in public housing redevelopment programs like HOPE VI (Housing Opportunities for People Everywhere). HOPE VI is a competitive grants program administered by the US Department of Housing and Urban Development (HUD) whose four main goals are: (1) to "improve" the living environment for public housing residents of "severely distressed public housing" projects through the demolition, rehabilitation, reconfiguration, or replacement of obsolete public housing projects; (2) to revitalize sites on which such public housing projects are located and contribute to the improvement of the surrounding neighborhood; (3) to provide housing that will avoid or decrease the concentration of very low-income families; and (4) to build sustainable communities.[1] In reality, such programs are part of a larger suite of urban policy decisions designed to strategize urban development, erase stigmatized structures from the landscape, and re-image the city as a safe zone (Crump, 2002). However, such strategies typically come at the cost of poor communities of color whose very existence challenges the formal representations of the city that are sought by decision-makers. Indeed, since the inception of the HOPE VI program in 1993, as many as 206 cities across the country have implemented 559 of these redevelopment projects, resulting in the demolition of tens of thousands of public housing units and the displacement of more than a hundred thousand poor households (Manzo, 2010).

This chapter examines the lived experience of place among immigrant residents of an ethnically diverse public housing site in the Pacific Northwest of the USA, which is undergoing redevelopment through HOPE VI, to understand the intersection of place meaning, cultural identity, and social support networks among and across diverse residents. This research demonstrates that such communities can be seen as microcosms of the transcultural city, particularly when mutual support networking traverses racial, ethnic, and linguistic boundaries to enhance quality of life and aids community members in the common project of living with limited resources. Together, residents of this neighborhood all, in their own way, struggled to survive economically, maintain a household, and participate in community life. These shared efforts and goals comprise "the common life" (Fullilove, 2004; Levy, 2005; Manzo et al., 2008). This common project of living transcends cultural barriers and provides important lessons for achieving the transcultural city.

Immigrants and Social Ties

Immigrants' ability to empower themselves through mobilization and solidarity may be made easier through spatial concentration (Mazumdar et al., 2000; Crump, 2002; Goetz, 2003; Lin, 2000; Lin & Dumin, 1986; Miller, 2011). For example, Hmong public housing residents in the Midwest of the USA felt that "spatial concentration facilitated childcare and promoted mutual support" (Crump, 2002, p. 589). Concentration also helps maintain identity and obtain services. Similarly, the spatial concentration of businesses, social gathering opportunities, and places of worship has also been critical to the success of West African cultures in immigrant-receiving cities like Seattle (Miller, 2011). Conversely, living among diverse neighbors may help immigrants adapt to a new country and navigate the ways of doing things in the adopted country (Sonn, 2002). In ethnically diverse communities there is also opportunity for transcultural mutual support networks to evolve (Manzo et al., 2008).

A special issue of *Housing, Theory and Society* (N.A., 2003, p. 1) on ethnic minorities and housing noted that more detailed-level studies of such communities offer:

> a much-needed refocus on qualitative research that begins to examine processes of labeling, interpersonal discrimination and exclusion, as well as seeing members of ethnic minorities as real live people with their own fears and hopes, whose situation can only be understood by examining the way the host country deals with them.

This chapter contributes to that body of knowledge by offering findings from a detailed case study of immigrants' experience of place in a diverse public housing community and offering insights for achieving the transcultural city.

A Closer Look at the Community

The public housing development that is the focus of this study is located in the Pacific Northwest of the United States. Originally built in 1943 as housing for defense industry workers, it was converted into a low-income housing development by the local Public Housing Authority in 1954. The site included 600 units of duplex, garden-style units (Figure 10.1), although at the time of the research 566 units were occupied.

Figure 10.1 Study site before demolition. Photograph by Jeffrey Hou.

The resident population was very ethnically and racially diverse with a total of 24 different languages spoken by residents. The predominant ethnicity of residents was Vietnamese (33 percent), with Somali (13 percent) and Cambodian (13 percent) the next most frequent (Figure 10.2). After English, the most frequently spoken languages were Vietnamese (36 percent), Somali (14 percent), and Cambodian (13 percent). The remaining quarter (26 percent) of household heads spoke one of 21 other languages (Arabic, Russian, Swahili, Ukrainian, Farsi, Spanish, Samoan, Amharic, Slavic Languages, Chinese, Polish, French, Italian,

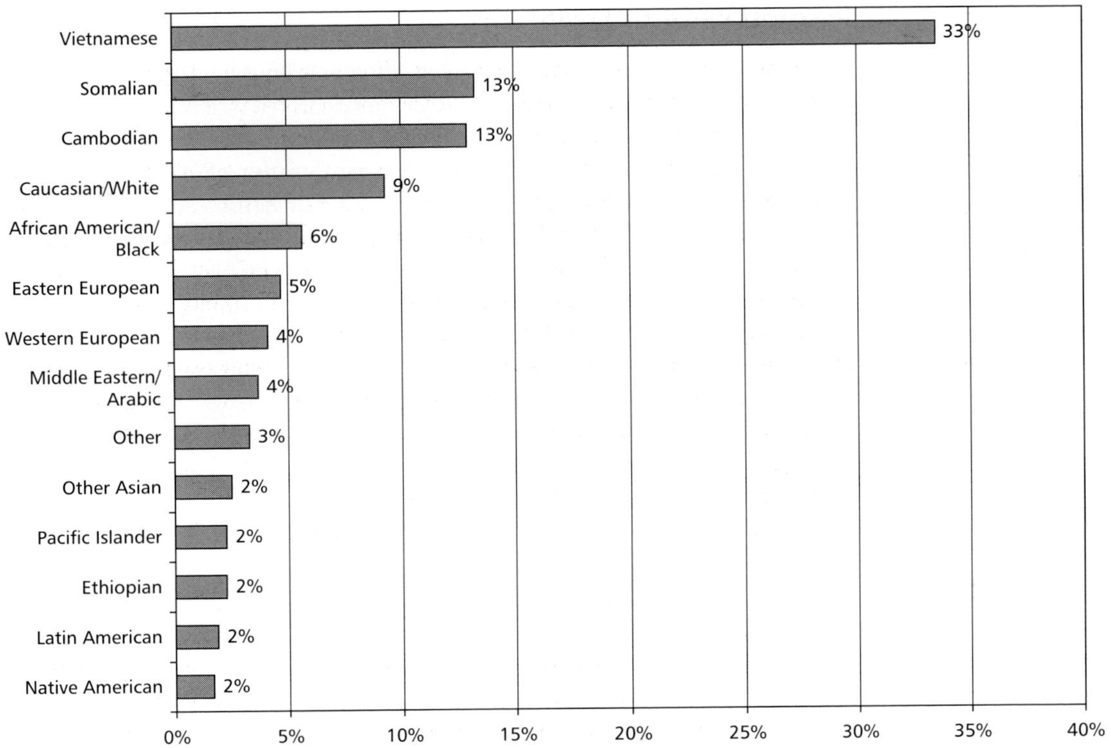

Figure 10.2 Ethnicity of household heads.

Khmer, Kurdish, Albanian, German, Punjabi, Tagalog, Thai, Tongan) revealing the enormous cultural diversity on site. This diversity reflects the role of this community as a haven to new immigrants; the majority of household heads were born outside of the USA, and 34 percent of the heads of household were not US citizens. Only about a quarter (23 percent) of household heads were born in the USA.

Over half of the households on site were families with children, while one in five housed elders, and a quarter had disabled residents. Sixty percent of all household heads were women and 40 percent were men. Many residents have limited formal education, with only about a quarter (28 percent) having a high school degree or the equivalent. About a third of the heads of household were employed and 41 percent of households had someone in them who received income from employment or a business. The average household income was about $12,000 a year. For those who worked, most of their occupations were in the service industry.

We studied this community in order to understand the role it played in residents' lives, and learn about their social ties, place attachments, and reactions to the redevelopment. This study took a mixed-method approach using both quantitative and qualitative strategies.[2] Quantitative data were collected via a survey with 512 of the 566 household heads living on site at the time of the research. Qualitative data were collected through in-depth interviews with a sample of English-speaking heads of households and focus groups with the four main ethnic groups on site—English, Vietnamese, Cambodian, and Somali.[3] These qualitative strategies were used to probe more deeply than the survey into residents' experience of living in the development.

A Good Place to Live

For residents, public housing meant stable, affordable housing. Despite the rhetoric of severe distress that surrounds HOPE VI sites, residents felt their housing community was a stabilizing force in their lives, enabling them to provide for their families and keep their families intact. This was reflected in both the length of residency of households and in their housing histories. The average household has lived on site for about 5 years although length of residency ranged from several months to 35 years. Notably, foreign-born residents were the least mobile, either not moving much—or at all—prior to coming to the USA. Indeed, immigrant interviewees mentioned a residence in their home country when identifying where they lived the longest. One Vietnamese resident explained that remaining in one family home was typical "in the Vietnamese culture . . . I always live [in] one house." Once they arrived in the USA, foreign-born residents experienced some mobility, but when considering residential history from the time of arrival in the USA, a full 82 percent of foreign-born interviewees lived at least half of their time in the USA at the study site, and within that group, 41 percent spent all or almost all of their time in the USA living there. This means that public housing provided them with critical stability.

Residents resoundingly thought their community was a good place to live. In fact, when asked in the survey to compare their housing development with the surrounding neighborhood, over half (60 percent) of residents thought that their community was a better place to live. This was certainly true of immigrant residents as the qualitative data showed. For example, Cambodian and Vietnamese focus group discussions illustrate how public housing provided a key stabilizing force compared to a tumultuous life in their home countries, particularly for refugees. One Cambodian resident explained that living at the study site was a vast improvement from the circumstances she escaped in her home country:

> I got bad, oh bad dreams before . . . like Pol Pot killing somebody . . . and I left from, I left my country to live here . . . it's better. It's peaceful.

Figure 10.3 A community meeting with the public housing residents. Photograph by Lynne C. Manzo.

One interviewee was a refugee from Iraq. He had spent time in an Iraqi prison for not voting for Saddam Hussein in a recent election before he fled to the USA with his family:

> They took me and put me in jail. And the first three months, they tortured me everywhere on my body, my chest, my ears, everywhere. And they put the rope in my hand, like, they tied it and they strap my body everywhere with the cable, wire cable. Was like that for three months. And then they, they let me go. After, week after, they come and take me again.

Later in the interview he explained that while he was struggling to adjust to life here, he felt he had a better quality of life now. He and his wife had recently become citizens and he was determined to make a good life here. Such accounts help us to appreciate the complexity of residents' lives, particularly immigrants and refugees, and the difficult pasts that have led them to the USA and to public housing in particular.

Overall, residents were also quite attached to the community. Residents reported significant attachment to the housing development as a whole and to their unit, with an average score of 8 for each on a scale of 1 to 10, with 1 being not at all attached and 10 being strongly attached. But it is in the qualitative data that we get a fuller picture of residents' lived experience of place. For example, when describing what they liked about the development in the in-depth interviews and focus groups, numerous comments focused on the physical layout of units and various qualities of the physical environment. A number of residents mentioned having a yard, a view, the opportunity to plant flowers, and a feeling that the units were "house-like" (because they had their own front door, a ground unit, and a porch) as aspects with which they were particularly satisfied. For some immigrant residents, yards, porches, and patios have particular cultural relevance. For example, members of the Cambodian focus group mentioned that according to the Khmer culture, outdoor activities were common and that every house should have a patio for barbequing, cooking, drying clothes, and to be outdoors in the good weather—"to breathe the air."

Safety and peacefulness was also an important source of satisfaction among interviewees and this was especially true for refugees. In these cases, residents enjoyed the fact that it was quiet, without trouble of crime or violence. As one resident put it, "This place is the best. It's peaceful. No one bothers you. It's the best." One non-immigrant resident, in describing what she likes about the housing development, mentioned peace and quiet. Interestingly, she thought that her refugee neighbors shared this value: "They are very tired because they come from countries with fights and they just want peace and quiet, too."

A Thriving Sense of Community

At the study site, a thriving social community enjoyed a long history, and residents had remarkably rich social relationships with their neighbors. Neighbors generally knew one another well and provided substantial support to each other. They went shopping together, gave advice to one another, shared food, helped each other with errands, borrowed small items from each other, and watched each other's children. They took out each other's garbage, watered their plants, and gave their grandchildren candy. Their kids played together and they went to each other's birthday parties. Residents told heartening stories about looking out for one another and helping each other when they could—moving an elderly neighbor's television for her, giving away unwanted furniture, or helping a wheelchair-bound neighbor make it up the hill—and all this occurring among diverse neighbors with notable cultural—and even linguistic—difference. Yet, for those with limited resources who struggle just to keep their households running, this mutual support made an important difference in the quality of their lives. It also demonstrates the power of human connection that forms the foundation of a healthy transcultural community.

Transcultural Mutual Support Networks

The fact that the housing development was comprised of people from many different places and ethnic groups emerged as an important theme in residents' experience of place. Fully 77 percent of residents discussed explicitly and without prompting how diversity was a positive feature of the community which they valued. Some residents also thought it was good for their children to be exposed to people of different backgrounds and cultures: "I think it's been good for the kids to grow up with such a variety of cultures. And they basically are kind of color-blind, which I think is positive."

Although such diversity also meant there were language differences, these did not preclude residents from engaging with their neighbors and developing positive relationships with them. On the contrary, there was a remarkable degree of mutual support networking across language barriers. As one native English-speaking white woman commented, "When I want to make a friend, I make a friend, I don't care who." Often residents described situations in which they figured out how to communicate non-verbally: "I helped a lady one night, in a domestic situation, and neither of us, we never did speak. It was completely non-verbal, hand gestures, and so on." In this case, a resident was able to help a neighbor in a domestic violence situation to get help, despite the fact that they did not share a common language. Being able to communicate across different languages and help in emergency situations was common in this community. Another interviewee described a time when she was able to help her neighbor who seemed to be having a heart attack:

> Yes, one of my neighbors here—they're Vietnamese and they don't know any English. I think he was having almost like a heart attack. I've learned on my own how to speak with your hands, even you don't know somebody's language. That's the way I was able to communicate to 911 what the problem was. I think it was great that I could help that day.

Even those neighbors who have difficulty communicating because of language differences still socialized with and provided mutual support to one another. For example, another resident explained that while she could not carry on full conversations with her neighbor who speaks only Arabic, they still brought food to one another. Another resident described a friendship that she developed with a neighbor who spoke a different language than her own:

> I had a neighbor who unfortunately moved away, who was just a doll, just would bring extra food from the Food Bank that she didn't need or want, bring us plants and flowers, and she was limited English-speaking, but there was still the, kind of, there was still a mutual, you know, affection between us.

Food, in fact, was an important commodity in this community often used in exchange for help, but also used as a way of sharing one's culture or expressing appreciation for neighbors. For example, a Filipino resident cooked one of her native dishes for a curious neighbor: "Like Filipino food, she smelled it, and she always wanted to taste it. So I let her taste it, and she asked me would I cook her some and [I did]." This kind of sharing seemed to bring about a better appreciation of residents' commonalities. Clearly mutual support crossed cultural boundaries and language "barriers" have not kept people from neighboring. Despite language and cultural differences, many residents recognized that they shared the common project of living.

Helping with Cultural Adjustments

Interviewees' descriptions of neighbor relationships suggest that there is an added value for immigrant residents in interacting with their American neighbors or those immigrants who have lived in the USA longer than they. As one Somali woman commented, "There's a lot of different cultures here. So, some of the people are from where I lived . . . at the same time, I am learning about American and other cultures." One resident who emigrated from Russia talked about how helpful her neighbor was when her family first moved into the housing development and were learning how to get by in a new country:

> We have a neighbor . . . she used to live here for several years before she moved away. Without her, I mean, she'd been supporting us from in and out, to cheer us up, to getting used to this whole place, and she helped watch the family, she came here to help, teach [us] how to do things, take care of little stuff that, that will make us happy.

Other immigrant interviewees also described how neighbors helped them to acclimate to life in the USA: "Around here very, very nice people. They help us a lot because before we live here, we don't know anything. We learn about life, too—through the people. So, yeah, they help."

Another American-born white resident explained how he believed word of mouth has spread among his immigrant neighbors about how he is willing to help them:

> I'm pretty sure they pass it down to everybody, like "That's the guy down there you should know. You should know him and ask him some things." I pretty much

help a lot. A lot of people may not know where things are around here and I can pretty much show them.

What is notable is that we did not really hear stories about cultural differences creating conflicts and significant barriers. Very few negative comments were made about neighbors except in cases where there were concerns about illicit activity, rowdy youth, or noise, and these concerns were not expressed in any way that suggested they were based on any racial or ethnic tensions.

Within Group Social Ties

While residents enjoyed the diversity of their neighbors and developed social relationships with people across different cultural backgrounds, immigrant residents also appreciated the presence of others from their homeland. Many found that being around people from their same background was a source of comfort and reassurance, and that it "felt good." As one Cambodian refugee interviewee declared laughingly when asked what it was like living in the development, "It's normal! Because there are so many other Cambodians around." As a Vietnamese interviewee put it: "I'm happy to see my country people. Because I never go back to Vietnam." So he was pleased that he could find a cultural connection to other Vietnamese while still making a new life for himself in the USA. Similarly, another Vietnamese interviewee explained, "Our friends are from Vietnam who came over here. Oftentimes, we meet a friend who lives in another part and they come over and we can talk." Another Vietnamese resident also said he had friends on site whom he knew from Vietnam and who shared his difficult past: "In Vietnam, we worked together. In 1975, we went to jail [as political prisoners together] for about six or seven years. Then they let us go to America and so we know each other."

Such social support among members of one's own ethnic group was evident across different ethnic groups. For example, one Somali woman mentioned that, "All of my friends speak my language and are originally from Somalia." She explained that she liked the area in general because it had a large Somali population. When she was planning to move to the USA, she called many people who had moved all over the USA whom she knew from Somalia, to inquire about what life was like there. When she came to visit in Seattle, she said of the Somalis here, "Everyone says welcome." They took her to restaurants and didn't ask her to pay. When she moved here, they shopped for her, they bought beds for her and her son, and one friend even got her a job.

The presence of ethnic-based services and amenities both in the area and on site strengthened feelings of belonging to a cultural community. For example, a number of people regularly went to the local supermarkets and social halls to see other co-ethnics and talk. Some residents also talked about the value of having bilingual service providers and staff on site, particularly for getting information on the relocation process that everyone could understand. So clearly there were strong cultural connections among immigrants within the same ethnic group. The sense of comfort and belonging offered by co-ethnics was unique. At the same time, there is no question that mutual support and networking transcended cultural and language groups, and were very much a part of life in this community.

Walking the Tightrope—Emergent Hybrid Identities

The stories residents told about their mutual support networks revealed that immigrants' experiences in the community added a new dimension to their cultural identity and experience of the city. For example, the Somali resident quoted earlier about her experience coming to the USA and integrating into the community described neighbors from her home country as coming from her "first culture." She explained that while she appreciates their presence, at the same time, she enjoyed learning about American and other cultures. The fact that she referred to her own culture as her "first culture" suggests that she saw herself belonging to several cultures at once, and that her cultural identity was evolving based on her newer experiences living in the USA.

Discussions from the Vietnamese and Cambodian focus groups also depicted how immigrants walk a tightrope between different worldviews, and how new cultural understandings emerged from a blending of experiences from one's home country and the accumulation of experiences in the USA. Interviews with immigrant residents reveal how they often relied on their children to serve as cultural translators, not merely in terms of literal translations of information but to understand how to get things done, for example, how to fill out forms or communicate with service agencies. Children in particular seemed to straddle two worlds—that of their immigrant limited English-speaking parents with their past experiences, values, and traditions, and the American world of school and their young peers. Housing Authority staff were also quite helpful with cultural translations as well, as many staff were multilingual and some were immigrants themselves. The idea of adjustment to a life in the USA and to public housing in particular, is well summarized by a participant in the Vietnamese focus group:

> For the Vietnamese, in general, we follow our ancestors' advice "If we live in a tube, we become thin and long. If we live in a gourd bottle, we become big and short." That's why we can adapt ourselves well to different circumstances.

Immigrant residents' hybrid identities were also manifest in the mix of feelings and attitudes they expressed as they negotiated the challenges and stigma of living in public housing. While redevelopment of the community was intended to be participatory, refugees in particular struggled to feel able to speak their minds to "government officials" and authority figures. As one Cambodian resident noted in the focus group discussion:

> They asked us once or twice . . . what did we need from them? We can't order them. It's up to them. Because we are average people, in whatever condition they allow us to live is up to them. It's up to people at the top, the high ranking officers.

Cambodian residents also struggled with the necessity for government assistance and prejudice from their fellow Cambodians who did not live in public housing:

P1: Some time ago, when I joined a ceremony at the [Khmer] temple, they said that "those people who live in the public housing meet dead end . . . they could find a way out."
P6: I fully heard from both of my ears . . . If we feel shameful, just walk out.
P8: Let them say whatever they want to say . . . they are not giving our rice to eat.

Later in that same discussion a resident noted, "That mentality has passed from one generation to the next, even after when we came here. It happens everywhere but it also depends on each family. But in general, it happens in our community."

Discussion

As Fainstein (2005) notes, "Diversity has become the new orthodoxy of city planning" (p. 3). It comes in many forms—cultural, racial/ethnic, income, uses, and architectural design. In the study site, the housing type and household incomes were homogeneous, but the population was enormously ethnically diverse. As a place of transcultural interaction, this housing community can be considered one of Al Sayyad's (2001) hybrid places, where cultural identities meet and transform each other, a place marked by vitality. Despite significant differences in culture and language, a strong sense of community and a remarkable degree of mutual support took place among diverse neighbors, as residents traversed differences that might otherwise be seen as barriers. Through the emic perspective of the residents, it is clear that "the projects" are not abstracted space—homogenized sites of distress that call for significant interventions—but rather they are homes where the tasks of everyday life and the common project of living unfold (Vale, 2002; Manzo et al., 2008).

Yet, as noted earlier, HUD has deemed this site "severely distressed" and slated it for demolition. Shortly after the study, the community was relocated and new mixed-income housing was built in its place through HOPE VI (Figure 10.4). Like many mixed-income housing programs, this one seeks economic diversity by deconcentrating poverty, begging the question of what kind of diversity is important. Like Crump (2002), I would argue that, in claiming to advance income diversity, the redevelopment program reclaims critical urban space from poor communities to reimagine the city as a safe zone for increasingly mobile capital to settle (Crump, 2002). Such a strategy reflects Lefebvre's (1991) notion of representations of place—the master narrative of bureaucrats constructed through formalized discourse (policy). Here, the rhetoric of distress is deployed to erase stigmatized structures from the landscape. This is also commensurate with older deficit models of community development.

However, residents' everyday spatial practices and lived experience of place show a very different picture than one of distress; they reveal critical community assets and provide a glimpse into what the transcultural city can achieve. Detailed case studies of people's place experiences, such as those presented in this chapter, are critical for shedding light on the contradictions in representations of the urban landscape. These contradictions must be interrogated if we are to achieve the transcultural city, and not merely orchestrate staged diversity. Only then can we move toward a more just city.

Figure 10.4
Newly completed redevelopment of the public housing site under HOPE IV. Photograph by Lynne C. Manzo

Notes

1 These goals were stated in: US Department of Housing and Urban Development, FY 2008, Notice of Funding Availability for HOPE VI Revitalization Grants.
2 This study was conducted as a formal evaluation of the HOPE VI redevelopment of this particular site. The research was conducted in collaboration with Dr. Rachel Garshick Kleit of the UW Evans School of Public Affairs.
3 Both the interviews and focus groups lasted about an hour and a half on average, and were audio-taped and transcribed. The non-English language focus groups were translated into English by bilingual interviewers before they were transcribed. All interviews and focus group discussions were coded and analyzed using the Atlas.ti 5.0 qualitative analysis software program that is based upon grounded theory (Glaser and Strauss, 1967; Strauss and Corbin, 1990).

References

Al Sayyad, N. (2001). Prologue: Hybrid Culture/Hybrid Urbanism: Pandora's Box of the "Third Place." In Nezar Al Sayyad (Ed.), *Hybrid Urbanism: On the Identity Discourse and the Built Environment*. Westport, CT: Praeger, 1–18.
Crump, J. (2002). Deconcentration by Demolition: Public Housing, Poverty, and Urban Policy, *Environment and Planning D: Society and Space*, 581–596.
Fainstein, S. (2005). Cities and Diversity: Should We Want It? Can We Plan for It? *Urban Affairs Review*, 41(1): 3–19.
Fainstein, S. (2010). *The Just City*. Ithaca, NY: Cornell University Press.
Fullilove, M. (2004). *Root Shock: How Tearing up City Neighborhoods Hurts America, and What We Can Do About It*. New York: Random House.
Glaeser, E. L. (2000). The Future of Urban Research: Nonmarket Interactions. *Brookings–Wharton Papers on Urban Affairs*, 101–149.

Glaser, B. & Strauss, A. (1967). *The Discovery of Grounded Theory: The Strategies for Qualitative Research*. Chicago, IL: Aldine.

Goetz, E. (2003). Housing Dispersal Programs. *Journal of Planning Literature*, 18, 3–16.

Hannerz, U. (1996). *Transnational Connections: Culture, People, Places*. London: Routledge.

Lefebvre, H. (1991). *The Production of Space*. Cambridge, MA: Blackwell. First published as *La Production de l'espace*, 1974.

Levy, D. (2005). Tending Home: Residents' Ambivalent Responses to Relocation from Public Housing Developments. Paper presented at the annual conference of the Urban Affairs Association, Salt Lake City, UT.

Lin, Nan. (2000). *Social Capital: A Theory of Social Structure and Action*. Cambridge: Cambridge University Press.

Lin, Nan & Dumin, M. (1986). Access to Occupations through Social Ties. *Social Networks,* 8: 365–385.

McCann, E. J. (1999). Race, Protest and Public Space: Contextualizing Lefebvre in the U.S. City. *Antipode*, 31(2): 163–184.

Manzo, L. C. (2010). Recognizing the Lived Experience of Place: Challenges to Genuine Participation in Redeveloping Public Housing Communities. In S. Sutton and S. Kemp (Eds), *The Paradox of Urban Space: Inequity and Transformation in Marginalized Communities*. New York: Palgrave Macmillan.

Manzo, L. C., Kleit, R. G., & Couch, D. (2008). "Moving three times is like having your house on fire once": The Experience of Place and Displacement among Public Housing Residents in the Pacific Northwest. *Urban Studies*, 45(9): 1855–1878.

Mazumdar, S., Mazumdar, S., Docuyanan, F., & McLaughin, C. M. (2000). Creating a Sense of Place: The Vietnamese-Americans and Little Saigon. *Journal of Environmental Psychology,* 20(4): 319–333.

Miller, R. (2011). Urban Design with Global Publics: West African Immigrants' Appropriation of Space in the Rainier Valley. Unpublished Master's Thesis. University of Washington.

N.A. (2003). Editor's Introduction. *Housing, Theory and Society*, 20(4): i–ii.

Pestieau, K. & Wallace, M. (2003). Challenges and Opportunities for Planning in the Ethno-culturally Diverse City: A Collection of Papers—Introduction. *Planning Theory and Practice*, 4(3): 253–258.

Rishbeth, C. (2004). Ethno-cultural Representation in the Urban Landscape. *Journal of Urban Design*, 9(3): 311–333.

Sandercock, L. (1998). *Toward Cosmopolis*. Chichester: John Wiley.

Sandercock, L. (2003). *Cosmopolis II: Mongrel Cities of the 21st Century*. New York: Continuum.

Sandercock, L. & Attili. G. (2009). *Where Strangers Become Neighbors: Integrating Immigrants in Vancouver, Canada*. Urban and Landscape Perspectives, 4. Dordrecht: Springer.

Sonn, C. C. (2002). Immigrant Adaptation: Understanding the Process through Sense of Community. In A. T. Fisher, C. C. Sonn, & B. J. Bishop (Eds), *Psychological Sense of Community: Research, Applications, and Implications*, pp. 205–222. New York: Kluwer Academic/Plenum Publishers.

Strauss, A. & Corbin, J. (1990). *Basics of Qualitative Research: Techniques and Procedures for Developing Grounded Theory*. Thousand Oaks, CA: Sage Publications.

Talen, E. (2006). Design That Enables Diversity: The Complications of a Planning Ideal. *Journal of Planning Literature*, 20(3): 233–249.

Vale, L. J. (2002). *Reclaiming Public Housing: A Half Century of Struggle in Three Public Neighborhoods*. Cambridge, MA: Harvard University Press.

CHAPTER 11

Spaces of Negotiation and Engagement in Multi-ethnic Ethnoscapes

The "Cambodia Town" Neighborhood in Central Long Beach, California

Felicity Hwee-Hwa Chan

> By ethnoscape, I mean the landscape of persons who constitute the shifting world in which we live: tourists, immigrants, refugees, exiles, guestworkers and other moving groups and persons constitute an essential feature of the world, and appear to affect the politics of and between nations to a hitherto unprecedented degree. This is not to say that there are not anywhere relatively stable communities and networks, of kinship, of friendship, of work and of leisure as well as of birth, residence and other filiative forms. But it is to say that the warp of these stabilities is everywhere shot through with the woof of human motion, as more persons and groups deal with the realities of having to move, or the fantasies to move.
>
> (Appadurai 1990, p. 297)

Globalization has ushered in a new era of reconsidering the claims and concerns of multicultural societies as we witness an increase of discourses on immigrant integration. A useful concept that illustrates the fluid movement of people-cultures across borders is the neologism of "ethnoscapes" introduced by Appadurai (1990, 1991, 1996). While Appadurai's intention appears to be communicating the high fluidity and seamlessness of flows and identities across borders, the phenomena of global ethnoscapes from the quote above implies three types of tensions. They are: *the tension between stable communities and those on the move*; *the tension between different cultures coexisting in space*; and lastly, *the tension between the material and the imagined*.

Central to these tensions, as Gupta and Ferguson (1997) remind us, is the role geographical space plays in the articulation and negotiation of identity among immigrants. Anecdotal evidence of growing tension, especially between immigrants and their receiving societies in cities, is on a rise regarding the sharing of urban space. In 2009, new construction of minarets was banned in Switzerland, followed by a spate of social disputes on the siting of mosques in New York and California in 2010. The German Chancellor Angela Merkel in 2010 announced that multiculturalism has "utterly failed!" At the neighborhood level, hostility and anger are displayed by the receiving societies over the changes in the use and character of neighborhoods that immigrants invariably bring and represent (Qadeer, 1997; Sandercock, 2003). In a survey study of forty-one communities in North America conducted in 2000, Putnam (2007) found that social solidarity, trust, and community cooperation are reduced in the presence of diversity. His findings show that inter-racial and intra-racial trust is the lowest in ethnically diverse cities like San Francisco and Los Angeles. He writes, "Diversity seems to trigger *not* in-group/out-group division, but anomie or social isolation. In colloquial language, people living in ethnically diverse settings appear to 'hunker down'— that is pull in like a turtle" (p. 149; author's emphasis).

Multi-ethnic settings thus appear to produce two perceivably interrelated effects: first, tensions arising from differences, and second, social isolation. These two effects can be mutually reinforcing: tensions arising from differences when tolerated but not addressed can produce a pulling-in effect to avoid face-to-face engagement. This lack of engagement may in turn create conditions conducive for entrenching stereotypes that lead to prejudice (Allport, 1954) and the perpetuation of fear and anxiety of the stranger (Kristeva, 1991; Sandercock, 2003). These circumstances of "parallel lives" (Cantle, 2005), where neighbors experience division on a daily basis, can lead to an insidious hardening of social exclusion that may already exist along socio-economic lines (Amin, 2002). Social contact, according to Allport (1954, p. 281), is a process that is capable of reducing prejudice, especially if contact takes place in conditions of equal status "in pursuit of common goals." This observation underscores the importance of the type and quality of contact also noted by Amin (2002) because repeated encounters are no guarantee for cultural exchange as they may in fact lead to entrenched prejudices.

Taking on this debate of social contact as a way to negotiate the three tensions from an urban studies viewpoint, engagement rather than non-engagement with difference is frequently favored as a process of public urban life as seen in the writings of Jane Jacobs (1961) and Richard Sennett (1970, 1974, 1989, 1990). Diversity is a value in the ideal of a good city (Lynch, 1981; Sandercock, 1998; Thrift, 2005; Amin, 2006; Talen, 2008), although it could become a distraction from the greater good of social justice (Fainstein, 2010). Sandercock's (2000, 2003) discussion about planning a "cosmopolis" shaped a new discourse about the impacts and implications of global immigration for planning multicultural cities. A review of the literature shows that the majority of the writings on planning multicultural cities is found in the following categories: adapting procedural planning by such means as increasing the sensitivity of planners to cultural difference (Sandercock, 2000, 2003; Umemoto, 2001; Pestieau & Wallace, 2003); revising regulations and guidelines to diversity (Zelinsky, 1990; Qadeer, 1997; Burayidi, 2000; Thompson, 2003; Germain & Gagnon, 2003); and making the planning process institutionally inclusive (Baum, 2000; Thomas, 2000; Reeves, 2005). Fewer works are found in the category of creating substantive planning and design outcomes such as intercultural urban programs and spaces (Amin, 2002; Woods & Landry, 2008), design diversity (Talen, 2008), and reconciliatory spaces (Bollens, 2006).

In Talen and Ellis's (2009, p. 192) recent discussion of the emerging agenda in American planning, they highlighted that cities of the future will need to negotiate the "two conflicting trends" of compactness and diversity. They emphasize the importance of studying existing diverse and dense communities and "the role that civic spaces—the provision of an ample and well-designed 'public realm' can play in supporting diversity," in addition to continuing research into how planning process and policies can support diversity.

This chapter thus proceeds to do the following: (a) explore and discuss the nested tensions in and through spaces of global ethnoscapes using the case of Cambodia Town, and (b) discuss the opportunities that existing civic spaces such as the neighborhood park, cultural center, and library offer for intercultural engagement between different ethnicities. The data for the discussion is drawn from preliminary observations and interviews with respondents who live and/or work in the dense, diverse, and low-income neighborhood in Central Long Beach, California, conducted between July 2010 and May 2011.[1]

Cambodia Town Neighborhood

In this chapter, the *Cambodia Town* neighborhood is partially an imaginary socio-spatial concept derived from its spatial adjacencies to Cambodia Town. As a concept, it highlights the inherent tensions between the imagined and material space in one of the most ethnically diverse districts in the United States. The official name of this neighborhood area on city records is "MacArthur Park/Whittier School Neighborhood Improvement Strategy Area," a name that perhaps reflects little of the current population represented in the neighborhood. According to the US Census 2010, this neighborhood has about 40,000 inhabitants of whom 43 percent are foreign-born immigrants from Central America (Mexico, El Salvador, Guatemala) and Southeast Asia (Cambodia, Vietnam, Thailand). Its ethnic composition is 58 percent Latinos, 20 percent Asians, 13 percent Blacks, and 7 percent Whites.

Geographically, the Cambodia Town neighborhood as referred to in this chapter includes the area between Pacific Coastal Highway and East 10th Street in the North–South direction bounded by Martin Luther King Jr. Avenue and Junipero Avenue along the East–West direction.[2] These boundaries include part of the official boundaries of *Cambodia Town*, a one-mile-long arterial corridor laced with a mix of Cambodian businesses (supermarkets, mom-and-pop grocery shops, jewelry retail, restaurants, cafés, video shops), Latino and Thai businesses along East Anaheim Street (between Atlantic Ave and Junipero Ave). In terms of location, Cambodia Town forms a small but central part of the neighborhood (Figure 11.1).

Tensions in and through Space

Diverse, low-income, and transient, according to residents and community organizers this neighborhood suffers from a high level of crime and is referred to as a "tough neighborhood" and a "war zone."[3] Gang-related shootings between Latino and Asian gangs proliferated in the 1990s, attributed to perhaps the sharp increase of the Cambodian community and the socio-economic displacement of the other existing ethnic groups such as Latinos and African Americans (Needham & Quintiliani, 2007). According to a Cambodian resident who has lived in the neighborhood since the early 1980s, Cambodians who came as refugees post-1975 experienced a very difficult period of settlement through the 1990s as they were bullied by Latinos and African Americans who regarded Cambodians as "the stranger" and "the newcomer" in the neighborhood. As a means of self-protection, some Cambodian youths banded together to form ethnicity-based gangs such as

Figure 11.1 Urban scenes and landscapes along East Anaheim Street. Photographs by Felicity Hwee-Hwa Chan.

Asian Boyz. Gangs provided refuge and a community for the many Cambodian youths who became school dropouts because of their academic difficulties in schools and the lack of support at home as their refugee families struggled with the transition to a new cultural way of life in America.

It is against this violent and scarred past that the tensions of the Cambodia Town neighborhood as a global ethnoscape must be discussed. The concept of Cambodia Town was proposed by a loose coalition of wealthy Cambodian businessmen who called themselves the "Cambodian Task Force" in 2000. Cambodia Town was to serve strategic purposes of branding local businesses and creating a space of identity for the Cambodian diaspora. When the proposal was introduced, the notion of Cambodia Town immediately unleashed undercurrents of the three nested tensions within the Cambodian community *between those on the move and those who are stable*, *between the different ethnic groups in the neighborhood*, and *between the real and the imagined*.

A Cambodian resident interviewee emphatically pointed out that Cambodia Town is proposed by a group of Cambodians who "don't live here" and it gives a wrong impression to neighbors that Cambodians who *actually* live in the neighborhood are trying to "claim the place." This resident elaborated that establishing Cambodia Town is akin to a gang-like act of drawing boundaries and territories. Cambodian residents feel that marking space this way creates divisions between neighbors living in the area because the neighborhood space is neither perceived nor lived in an ethnically physical contiguous way, unlike in the case of ethnic enclaves such as Chinatown. Further, the name "Cambodia Town" imposes a contentious singular national identity in what is in fact a shared space of multiple nationalities. A less fractious name, such as "Little Phnom Penh" that focuses on the city rather than a culture or a group, according to the interviewee, can still fulfill the same intent without ruffling the feathers of other groups. The unease is prompted by fears that the Cambodian resident community will bear the brunt of violent reprisals from other ethnic groups over encroachment of territories. As expected, the proposal elicited immediate unease amongst the Latino and African American small business owners who did not see their identity fitting into the concept of Cambodia Town. The proposal was swiftly rejected by the City of Long Beach.

The rejection in 2000 made the proposal's proponents realize that support from the local community was critical. Between 2001 and 2006, husband and wife team Sithea and Richer San (founders of Cambodia Town, Inc.), with the assistance of the third party California Conference of Equality and Justice (CCEJ), conducted a survey with major stakeholders along East Anaheim Street to seek support from the Latino, African American, and Asian business owners. The survey asked if business stakeholders would support a cultural and business district. Of the stakeholders interviewed, 90 percent were supportive. In 2007, the concept of Cambodia Town was finally approved by the city council with near unanimous support.

However in order for the district to be formalized in space with official road signs to demarcate the boundaries of Cambodia Town, the city requires Cambodia Town to become a Business Improvement District (BID). However, resistance against the BID was evident amongst the businesses as only less than half of the two hundred businesses gave immediate support for the BID. The resistance came as somewhat of a shock to the organizers of Cambodia Town, Inc., given the initial positive feedback for a cultural and business district. Mom-and-pop shop owners found it economically hard to commit financially to a BID, perhaps as a result of the economic downturn beginning in late 2007. Further, as most of the businesses in the neighborhood are ethnic-based (Thai, Vietnamese, Mexican, African American), a Cambodia Town identity BID might not promote their business interests and identity. It was decided that until Cambodia Town, Inc. received more support from local business owners, the official city sign for Cambodia Town would have to be placed on hold.[4] In a surprising turn, in July 2011 after negotiating through "a complicated web of issues of liability, policy and council votes," the first two boundary markers for Cambodia Town were installed with the support of the city amidst debates over the final name of the district. The district is now officially recognized as "Cambodia Town Cultural District" (Long Beach Press Telegram, July 16, 2011).

The city's original approval of the concept, subject to the implementation of BID that is focused on commerce and local business buy-in instead of culture per se, is worth noting. By prioritizing local economic development instead of ethnic identity, the city minimizes the racial overtones of the proposal in a setting of diverse vested interests and tactically leverages the interest of Cambodian businessmen for Cambodia Town to bring about tangible infrastructure investment for local economic development. However, the recent twist of development to rename Cambodia Town as a cultural district instead, and the conspicuous silence about the business component, points to the power culture as a broker of spatial negotiation when combined with political finesse in a diverse setting.

Urban Interculturalism and Civic Spaces

While inter-ethnic gang shootings have declined in the neighborhood over the last decade because of a variety of interventions including gang intervention in schools, police enforcement, the moving on of Asian gangs to more lucrative nationwide interests, and the city's initiatives to emphasize unity in diversity,[5] the Cambodian residents remain cautious and avoid the streets and public spaces after dark. Many Cambodian parents would not allow their children to walk even a short distance home from school. In fact, most shops close by seven in the evening. Community activities and meetings end no later than eight in the evening so that residents will be encouraged to attend. Although the neighborhood is no longer as violent as it used to be, every so often there is a shooting incident that brings back old fears, albeit not necessarily racially motivated. A drive-by shooting in January 2011 of a policeman at the peripheries of the MacArthur Park by an African American man has stopped a long-time African American lady resident from bringing her children to the park for recreation. Perception of a lack of a sense of safety in the public spaces of Cambodian Town neighborhood has restricted social life in the neighborhood, and social anomie is commonly experienced even in a dense neighborhood like this.

In the face of these socio-spatial tensions, inherent in the process of global ethnoscape formation, and of the tendencies for anomie in diverse settings as outlined by Putnam (2007), planners and policy-makers will increasingly be called on to address these tensions in their plans and policies through answering such questions as: what are the infrastructure (programs, policies, institutions, spaces) and conditions that would make productive engagement between different cultures possible? And how can the rational model of planning that has relied predominantly on planning standards of "universal needs" for a homogeneous population be adapted to a more diversified public who may have competing needs (Sandercock, 2000, 2003)?

I am proposing that we begin from existing everyday public infrastructure of the neighborhood such as the civic spaces (park, library, and community center). Although the efficacy of contemporary public spaces as sites of multicultural engagement is questioned by Amin (2002), the role of civic spaces in neighborhoods as everyday zones of encounter (Woods & Landry, 2008) makes research about these spaces in multi-ethnic settings necessary.

Three civic amenities in the Cambodia Town neighborhood, namely MacArthur Park, Homeland Cultural Center, and Mark Twain Library, are chosen because of their physical and social centrality to the residents and their official recognition by the city as a center of significant social amenity in an impoverished neighborhood. Thus, a study of these spaces can offer valuable insights into whether and how these civic spaces support opportunities for productive intercultural engagement that could address the socio-spatial tensions and anomie in multi-ethnic neighborhoods.

To conceptually ground the discussion and analysis, I would like to borrow the concept of *urban interculturalism* from Amin (2002). The concept advocates proactive intercultural engagement, dialogue, and understanding as means of addressing social tension in ethnically diverse cities (Sandercock, 2003; Woods & Landry, 2008). According to Amin, urban interculturalism stresses the importance of the prosaic and the everyday "negotiation of difference" in the "local micropublics of everyday interaction." The emphasis here is on creating opportunities and spaces within the fabric of the modern city for *regular sociality*[6] in the city. Amin views good intercultural spaces as "spaces of cultural displacement," where individuals are given opportunities to "break out of fixed relations and fixed notions." For Amin, these opportunities must be combined with "joint projects across ethnic divisions" such as in communal gardens, community centers, and neighborhood-watch schemes. Central to Amin's discussion is the importance of the *value of intentionality* in the crafting of these spaces.

Expanding beyond Amin, Woods and Landry (2008) describe good intercultural spaces by values of "regular contact" (p. 158), "safe and neutral" (pp. 188, 191), "well managed and creatively programmed" (p. 183). This *value of neutrality* of space from any one particular group is also stressed by Bollens (2006) in his synthesis of the role and scope of urban planning as peace-making efforts in previously divided cities in Spain and Bosnia. For Bollens, sensitivity to the tension lines between communities is critical when designing spaces of interaction. These spaces of interaction crafted to facilitate contact and exchange between previously warring groups need to pay special attention to maintaining the neutrality of the location, the symbols, and the functions of new intercultural spaces.[7] These insights corroborate with findings from preliminary interviews in the neighborhood where the neutrality of spaces against ethnic territorialization was highlighted by those who live and work there as an important criterion to encourage the use of a civic space.

Good spaces of intercultural understanding are hence spaces where stereotypes can be intentionally questioned and dismantled on safe and neutral grounds, where prejudice is hence intentionally challenged, and cultural diversity can be learnt and appreciated. Using the twin qualities of *neutrality* and *intentionality* drawn from these writings, I analyze the civic spaces in the Cambodia Town neighborhood for these qualities that are conducive for urban interculturalism. The working definition for the *value of neutrality* in this chapter refers to the physical and social accessibility of the civic space by all ethnic groups. To assess this, I propose to use *access, location,* and *perception of accessibility* as criteria. The *value of intentionality* refers to the capacity of the space to create opportunities that can displace stereotypes. This capacity is assessed by the vision for the civic space and the programming (programs and staffing) of spaces for intercultural engagement. My analysis draws on participant observations and interviews.

MacArthur Park

MacArthur Park is an open, visible, and publicly accessible civic space located along East Anaheim Street adjacent to commercial and residential areas (Figure 11.2). The park, together with the community library and recreation building, was established in 1943. Over the years, basketball courts, tennis courts, a playground, picnic benches, and a public restroom were added. In 1994, the community recreation building was replaced by the Homeland Cultural Center and the library (also known as the Mark Twain Library) was relocated across the street in 2007 to a new and bigger building.

Figure 11.2
MacArthur Park, located across the street from Mekong Mart. Photograph by Felicity Hwee-Hwa Chan.

As the largest green space in the dense neighborhood, MacArthur Park is a well-used neighborhood space during the week. The park, according to a community organizer, is regarded as "very, very important because this neighborhood is very crowded." From site visits, I observe that the users of the park are made up mostly of Latinos, African Americans, and Asians, in descending order, many of whom walk to the park from their nearby homes. Amongst the Cambodians I interviewed, MacArthur Park is perceived as unsafe and is associated with the fear of street violence that was targeted at Cambodians in the past. The park is also considered rather small and poor in amenities in comparison to the surrounding parks outside the neighborhood, especially for teenagers and young adults. Many of them (Cambodian, Latino, White, African American) would travel to other parks in the vicinity of the neighborhood because they were larger and there were more amenities.

Fieldwork observation indicates that the park appears to have a sustained level of *neutrality* based on its use by different ethnic groups from the neighborhoods, albeit not having as much popularity with Cambodians in general. However, the degree of its neutrality is differentiated within the park grounds. The children's playground is observed to have the most open access and be used by all ethnicities throughout the day. There was, however, little visible interaction between adults of different ethnicities even though their children might play together. The basketball court seems to be regularly occupied by African American youths and the open field is used by Latino boys to play soccer, albeit more frequently in the past according to the park management. Of note here is that while the park is generally open to all groups, activities and interaction are observed to be mostly within specific ethnic groups.

Based on these findings, the park lacks *intentionality* in creating opportunities for intercultural contact and exchange. It does not have a vision and programming of its spaces to encourage the cultural displacement that Amin discusses. Its function is largely as a green space for recreation. However, this lack of intentionality has not prevented the park from being used and read as a common social space of the neighborhood where community events are held and where children of different ethnicities can socialize through play. In fact, for the city and some members of the community, MacArthur Park is identified as the socio-spatial center for the East Anaheim corridor and an anchor space for the neighborhood into the future. Thus, its perceived openness holds potential for the park to become an active intercultural space for the community if it can be harnessed through programs and events; for example, organizing sports activities where youths of different ethnicities teach basketball and soccer to children, or sports competitions that encourage mixed-ethnic teams from the neighborhoods.

Homeland Cultural Center

While the MacArthur Park currently lacks vision and programming of its spaces for intercultural connection, the Homeland Cultural Center is a programmed social space, located at the southern edge of the park along East Anaheim Street, that was set up with a clear mission for engagement (Figure 11.3).[8] Its mission is "to provide a safe place for the cultivation, preservation and promotion of art forms, both traditional and contemporary, representing the diverse array of cultures that make up Long Beach today, and in the process to bring these cultures together" (Information Sheet of Homeland Cultural Center). It partners with different grassroots and ethnic organizations and offers many different programs like break-dancing, graffiti murals, pop-locking, writing classes for children and adults, music and dance classes, and the ethnic arts program, including Khmer Martial Arts, African Dance, Cambodian Music, Micronesian Music and Dance, Hmong Arts.

Intentional in its vision and programming to promote cultural arts, the Homeland Cultural Center's focus on the arts offers the possibility for participants keen in the arts to engage in joint projects that can result in what Amin terms as cultural displacement. For example, the break-dancing and pop-locking programs attract enthusiasts of different ethnic groups and bring visitors from different parts of Los Angeles as well as from around the world. These programs provide the conditions for regrouping based on interests and hence become a platform for engagement. The center's ethnic arts program, purposeful in its representation of the major culture groups living in the area, offers opportunities for intercultural engagement and understanding. A new adjoining 66-seat theater hosts weekly open-microphone nights and poetry readings and expands opportunities for local artists of different ethnicities from Long Beach to engage with the youths and children.

Figure 11.3 Homeland Cultural Center. Photograph by Felicity Hwee-Hwa Chan.

From the perspective of *neutrality*, its open accessibility for all ethnic groups and geographical visibility (being located in a public park and next to a bus stop) make Homeland Cultural Center a fairly neutral space. However if we consider the perception of accessibility as a criterion, its lack of social presence and significance for some ethnic and age groups as well as for its immediate neighborhood residents indicate that its value of neutrality is compromised. For example, the Hmongs use the cultural center as their cultural space on Sundays from morning to mid-afternoon, but the Cambodian residents interviewed express that they do not quite relate to the activities in the center. In addition, activities of the Homeland Cultural Center attract only a selected segment of the local community, i.e., those 18–30 years old of Latino and African American descent. It functions like a youth center with volunteers who are largely made up of Latinos and African American youths and adults.

Further, its activities appear to be attracting more people from outside the neighborhood than from within. Amongst the residents (African American, Anglo, Latino, and Cambodian) I spoke to, many of them were oblivious to the activities and opportunities at the cultural center. In fact, during the interviews I found myself informing them of the center's location and activities. A possible reason for its lack of visibility and popularity with the neighborhood's residents could be due to the timing of the activities at the center, which begin mostly in the late afternoon and early evening. These are times when life in the public spaces of the neighborhood is quickly winding down due to safety concerns.

While there is indeed opportunity and intentionality for the community to engage, these findings suggest that the scope of activities at the cultural center could be too narrowly defined to certain art forms that do not, and may no longer, reflect the interests and needs of its immediate resident population. The value of intentionality for intercultural (and intergenerational) engagement is undermined, which in turn compromises the neutrality of the space.

Mark Twain Library

Mark Twain Library is located across the street from MacArthur Park in a brand new building that opened in 2007 (Figure 11.4). Accessible and well located along East Anaheim Street, the library serves an average of 900 visitors per day from the surrounding neighborhood who walk to the library. It is well patronized by all ethnicities. The library is particularly popular with the Cambodian community in contrast to the Homeland Cultural Center. It boasts the largest collection of Khmer books in the United States. According to the librarian, the library is perceived as a peaceful and safe place that is meant for everybody, evident by only one case of graffiti gang-tagging thus far. In essence, quoting the librarian, "It's kind of a *neutral* territory."

While the library's functions are not *intentionally* set up as a space for active intercultural interaction and learning, its facilities and program, with a clear focus on learning for all, provide the pre-conditions that encourage values of openness and understanding between visitors. In terms of facilities, it has a multilingual collection of Khmer, Vietnamese, Spanish, and English books. The library offers trilingual story time with stories in English, Khmer, and Spanish during the weekdays, free Khmer language classes, as well as a new section of language learning tapes and videos. In addition, the library has a community room that is very well used by events including neighborhood meetings, US Census training sessions, early childhood development courses, and more. Most significantly, through its multilingual

Figure 11.4 Interior of the Long Beach Public Library Mark Twain Branch. Photograph by Felicity Hwee-Hwa Chan.

staff of homework helpers and library workers who speak Khmer, Vietnamese, Tagalog, and Spanish, the library attracts people of different ethnicities. It has also become a space where residents have established a sense of familiarity and comfort with the librarians. The library is especially popular as an after-school destination for mothers and their children who need a space to get homework help and recreate, since most of them live in crowded quarters that average five to six people per room. In fact, the after-school visitors to Mark Twain Library typically stay in the library from 3 p.m. to 7 p.m. until the library closes.

Overall, the library's program creates a supportive environment for urban interculturalism to grow. The library hosts the regular Neighborhoods Services Bureau meetings when the community gathers to discuss common problems in the neighborhood in Spanish, Khmer, and English. In addition to hiring staff workers who speak the different languages used in the neighborhood, the library also celebrates all major ethnic holidays with special programs and displays. Its limited access card, a policy of the City of Long Beach, allows residents of the neighborhood without official identification to borrow books.

Read as a space of investment for the future both metaphorically and physically, it is a strong and tangible signal to the resource-poor neighborhood that the library is a space of change. The local residents' perception of the library as a place of comfort and safety and its increasing utility as a space for community meetings makes it a space that brims with potential for urban interculturalism. It can serve as a seeding ground to nurture intercultural engagement and understanding in the neighborhood.

The above discussions about the value of intentionality and neutrality of civic spaces for intercultural engagement illustrate that these two values can be reinforcing in both positive and negative directions as seen in the case of the library and the cultural center respectively. Figure 11.5 maps out a conceptual comparison of the quality of the three spaces as intercultural spaces based on the values of neutrality and internationality.

Neutrality is a good starting point to design a civic space, but without an explicit intentionality for intercultural programs that engage the current needs of its neighborhood population, its capacity to be a good intercultural space where interaction can lead to dismantling of stereotypes and new appreciation of cultural diversity is undermined, as at MacArthur Park. The case of Homeland Cultural Center illustrates that there is a nuanced tension between intentionality and neutrality and challenges a more robust understanding of the value of intentionality in the context of intercultural engagement.

Akin to Amartya Sen's (1992) "Equality of What?" the relevant question here is "Intentionality of What?" Is the intentionality for neutrality more important than intentionality for intercultural engagement and learning or vice versa? My sense is that intentionality for intercultural engagement must precede neutrality. Starting with the intentionality for intercultural engagement, i.e. the "negotiation of differences" in the "local micropublics of everyday interaction" (Amin, 2002), the presence of difference is made explicit and necessary. A civic space that prioritizes a value of intentionality for intercultural engagement and that continuously evaluates

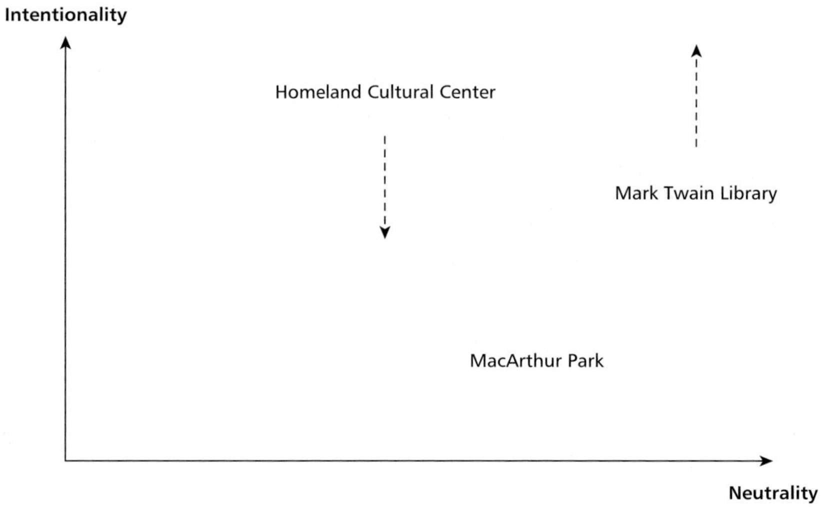

Figure 11.5 Conceptualizing the qualities of good intercultural spaces in "Cambodian Town" Neighborhood. Diagram by Felicity Hwee-Hwa Chan.

the changing needs of the community will be a space that serves as a robust neutral ground for engaging with differences and that can lead to building good intercultural understanding and relations.

Concluding Thoughts

The concept of global ethnoscapes by Appadurai (1990, 1991, 1996) provides a useful starting point to frame the discussions about the socio-spatial tensions in globalizing multi-ethnic settings. Using preliminary fieldwork from the case of the Cambodia Town neighborhood, this chapter has attempted to flesh out the nested socio-spatial tensions *between communities who are stable and those on the move, between the co-habiting of different cultures in space,* and the tension *between the material and the imagined space* that are proving to be endemic in the formation of global ethnoscapes. I have shown how the proposal for a coherent homogeneous identity encapsulated in the concept of Cambodia Town exists as a point of tension in a landscape of different cultures and belongings. It illustrated how the imaginary can become materially salient and fractious when articulated in and through the space of everyday experiences made up of practical and symbolic meanings.

The discussion of the Cambodia Town neighborhood highlights the negotiation of coexistence in global ethnoscapes that is marked by struggles with social anomie and conflicts within the shared spaces of the neighborhood. Agreeing with Amin (2002), the everyday civic spaces in a local neighborhood are where efforts for intercultural engagement against stereotypes can be leveled most effectively. How these spaces are used and can be used to productively transform human relations lies not outside but within the professional realm and responsibility of urban planners and policy-makers! Human flourishing in globalizing cities of diversity and transculturalism requires tangibly re-thinking and re-shaping urban interventions with a new consciousness for multilevel processes and a heart for transformative intercultural human relations.

Notes

1 The names of the interviewees are kept anonymous.
2 The boundaries were derived from asking respondents who work and live in the area. They are not the official boundary lines but the boundaries as perceived by its users.
3 Private communication, November 2010.
4 Private communication, January 2011.
5 In addition to these interventions to improve relations between the ethnic groups, the city organizes community events such as neighborhood clean-ups, ethnic festival parades, a unity festival, and potluck parties, and creates outreach programs in schools and in the local library to educate the multi-ethnic population about the services of the city. In 2003, the city commissioned a mural titled "At the Close of Day" by Long Beach resident artist Guillermo Avalos to formally welcome the Cambodians to Long Beach.
6 Sociality, according to Malpas (1997, p. 61) is awareness and "a grasp of one's own perspective on the world as one perspective amidst a multiplicity of other possible perspectives." This grasp of multiplicity of creatures other than oneself necessitates "a grasp of objective space and vice versa" (p. 70).
7 Private communication with Scott Bollens, September 2009.
8 The Homeland Cultural Center is managed as part of the MacArthur Park. In fact, the onsite park manager also directs and manages the programs at Homeland Cultural Center.

References

Allport, G. W. (1954). *The Nature of Prejudice*. Cambridge MA: Addison-Wesley.

Amin, A. (2002). Ethnicity and the Multicultural City: Living with Diversity. *Environment and Planning A*, 34, 959–980.

Amin, A. (2006). The Good City. *Urban Studies,* 43, 1009–1023.

Appadurai, A. (1990). Disjuncture and Difference in the Global Cultural Economy. *Public Culture,* 2, 1–24.

Appadurai, A. (1991). Global Ethnoscapes: Notes and Queries for a Transnational Anthropology. In Richard Fox (Ed.), *Recapturing Anthropology: Working in the Present* (pp. 188–210). Santa Fe, CA: School of American Research.

Appadurai, A. (1996). *Modernity at Large: Cultural Dimensions of Globalization.* Minneapolis: University of Minnesota Press.

Baum, H. S. (2000). Culture Matters—But it Shouldn't Matter Too Much. Urban Planning in a Multicultural Society. In M. A. Burayidi (Ed.), *Urban Planning in a Multicultural Society* (pp. 115–136). Westport, CT: Greenwood.

Bollens, S. A. (2006). Urban Planning and Peace Building. *Progress in Planning,* 66(2), 67–139.

Burayidi, M. (2000). *Urban Planning in a Multicultural Society*. Westport, CT: Praeger.

Cantle, T. (2005). *Community Cohesion: A New Framework for Race and Diversity*. New York: Palgrave Macmillan.

Census 2010, from: http://factfinder2.census.gov/faces/nav/jsf/pages/index.xhtml

City of Long Beach, from: http://www.longbeach.gov/park/parks_and_open_spaces/parks/macarthur_park_history.asp

Fainstein, S. S. (2010). *The Just City*. Ithaca. NY: Cornell University Press.

Germain, A. & Gagnon, J. E. (2003). Minority Places of Worship and Zoning Dilemmas in Montreal. *Planning Theory & Practice*, 4(3), 295–318.

Gupta, A. & Ferguson, J. (Eds) (1997). *Anthropological Locations: Boundaries and Grounds of a Field Science.* Berkeley and Los Angeles: University of California Press.

Jacobs, J. (1961). *The Death and Life of Great American Cities*. New York: Vintage Books.

Kristeva, J. (1991). *Strangers to Ourselves*. New York: Columbia University Press.

Long Beach Press Telegram, Long Beach Cambodia Town signs recognize ethnic neighborhood, by staff writer Greg Mellen 7/16/11 (accessed August 1, 2011, from: http://ki-media.blogspot.com/2011/07/long-beach-cambodia-town-signs.html).

Lynch, K. (1981). *Good City Form*. Cambridge, MA: MIT Press.

Malpas, J. (1997). Space and Sociality. *International Journal of Philosophical Studies*, 5(1), 53–79.

Needham, S. & Quintiliani, K. (2007). Cambodians in Long Beach, California. *Journal of Immigrant & Refugee Studies*, 5(1), 29–53.

Pestieau, K. & Wallace, M. (2003). Challenges and Opportunities for Planning in the Ethno-culturally Diverse City: A Collection of Papers—Introduction. *Planning Theory & Practice*, 4(3), 253–258.

Putnam, R. D. (2007). E Pluribus Unum: Diversity and Community in the Twenty-first Century: The 2006 Johan Skytte Prize Lecture. *Scandinavian Political Studies*, 30(2), 137–174.

Qadeer, M. A. (1997). Pluralistic Planning for Multicultural Cities. *Journal of the American Planning Association*, 63(4), 481–494.

Reeves, D. (2005). *Planning for Diversity: Policy and Planning in a World of Difference*. London: Routledge.

Sandercock, L. (1998). *Towards Cosmopolis: Planning for Multicultural Cities*. Chichester: John Wiley.

Sandercock, L. (2000). When Strangers become Neighbors: Managing Cities of Difference. *Planning Theory and Practice,* 1(1), 13–30.

Sandercock, L. (2003). *Cosmopolis II: Mongrel Cities of the 21st Century*. New York: Continuum.

Sen, A. (1992). *Inequality Reexamined*. New York: Oxford University Press.

Sennett, R. (1970). *The Uses of Disorder: Personal Identity and City Life.* New York: W.W. Norton.

Sennett, R. (1974). *The Fall of Public Man*. New York: W.W. Norton.

Sennett, R. (1989). The Civitas of Seeing. *Places*, 5(4), 82–84.

Sennett, R. (1990). *The Conscience of the Eye: The Design and Social Life of Cities*. New York: Alfred A. Knopf.

Talen, E. (2008). *Design for Diversity: Exploring Socially Mixed Neighborhoods*. Oxford and Burlington, MA: Architectural Press.

Talen, E. & Ellis, C. (2009). Compact and Diverse: The Future of American Urbanism. In Blanco et al., "Hot, Congested, Crowded and Diverse: Emerging Research Agendas in Planning." *Progress in Planning* 71, 153–205.

Thomas, H. (2000). *Race and Planning: The UK Experience*. London: UCL Press.

Thompson, S. (2003). Planning and Multiculturalism: A Reflection on Australian Local Practice. *Planning Theory & Practice*, 4(3), 275–293.

Thrift, N. (2005). But Malice Aforethought: Cities and the Natural History of Hatred. *Transactions of the Institute of British Geographers*, 30, 133–150.

Umemoto, K. (2001). Walking in Another's Shoes: Epistemological Challenges in Participatory Planning. *Journal of Planning Education and Research,* 21, 17–31.

Woods, P. & Landry, C. (2008). *The Intercultural City: Planning for Diversity Advantage*. London: Earthscan.

Zelinsky, W. (1990). Seeing beyond the Dominant Culture. *Places*, 7, 33–35.

CHAPTER 12

From a Neighborhood of Strangers to a Community of Fate

The Village at Market Creek Plaza

Michael Rios

The following chapter introduces the concept of cultural citizenship to demonstrate how marginalized groups use place to negotiate competing imaginaries of culture and, in the process, create new physical and political spaces that bridge differences among diverse ethnic groups and between these groups and community-serving institutions. Moreover, I argue that place rights ultimately reflect, and are the outcome of, various negotiations—of belonging to a territorial community, in participating in local decisions, and navigating the political landscape at multiple scales. As an illustrative example that draws attention to the link between place and cultural forms of citizenship, I describe the Village at Market Creek, a newly imagined community in San Diego, California.[1] Using the analytical lens of cultural citizenship, I explain how places provide an important location for cross-cultural exchange and brokering to occur, and where negotiations of belonging, authorship, and, ultimately, power can be fully realized.

Place, Culture, and Citizenship

From a scholarly perspective, "place" does not represent a coherent single idea. Multiple disciplinary threads ranging from philosophy, geography, and architecture, to cultural anthropology and environmental psychology, have defined the concept. Broadly defined, place refers to: territorialized local communities, collective memories associated with territory, claims of authenticity by local actors, phenomenological associations with locales, and social relationships among people in territorial communities (Casey, 1993; 1997; Relph, 1976; Norberg-Schulz, 1980; Tuan, 1977). In other words, place is a simultaneity of setting, identity, ideas, and practice (Agnew, 1987). Similarly, placemaking can be described as a historically contingent process that holds together the tensions and contradictions inherent in place construction (Pred, 1986). This process involves participation in both "production of meaning and in the means of production of a locale" where the symbolic and spatial are simultaneously expressed, whether imagined from within a particular territorial community by place dwellers, or imposed by external forces of the state or market (Lepofsky & Fraser, 2003).

Whereas general processes exist in and through space, places consist of territories that have specific histories, myths, and meanings that are situated within power relations (Creswell, 1991; Shields, 1991). This helps explain why placemaking is often used as a defensive reaction to the forces of globalization and as a way to avoid political economic concerns through nostalgia, essentialism, and a fetish for historical forms of communal life (Harvey, 1993). While this is an important critique, less attention has been paid to place as the setting to imagine territorialized identities, create narratives of meaning, negotiate between larger processes and the particular, and produce new political projects from the bottom up.

Place as a setting for rights claims draws attention to the types of political communities available for practice. However, in this case, *practice* does not concern general statements of social or spatial behavior (Liggett & Perry, 1995), but rather particular visions, enactments, and intersections of culture, power, and place (Gupta & Ferguson, 1997). This type of "emplaced" practice makes explicit the connection between material and political relations, and, in particular, the relationship between political efficacy and practices on the ground. Some have argued that the "power to act" resides locally, but that this power also incorporates extra-local forces aimed at producing a progressive or extroverted sense of place between global, local, and trans-local processes (Agnew, 1987; Amin, 2004; Massey, 1991). As Doreen Massey (2005) has pointed out, places necessitate invention and negotiation to make sense out of the complex power-geometry of space and time. This "practicing of place" is inherently political and requires action between people—re-acting and pro-acting, drawing together local circumstance with external forces at play. Challenging but important roles for practice are to reveal knowledge and to link, bridge, and mobilize multiple imaginations and narratives to create a *place* for transformative politics.

Despite a focus on politics, space, "the right to the city," and "insurgent citizenship," planning and geography scholars have paid little attention to culture in relation to these topics (Brenner, 2000; Douglass & Friedmann, 1998; Friedmann, 2002; Holston & Appadurai, 1999; Mitchell, 2003; McCann, 2002; Purcell, 2003; 2008). The use of cultural resources by marginalized groups to claim rights and create new physical and political spaces has often been described as cultural citizenship. Cultural forms of citizenship represent "the ways people organize their values, their beliefs about their rights, and their practices based on their sense of cultural belonging rather than on their formal status as citizens" (Silvestrini, 1998, p. 44). Additionally, cultural citizenship takes place in "zones of difference within and between cultures" (Rosaldo, 1989, p. 28) and involves negotiations of belonging, negotiations of authorship, and negotiations of power. Negotiations of belonging center on the exertion of citizens' rights to both be different and command a sense of belonging in the nation state, while also having the ability to participate in the democratic process (Benmayor et al., 1997; Rosaldo & Flores, 1997). Negotiations of authorship concern how individuals and groups claim ownership of the specific cultural context that they inhabit and participate in, and is inclusive of the right to challenge cultural consensus and hegemony (Boele van Hensbroek, 2010). Cultural practices and political rights are not mutually exclusive but intersect in ways that constitute one another. However, cultural forms of citizenship do not necessarily guarantee other entitlements (Pakulski, 1997). The prospects for securing place rights lay in the ability of cross-cultural coalitions to re-imagine the production of space toward political ends that has economic and legal benefits. It is with this insight—that rights are gained through various and overlapping forms of cultural negotiation *in* a place—that I now turn to the case of the Village at Market Creek.

The Village at Market Creek

The Village at Market Creek is a newly envisioned place in Southeast San Diego on a 45-acre site situated at the boundary of four different neighborhoods at the intersection of Market Street and Euclid Avenue. When fully built-out, the Village will include a mix of commercial, residential, and community-serving uses such as cultural facilities, parks, and the expansion of public open space around Chollas Creek, an arroyo that meanders through the entire Market Creek site (Figure 12.1). The four neighborhoods that converge around the Village are part of the larger Diamond District that includes close to 90,000 residents with a Latino majority. According to the 2005–2009 American Survey, 41 percent of the population in the four census tracts surrounding the Village is white, 27 percent is black, and 14 percent is Asian. For the same geographical area, 51 percent of the total population is Latino, primarily of Mexican descent. These multi-ethnic neighborhoods are rather diverse and also include populations from the Philippines, Laos, Vietnam, Somalia, Nigeria, Sudan, Somoa, and Guam, among other sending regions.

A Village Takes Shape

The story of the Village at Market Creek begins in the mid 1990s when residents and a handful of organizations participated in a community-visioning exercise sponsored by the CA Energy Commission, an unlikely convener due to its mandated focus on statewide energy policy. The Commission was interested in ways local communities could physically develop to reduce energy consumption. For the Commission and its consultants, this included transit-oriented development strategies that would increase housing densities around a trolley station that was built in 1986 near the corner of Market Street and Euclid Avenue.

Building on this early effort, the Jacobs Family Foundation (JFF) made an initial commitment to focus their mission on neighborhood revitalization in the Diamond District after funding several projects in the area. At the time, the area was known as the "Four Corners of Death," as it was home to 42 gangs (Wittenberg, 2009). A "convergence moment," as recounted by JFF President and CEO Jennifer Vanica, was during a community meeting when residents were complaining about a 20-acre lot with an abandoned aerospace factory (known as the Langley site) where much crime and gang violence took place (Korten, 2009). Moreover, parents and grandparents were upset about the 300 kids who walked beside the lot each day to and from school (p. 70). Joseph Jacobs (the founder of the foundation) responded by buying the property and committing to work with residents to plan for its redevelopment. In 1997, the Jacobs Center for Neighborhood Innovation (JCNI), the newly formed non-profit community development arm of the JFF, bought the 20-acre lot for $3.1 million and hired Roque Barros, a seasoned community organizer, to engage

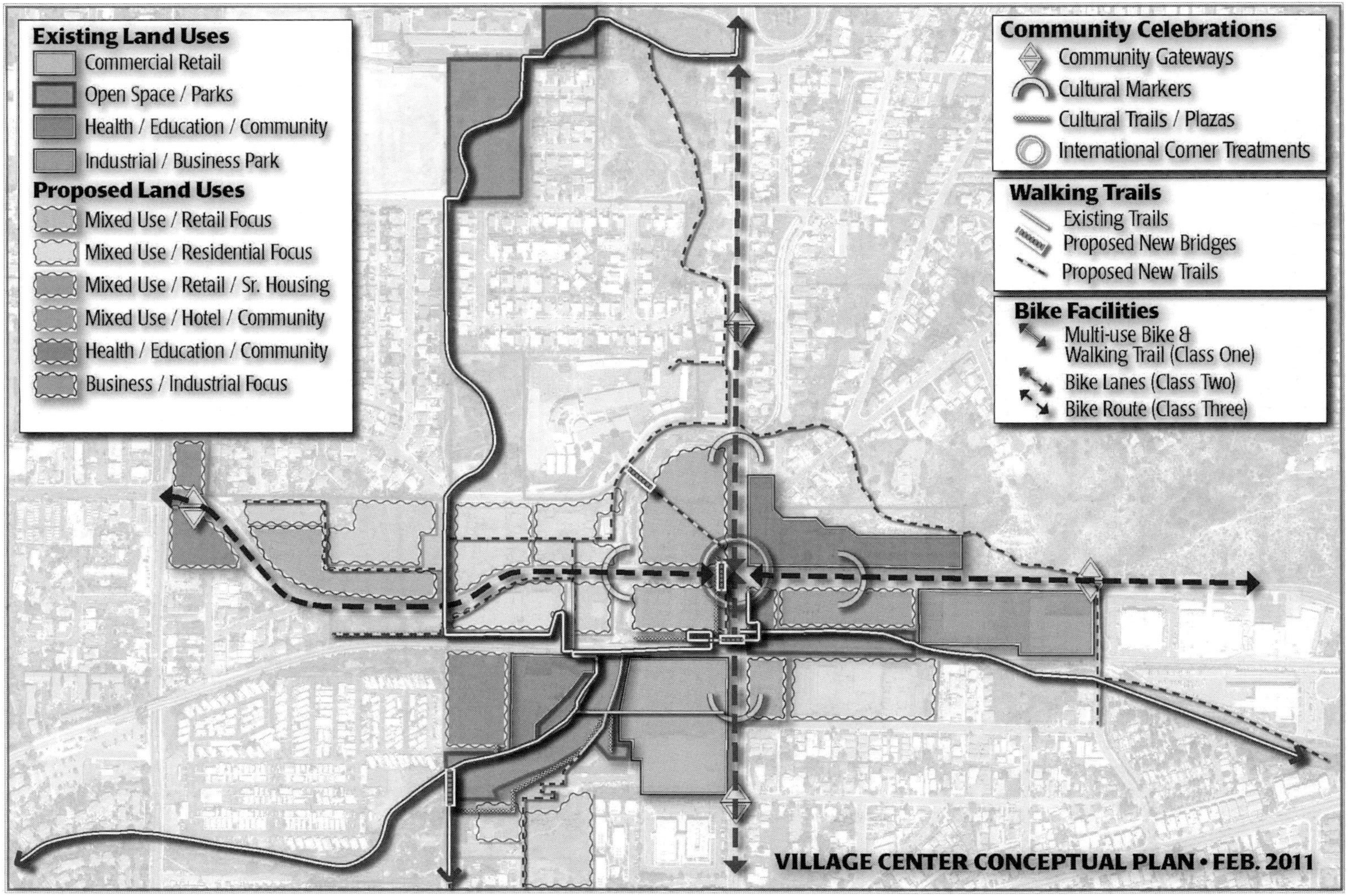

Figure 12.1 The Village at Market Creek is envisioned as a transcultural center that will include mixed-use development, parks and trails, and cultural plazas, gateways, and totems. Courtesy of KTU+A.

local residents in a participatory planning process. This was a strategic move as there was significant mistrust in the community, much of it divided along ethnic and racial lines. To begin, Barros organized a Halloween carnival at the old factory; thousands of people came and shared stories about the site and their feelings about it.

In order to survey community residents about their needs and desires for the site, Barros helped form International Outreach Teams made up of residents. To address racial tensions that became apparent during the surveying, he recruited and paid stipends to individuals from different cultural groups living in the community who would later serve as "community coordinators." For example, there were historical tensions between a group of Latinos and Samoans living on the same street. Many did not know why there was conflict, but acknowledged it had existed for some time. Working with Barros, coordinators representing the two groups organized a block party to identify issues that the Latinos and Samoans had in common. One issue was the lack of lighting on the cul-de-sac. A media event was orchestrated to draw attention to the issue and to demonstrate that they could work together. One outcome was the initiation of cross-cultural exchanges. For example, many Latino families were interested and participated in Samoan dance classes taught by their neighbors.

Coordinators were also recruited from seven other cultural groups including African Americans, Chamorros (from Guam), Kumiai Indians (San Diego's original inhabitants), Filipinos, Latinos, Laotians, Samoans, Somalis, and Sudanese. Further cross-cultural exchanges ensued leading to a series of "ethnic nights" that highlighted the cultural heritage of each group. These events culminated in a "Unity Day" featuring different food, clothing, and dance traditions. The Langley site was transformed into a series of cultural villages with individual cultural houses for each group. Several thousand people turned up for the event. A "Spirit of Unity" event now takes place each year as well as an annual "Arts and Culture Fest" celebrating the diverse cultural traditions of the community.

Representatives from different cultural groups were also involved in decisions concerning development of the Langley site. Working with JCNI staff and a local architect, they developed proposals for a commercial center, including a new grocery store (given that there were no grocery stores in the immediate area). In January 2001, Market Creek Plaza opened with Food-4-Less as the anchor grocery store tenant, as well as other commercial tenants, including restaurants, a Wells Fargo Bank, and a Starbucks café (Figure 12.2).

In the same year, the community also formed the Euclid Market Action Team (EMAT) in response to the city's decision to relocate produce distribution warehouses to the Market Creek area due to the displacement of warehouses caused by the development of the $450 million PETCO Park baseball stadium. The relocation of the warehouses and the expected truck traffic was a particular threat to existing community institutions such as the Tubman-Chavis Cultural Center, the Elementary Institute of Science, and Malcolm X Library.

Figure 12.2 Residents worked with a local architect to design buildings expressive of the forms and colors from sending regions. Photograph by Michael Rios.

A Vision for a Cultural Village

Organized by JCNI and the Coalition of Neighborhood Councils, several hundred angry residents went down to city hall to protest the proposed relocation of produce distribution warehouses; they were upset with the local councilman and the Southeastern Economic Development Corporation, the non-profit redevelopment organization in the area, for approving the relocation. At a tense meeting, the councilman challenged the community to come up with an alternative plan. Shortly thereafter, JCNI hired Niko Calavita, an urban designer and professor of planning at San Diego State, to develop the EMAT plan that provided a framework for "equitable development."

The kick-off of the EMAT planning process finally began in August of 2001, where nearly 200 participants showed up to develop the plan. Maps of the four neighborhoods surrounding the Euclid and Market Street intersection were used to reconfigure historical boundaries and analyze possible linkages between the neighborhoods. The initial workshop was followed by a series of "planning circles" in peoples' homes to solicit ideas about their vision for the community. Results of planning circles were used to develop a framework for the rest of the planning process and subsequent design workshops were then used to prepare the plan. A total of five workshops took place over a two-month time period. As part of the

EMAT planning process, community coordinators also conducted 800 neighborhood surveys in multiple languages within a one-mile radius of the corner of Euclid Avenue and Market Street (Robinson, 2005).

After the results from the surveys were gathered, a clear vision emerged. This was the first time that the concept for a "Cultural Village" was identified; it was envisioned as a multicultural and commercial hub that would connect the four adjacent neighborhoods, be joined by a proposed pathway, and include improvements along Chollas Creek. Once a plan was created on how to proceed, JCNI staff (now including the community coordinators as paid staff) in 2002 helped organize over 2,000 adults and 1,000 youth as part of three primary working teams focused on Art & Design, Business & Leasing, and Construction & Ownership Design. One of the outcomes of the working teams was the idea for an initial public offering (IPO) where residents could invest and own shares in the Market at Village Creek, the commercial portion of the Village. In 2006, 416 residents purchased ownership units in a Community Development IPO, the first of its kind in the United States. At that time, 20 percent of the project was owned by residents and 20 percent by a JCNI-fostered community foundation, with the assets of the remaining 60 percent held by JCNI. Two years later 40 percent of the plaza was owned by residents (Korten, 2009). According to JCNI staff, the goal is to eventually transfer complete ownership and control of Market Creek Plaza to a non-profit organization managed by residents.

Six years after the vision for the Cultural Village was created, a new and active multi-ethnic community has taken shape (Figure 12.3). As a result of their accom-

Figure 12.3 View of the Village at Market Creek showing Chollas Creek, the community amphitheater, cultural houses, and commercial center. Photograph courtesy of Jacobs Center for Neighborhood Innovation.

plishments, the City of San Diego finally changed the zoning designation for the Village at Market Creek in 2008, which provided the legal basis to fully implement the plan for the Cultural Village. Prior to the update, Southeast San Diego was also the only major area of the city without mixed-use zoning.

Securing Rights through Community Control

Beginning in 2010, JCNI staff invited approximately 40 residents of the community to form the VOCAL network which stands for "*V*oices *O*f the *C*ommunity at *A*ll *L*evels." For six months, weekly evening workshops were held to build the group's capacity. A core focus of the weekly gatherings was a series of "cultural workshops" where VOCAL members, led by design consultant Bennett Peji, brought in cultural artifacts, described these items, shared their cultural histories, and entered into a deeper cross-cultural dialogue. One commonality that emerged from these learning sessions was the recognition that the different cultural groups experienced oppression in one form or another as a result of Colonialism and Empire, and that it was important not to lose sight of this common experience. During the first workshop, an older African American male commented about the blues and the persistence of culture through music despite tragedy. He spoke about the persistence of pan-African culture even though more than 2 million Africans perished in the Atlantic Ocean (also known as the Middle Passage) during the slave trade and the separation of hundreds of thousands families. Similarly, other participants made reference to the violent histories of Colonialism, whether it was French occupation in Laos or British rule in Sudan. A Kumiai Indian artist shared a photograph art-piece depicting the mission wall where indigenous babies were buried, commenting, "We were also slaves," making reference to genocide inflicted by Franciscan missionaries at San Diego de Alcala.

The deeply personal and meaningful sessions were then followed by a series of VOCAL retreats where specific aspects of the Village were discussed with the larger community. The first of four retreats took place in November of 2010.[2] Over 100 people participated in the discussion that focused on urban design and land use elements of the Village plan. Subsequent VOCAL retreats drew attention to proposals for specific development sites, strategies to increase physical accessibility and connectivity, programming for new and existing parks and open space, and possibilities to address environmental sustainability through the Leadership in Energy and Environmental Design Neighborhood Development (LEED-ND) standards. In each of the VOCAL retreats, a team of planning and design consultants used methods and techniques to increase participation and solicit feedback.[3]

A major challenge during the planning and design process was keeping focus on the group's prior decisions as new participants showed up at subsequent retreats. Some residents expressed frustration with the technical nature and quick speed at which issues were being discussed. Others were skeptical and opined that JCNI staff and their consultants were making important decisions in private meetings. Tension also existed with some individuals, such as an African American senior, who felt that blacks should have a larger presence among JCNI staff and a more prominent voice in decision-making given their historical presence in the neighborhood, despite its growing multi-ethnic character.

Despite these grievances the retreats were generally successful, as measured by the level of resident participation and engagement with the planning and design consultants on specific issues. With general agreements reached about the size, scope, and character of the Village, residents and JCNI staff have begun to focus on the difficult tasks that lie ahead—to complete the build-out of the Village, transfer ownership of the foundation's assets and holdings to residents, and develop a collaborative leadership structure so that all decision-making will be in the hands of resident leaders by 2020. The story of the Village at Market Creek continues . . .

Discussion: Negotiations of Belonging, Authorship, and Power

As the case of the Village at Market Creek illustrates, place is a location where cross-cultural exchange occurs and it is the acquisition of rights through the claiming of space that cultural awareness and understandings, new identities and meanings begin to take shape. The point here is that rights, although not the only source of identity formation, are an important means through which marginalized groups negotiate different imaginaries of culture. Moreover, place ultimately reflects and is the outcome of these various negotiations—of belonging to a community, in participating in local decisions, and struggling to gain power. The multi-ethnic community that coalesced around the Village at Market Creek evolved out of these negotiations. It began with creating a vision for a cosmopolitan village that impelled local citizens to engage in deep dialogue about past and present conditions in order to secure the right to envision a place that could belong to everyone. The case demonstrates the importance of cultural citizenship in the formation of place rights inasmuch as it involved tensions and negotiations between group histories, recognition in decisions, and community control.

As described in this chapter, negotiations of belonging center on citizens' rights and the ability to be different-but-equivalent in terms of associating with, and participating in, a larger political community. At the Village at Market Creek these negotiations took several forms, including cross-cultural learning and exchange and envisioning of place through the built environment. A critical aspect of this was the brokering between and among the different cultural groups as well as with JCNI staff, project consultants, private developers, and elected officials. An explicit strategy was the identification of cultural diversity in the surrounding area and the creation of a safe and respectful space in which different worldviews could be expressed and shared (Figure 12.4). The result of the various cultural exchanges (such as festivals, workshops, visioning forums, etc.) was a deep respect for cultural difference and heritage, which enabled participating groups to facilitate intercultural communication after identifying common struggles of oppression.

This shared experience became a foundation from which to build solidarity and a dynamic multi-ethnic polity that is still evolving today. From this newly formed political space, different cultural groups made claims about their right to live in coexistence with other cultures and flourish together in a place that had been once identified as "the four corners of death." Members of the community practiced their cultural citizenship through negotiations of authorship concerning whose vision would be implemented, what constituted participation, and which decisions were

Figure 12.4 Sudanese women gather in their "cultural house," one of a number of structures that highlight the heritage of different cultural groups. Photograph courtesy of Jacobs Center for Neighborhood Innovation.

in the community's best interests. Some examples included demanding to be equal partners with JCNI staff on Village projects, approving or rejecting proposals from private developers and businesses, and selecting consultants for the planning, design, and implementation of the Village plan. In some instances this included firing project consultants that did not carry out agreed-upon goals, objectives, and requirements. Collectively, these activities do not represent one-time decisions or a limited sense of authority, but have expanded in scope through on-going resident education and training, place-to-place peer exchanges, and the goal of collaborative leadership through the efforts of VOCAL (Figure 12.5).

The desire to create collaborative leadership capacity is perhaps one of the greatest challenges for the Village as it ultimately raises the issue of power—among the different cultural groups, between these groups and JCNI staff, and between different scales of decision-making ranging from the local to the regional. While participation in local decisions is evidence that residents are taking ownership over their community's fate, there is no guarantee that the delicate balance of power will continue as the Jacob's Foundation divest all of their assets to community control by 2020. However, if the community development IPO pioneered by residents and JCNI staff that requires changes to state law is any indication, it would appear that resident leaders have the capacity to engage in future negotiations of power and are resilient enough to anticipate unforeseen political challenges.

Negotiations of belonging, authorship, and power are important inasmuch as negotiations establish what groups can expect of one another. Some communication scholars have argued that negotiation is the basis for identity construction, the bridging of cultural difference, and the precondition for successful intercultural communication (Berger & Calabrese, 1975; Gudykunst & Ting-Toomey, 1988; Jackson, 1999; Ting-Toomey, 1999). Negotiations are also the basis for agreements and provide shared experiences that maintain relationships into the future. These agreements, or "cultural contracts," measure the degree to which cultural values

Figure 12.5 Residents participate in a group exercise in a 2011 resident-to-resident learning exchange. Photograph courtesy of Jacobs Center for Neighborhood Innovation.

and commitments are exchanged between groups (Jackson, 2002). While cultural contracts often focus on exchanges between traditionally marginalized groups and majority groups in power, the case of Market Creek demonstrates that agreements can be utilized in transcultural processes that involve conflictive and non-conflictive, equal and unequal social relationships, and between individuals, groups, and institutions. In the case of the Village at Market Creek, cultural contracts required the adaptation of cultural values through negotiations between marginalized groups and institutions such as JCNI, as well as the co-creation of mutually valued agreements between the marginalized groups living in the neighborhood.

Key to these various forms of negotiation that resulted in agreements is the brokering between various identities. Brokering is a transactive process where someone acts as an agent for others. In cross-cultural exchanges an agent facilitates the communication and transaction of customs and values toward a shared goal that is codified (explicitly or implicitly) in an agreement, whether short or long term. In the Village at Market Creek there was no single broker but rather a fluid group of agents that were called to facilitate cross-cultural exchange—from the early ideas of the Village and throughout the process. At the beginning this included JCNI staff such as Roque Barros, who reached out to the Diamond District's multi-ethnic communities. The formation of an international outreach team that would evolve into community coordinators also enabled cross-cultural dialogue between different cultural groups and with institutions such as JCNI. Even the planning and design consultants facilitated dialogue and agreement between participants, as well as finding common ground between technical understandings of project development and the specific values of community members centered on cultural expression, quality of life, and local economic development.

Without the skilled brokering between the various identities it is doubtful that agreements would have been reached. Moreover, the transcultural coalition that has emerged over the past decade is testament to the successful negotiations that occurred over the same time period. The Village at Market Creek demonstrates how transcultural agency and capacity can be developed, and is evidence of the hopes and struggles that multi-ethnic communities encounter in creating a place of their own.

Notes

1 A mixed method approach was used for the case study. After initial interviews with JCNI staff and community leaders in February 2010, data were collected through a focus group with eight Community Coordinators in December 2010; formal interviews with staff and residents in November 2010, December 2010, and January 2011; and personal observations at three retreats on November 13, 2010, December 11, 2010, and January 22, 2011. Archival data were also collected including newspaper articles; staff and consultant reports; Internet sources; and video recordings of cultural workshops.

2 The four retreats were led by a multidisciplinary team of architects, planners, and landscape architects. The consultant team was led by MIG, a planning and communications firms based in Berkeley, California. Other consultants included KTU+A, a planning and landscape architecture firm from San Diego, and Niko Calavita, a retired planning professor from San Diego State, who had been working with the Jacobs Center since 2001.

3 The consultant team used a variety of techniques to solicit feedback from retreat participants. This included a visual primer on urban design concepts and precedents, visual preference surveys, small group discussions, gaming techniques, and validation of participant preferences through voting on specific options presented by the consultants. Multilingual translators assisted non-English participants at each of the retreats.

References

Agnew, A. (1987). *Place and politics: The geographical mediation of state and society.* London: Allen and Unwin.

Amin, A. (2004). Regions unbound: Towards a new politics of place. *Geografiska Annaler, Series B: Human Geography* 86(1), 33–44.

Benmayor, R., Torruellas, R., & Juarbe, A. (1997). Claiming cultural citizenship in East Harlem: "Si esto puede ayudar a la comunidad mia." In W. V. Flores & R. Benmayor (Eds), *Latino cultural citizenship: Claiming identity, space and rights* (pp. 152–209). Boston, MA: Beacon Press.

Berger, C. & Calabrese, R. (1975). Some explorations in initial interactions and beyond. *Human Communication Research* 1, 99–112.

Boele van Hensbroek, P. (2010). Cultural citizenship as a normative notion for activist practices. *Citizenship Studies* 14(3), 317–330.

Brenner, N. (2000). The urban question as a scale question: Reflections on Henri Lefebvre, urban theory and the politics of scale. *International Journal of Urban and Regional Research* 24(2), 361–378.

Casey, E. (1993). *Getting back into place: Toward a renewed understanding of the place-world*. Bloomington: Indiana University Press.

Casey, E. (1997). *The fate of place: A philosophical history*. Berkeley: University of California Press.

Creswell, T. (1991). *Place: A short introduction*. Oxford: Blackwell.

Douglass, M. & Friedmann, J. (1998). *Cities for citizens: Planning and the rise of civil society in a global age*. New York: J. Wiley.

Friedmann, J. (2002). *The prospect of cities*. Minneapolis: University of Minnesota Press.

Gudykunst, W. & Ting-Toomey, S. with Chua, E. (1988). *Cultural and interpersonal communication.* Newbury Park, CA: Sage.
Gupta, A. & Ferguson, J. (1997). *Culture, power, place: Explorations in critical anthropology.* Durham, NC: Duke University Press.
Harvey, D. (1993). From space to place and back again: Reflections on the condition of postmodernity. In J. Bird, B. Curtis, T. Putnam, G. Robertson, & L. Tickner (Eds), *Mapping the futures: Local cultures, global change* (pp. 3–29). London: Routledge.
Holston, J. & Appadurai, A. (1999). Cities and citizenship. In J. Holston (Ed.), *Cities and citizenship*. Durham, NC: Duke University Press.
Jackson, R. (1999). *The negotiation of cultural identity*. Westport, CT: Praeger.
Jackson, R. (2002). Cultural contracts theory: Toward an understanding of identity negotiation. *Communication Quarterly* 50(3), 359–367.
Korten, A. (2009). *Change philanthropy: Candid stories of foundations maximizing results through social justice.* San Francisco, CA: Jossey-Bass.
Lepofsky, J. & Fraser, J. C. (2003). Building community citizens: Claiming the right to place-making in the city. *Urban Studies* 40(1), 127–142.
Liggett, H. & Perry, D. (1995). *Spatial practices: Critical explorations in social/spatial theory*. Thousand Oaks, CA: Sage.
McCann, E. J. (2002). Space, citizenship, and the right to the city: A brief overview. *GeoJournal* 58(2–3), 77.
Massey, D. (1991). A global sense of place. *Marxism Today*, 24–29.
Massey, D. (2005). *For space.* London: Sage.
Mitchell, D. (2003). *The right to the city: Social justice and the fight for public space.* New York: Guilford Press.
Norberg-Schulz, C. (1980). *Genius loci: Towards a phenomenology of architecture*. New York: Rizzoli.
Pakulski, J. (1997). Cultural citizenship. *Citizenship Studies* 1(1), 73–86.
Pred, A. (1986). *Place, practice and structure: Social and spatial transformation in southern Sweden, 1750–1850.* Cambridge: Polity Press.
Purcell, M. (2003). Citizenship and the right to the global city: Reimagining the capitalist world order. *International Journal of Urban and Regional Research* 27(3), 564–590.
Purcell, M. (2008). *Recapturing democracy: Neoliberalization and the struggles for alternative futures*. New York: Routledge.
Relph, E. (1976). *Place and placelessness*. London: Pion.
Robinson, Lisa (2005). *Market Creek Plaza: Toward resident ownership of neighborhood change*. Oakland, CA: Policylink.
Rosaldo, R. (1989). *Culture and truth: The remaking of social analysis*. Boston, MA: Beacon Press.
Rosaldo, R. & Flores, J. (1997). *A companion to Latina/o studies*. Oxford: Blackwell.
Shields, R. (1991). *Places on the margin*. London: Routledge.
Silvestrini, B. (1998). The world we enter when claiming rights. In W. Flores & R. Benmayor (Eds), *Latino cultural citizenship: Claiming identity, rights, and space*. Boston, MA: Beacon Press
Ting-Toomey, S. (1999). *Communicating across cultures*. New York: Guilford Press.
Tuan, Y. (1977). *Space and place: The perspective of experience*. Minneapolis: University of Minnesota Press.
Wittenberg University. (2009, December 18). Jennifer Vanica '76 is transforming the national conversation about community development. *Wittenberg Magazine*. Retrieved April 11, 2009, from: http://www9.wittenberg.edu/magazine/features/leading-change/

CHAPTER 13

"Dumb White Kids" and "Asian Nerds"

Race and Ethnic Relations in Silicon Valley Suburban Schools

Willow Lung-Amam

Fremont is a window into a new phase in race relations in America.

(Claudia Sanchez, 2004, *National Public Radio*)

Nestled in the foothills of Fremont, California, is Mission San Jose, a neighborhood that only twenty years ago was known primarily for its historic landmark and namesake, an eighteenth-century Spanish mission. More recently, this Silicon Valley suburban community has become internationally recognized for another landmark—Mission San Jose High School (Figure 13.1). Up until the mid-1990s, Mission San Jose High was a "typical" suburban American high school, made up of a largely white, middle-class student body. Today it is a premier destination for high-income, highly educated families from all over the world, especially Asia.

In a case study of Mission San Jose High (herein Mission High or Mission), this chapter investigates the impacts of Asian immigration on the social and academic life of the school and racial and ethnic relations among students. I argue that demographic shifts and the changing culture of the school have resulted in social tensions over disparities in academic achievement and a sense of academic competition, particularly between Asian American and white students. Paradoxically, Mission's increasing diversity and changing academic culture have helped to neutralize racial and ethnic conflicts and normalize social differences of all kinds.

This study brings attention to the everyday suburban landscape as a locus of interracial and interethnic conflicts and interactions. Currently, most studies on urban diversity focus on urban centers and traditional public spaces as the supposed centers of contemporary cosmopolitan life. Environmental design studies aimed at better understanding the spatial needs and desires of different ethnic groups and strategies for "designing for difference" have largely focused on urban parks and plazas (Jacobs, 1961; Loukaitou-Sideris, 1995; Rishbeth, 2001; Low et al., 2005). So too has much of the multi- and intercultural scholarship that looks at cities as spaces to foster respect and tolerance for ethnic and cultural differences (Burayidi 2003; Sandercock, 2003; Amin, 2002).

The limited geographical focus of extant studies contributes to the common assumption that most people encounter social difference and diversity in their everyday lives mainly, if not exclusively, within central cities and traditional public spaces. But, in fact, over half of all Americans, including the majority of immigrants and minorities, live in suburbs (Singer et al., 2008; Frey, 2011). The average

Figure 13.1 Mission San Jose High School in Fremont, California, is an internationally renowned public high school, especially among Asian immigrants settling in the Silicon Valley. Photograph by Willow Lung-Amam.

American today is just as, if not more, likely to meet people of different ethnic and cultural backgrounds in suburban shopping malls, office parks, big box stores, and in their neighborhood schools as in downtown public parks or squares. In a rapidly globalizing and decentered metropolis, I argue that the current scope of scholarly discourse on urban diversity needs to expand to include places, people, and problems that do not fit the prototypical urban imaginary—places like Mission San Jose High. Mission San Jose High exemplifies both the challenges of fostering diverse urban schools and neighborhoods as well as the possibilities inherent in the banal spaces of everyday suburban landscape for learning to live together amidst our differences.

Methods

This chapter is based on interviews, participant observations, and archival research on the Mission San Jose neighborhood and its schools collected between 2010 and June 2011. I conducted in-depth interviews with 22 current and former Mission High parents, 12 students, and 10 school administrators. Among parents, 14 were Chinese immigrants, 2 were Asian Indian immigrants, and 6 were native-born whites. Among students, 11 were American-born and 1 was a foreign-born student raised in the USA. Three students identified as white, 3 as Chinese American, 2 as Indian American, 1 as Japanese American, 1 as Korean American, and 2 as having mixed ancestry (Caucasian and Afghani, and Caucasian and Chinese). School administrators included 5 high-ranking Mission High staff, 4 teachers, and 1 Fremont Unified School Board member.[1]

To investigate demographic changes in the neighborhood and school, I analyzed US Census data for Fremont and Mission San Jose from 1970 to 2010, and demographic data on Mission San Jose High from the California Department of Education (1980 through 2011) and the school administration (2009 through 2011).

For insights on Mission High's changing school and academic culture, I reviewed archives of its student newspaper, *The Smoke Signal,* from 1974 to 2011, as well as local, regional, and national newspaper databases. And over the course of about a year and a half of research, I sat in on several classes and otherwise informally observed students in their daily lives at Mission High.

From White to Asian American Schools and Neighborhoods

Immigration reform, globalization, and economic and political restructuring in the later half of the twentieth century has changed the face of many neighborhoods throughout Silicon Valley. Since the dot-com boom (1995–2000), the Silicon Valley has become a hotbed of foreign-born engineers and researchers, especially those from China, Taiwan, and India. In 2000, 53 percent of Silicon Valley engineers were foreign-born (Saxenian, 2006).

Fremont is a popular destination for many Silicon Valley immigrants who have clustered within neighborhoods like Mission San Jose with high-end new housing and high-performing public schools. Among high-tech, Asian immigrant families, schools are often the number one priority in their residential location (Chang & Lung-Amam, 2010).

Asian immigration in Mission San Jose has completely changed its schools. Paula, who grew up in the area, recalled that Mission High was referred to as "Little Scandinavia" for its predominance of blond-haired, blue-eyed students well into the 1980s. When the district first began recording racial demographics in 1981, Mission High was 84 percent white. Mexican and Japanese families who had traditionally lived on and worked the local farms, nurseries, ranches, and orchards made up the vast majority of its minority residents. But as Maria, a longtime resident of Mission San Jose, recalled, "The 1990s marked the end of the dominance of the white, blond-haired group at Mission High." Between 1981 and 2009, Mission High's white population dramatically dropped to 14 percent, while its Asian American population grew from 7 percent to 83 percent. Growth among students of Chinese and Indian descent far outpaced that of other Asian groups. In 2009, Chinese Americans made up 49 percent of Mission's Asian students, Indians 17 percent, with another 25 percent of Asian students identifying their ethnicity as "other." The latter likely includes the large number of students of white-Asian ancestry, also referred to as "Wasians" or "Hybrids," at Mission. Although neither Mission High nor the district records parental country of origin, most Asian students are reportedly from Taiwan. Latino and African American students make up only 2 percent and 0.5 percent of the population, respectively (Table 13.1). In 2010, 76 percent of Mission High's Asian students were American-born and 69 percent spoke a

Table 13.1 Mission San Jose High's student population went from predominantly white to Asian American in only a few decades. At the same time, the academic standards of its students have increased.

		1981*	1985	1990	1995	2000	2005	2010
Race	White	84%	81%	71%	53%	39%	19%	12%
	Asian	7%	10%	25%	41%	57%	58%	84%
	Hispanic	7%	6%	3%	4%	3%	2%	2%
	Black	1%	1%	1%	2%	1%	1%	1%
	Other	1%	1%	1%	0%	0%	20%	1%
API Scores**		—	—	—	—	882	935	951

*1981 is the first year that the state of California began recording racial data in schools.
**California's current Academic Performance Index (API) standardized testing system began in 1999.

non-English language at home. The dominant profile of today's Mission High student is an American-born Asian student with immigrant parents.

Asian immigrant families, who are often employed as Silicon Valley engineers and researchers, have raised the class status of Mission High students. Many arrive at school in Lexuses, Audis, and BMWs, otherwise known as "Basic Mission Wheels." Joseph, who graduated from Mission in the early 1980s, said that in his day everyone just drove "a car to get around in," whereas today the majority of students have fancier cars than he does. In 2009, less than 4 percent of Mission High students qualified for free and reduced lunch, compared to 19 percent district-wide. And in 2005, Mission San Jose was ranked on *Forbes* magazine's list of the 500 most affluent communities in the United States, with a median income of over $114,000.

The demographic shift at Mission High has corresponded with a significant change in its academic culture. In 1974, students averaged GPAs between 2.0 and 3.0 and often took classes in wood, auto, electric, and metal shop (Staff Writers, 1974). In 1987, 65 percent of students went to college, 40 percent to 4-year institutions and 25 percent to a community college (Walter, 1987). Today Mission High and its feeder schools are internationally recognized as world-class academic institutions. In 2008, 2009, and 2010, Mission was the number one comprehensive high school in California, based on its standardized test scores. In 2009, *U.S. News and World Report* rated Mission High as the 36th best academic school (among both public and private schools) and 4th best public open-enrollment high school in the nation. William Hopkins Junior High, its feeder school, had the highest standardized test scores among public junior high schools in California in 2005 and 2007. Its four elementary schools all rank amongst the top performing schools in the state. Since 2000, Mission High has maintained a near 100 percent graduation rate and in 2010 graduated 31 (out of 512 total graduates) valedictorians, all with grade point averages exceeding a 4.0. Eighty-one percent of students were on the honor roll and 94 percent of the 2010 graduating class enrolled in college. Sixty-four students went on to study at the University of California, Berkeley, and several enrolled at other prestigious institutions like Stanford, Princeton, Harvard, Cornell, and MIT. "All the Ivy Leagues know about Mission," explained Annie, a parent of two Mission High students.

Driven largely by Asian students' and parents' demands for more and harder classes, Mission High now offers primarily honors and advanced placement (AP) courses, particularly in the math and sciences. In the 2009–2010 academic year, 77 percent of juniors and seniors completed one of Mission's 52 AP sections, 85 percent of which were in math and science-related fields. In 2005, Mission had the highest AP statistics exam pass rate of any school of its size in the world (Aboumrad, 2005). Ohlone Community College is popularly known as "Mission on the Hill" because of the large number of Mission High students who attend classes there on weekends and over the summer. SAT prep classes, professional tutors, and other academic services proliferate throughout the Mission San Jose neighborhood. Students often attend afterschool and weekend Chinese classes and academic summer camps, and are even known to study their textbooks and get tutored on coursework the summer before classes begin. Long gone are reports of girls' locker room break-ins and wild homecoming parties. Instead *The Smoke Signal* dishes out advice on managing academic stress, getting enough sleep, keeping up grades, and selecting the right college (Figure 13.2). A popular Mission High cheer reaffirms the school's reputation: "Cosine, sine—cosine, sine—3.14159—2400s on SATs—and yes, we all take five APs."

Figure 13.2 Mission High students often suffer from high levels of stress over their grades. Drawing by Cassie Zhang, published in *The Smoke Signal.*

The social life of the school has also been transformed by changes in its demographics and academic culture. In the 1970s, administrators complained about the lack of student involvement in clubs. One student joked that the most popular student clubs were those with "no constitution, no officers, no dues, and no meetings" (Rosen, 1978). Today, administrators debate whether students start too many clubs to pad their college resumes. Stressed Out Students, a nationwide program started by a former Mission teacher that instructs students and parents on managing stress, is one of Mission's most active student clubs.

Sports serve as another indicator of social and cultural change. Up until the 1980s, Mission High was largely known as a football school and ranked among California's top teams for several years running. But for the past couple of years, it has struggled to even field a varsity football team. And in 2002 coaches canceled the season for lack of interest among upper classmen. According to Coach Kevin Lydon, trying to muster enthusiasm for football on the Mission campus was "like trying to sell electricity to the Amish" (Somashekhar, 2003). Students now joke that the only reason to go to a football game is to get physics extra credit (as the physics teacher is also the football coach). Meanwhile, the badminton team is larger than the football team and, like Mission's chess and debate teams, is highly competitive regionally and nationally.

In only a few short decades, Mission High has been transformed into an internationally renowned institution of predominantly Asian American students from upper-middle class immigrant families. The concentration of immigrants with economic resources and a new set of educational values, expectations, and standards has redefined the academic and social life of this once "typical" suburban American high school.

Academic Competition, Stress, and Social Segregation

Academic stress, competition, and changes in the social and academic culture at Mission impact everyday social relations. Racial and ethnic segregation is an everyday part of life at Mission High. A 2010 survey found that 72 percent of students thought that ethnicity played at least a "somewhat important" role in social relations on campus (Bernstein et al., 2010). Social groups tend to segregate themselves primarily along racial and ethnic lines, between whites, Chinese, and Indians. Indian students are sometimes accused of thinking of themselves as white and more assimilated than the Chinese students and, according to many, mix better with the white kids. Immigrant students are often labeled FOB ("fresh off the boat") or "fobby," suggesting they are non-assimilated and thus uncool. Among immigrant students, the social lines are often further delineated based on familial histories in different regions, social castes, and language groups. While students from mainland China sometimes refuse to work with students from Taiwan, students from Hong Kong sometimes reject Chinese mainlanders. Cindy quipped that at Mission High, "instead of the Bloods versus Crips, we have Chinese versus Taiwanese." Among Indian students, differences in familial castes, languages, regional origins, and religions can determine whom kids will work with in class and are critical identity markers (Shankar, 2008).

While racial and ethnic divisions are common to most schools and stem from several factors including peer and parental pressures and cultural differences, at Mission High, racial and ethnic divisions often center on students' academic performance. Competition over grades and cultural stereotypes about academic intelligence drive wedges between and among different groups. "Mission High is made up of two student bodies," explained *Smoke Signal* reporter Jennifer Kao (2005), "those in Honors and those in non-Honors classes." At Mission, these are often racial and ethnic dividing lines as well.

The academic disparities are most evident between white and Asian American students (considered by the California Department of Education to be the only "statistically significant" racial groups at Mission High). In 2009, Mission High's Asian American students, including students of both Chinese and Indian descent, had a 966 base Academic Performance Index (API) score compared to 890 for white students. While white students at Mission generally do well, they are disproportionately represented in the lower-division and special education courses, while Asian American students are overrepresented in the honors and advanced placement courses, particularly those in math and science.

The divide has generated crude stereotypes about students' intelligence and work ethics that reinforce social divisions. White students are often labeled the "dumb white kids," "blondes," "jocks," "rah-rahs," or "theatre kids" while Asian American students are sometimes referred to as "curve busters," "nerds," and "grade robots." The racial labels extend to all kinds of social actions—those perceived as being studious and academically oriented are called "Asian," whereas those considered non-academic are termed "white" and those Asian American students who follow them "white-washed." These derisive racial labels reflect tacit reinforcement of the model minority myth about Asian American academic success and, in contrast, a prevalent assumption that white kids, especially white girls, do not get good grades or study hard. "My best friend and I are blonde, light-eyed and in honors classes. When we walk into the room, you can tell from the body language [Asian students are] thinking 'Why are you in this class'?" said Lindsay (Brown, 2001). In a controversial article on reverse racism, *Smoke Signal* staff writer Anamarie Farr (2002) wrote, "I am part of a minority that is the object of discrimination at [Mission High]. No, I'm not a Gupta, Chan, Chen, Wu, or Wong. I am Farr and non-Asian . . . Just because I don't weigh myself down with 4 or 5 or even 6 AP classes does not mean that I lack intelligence." Alice recalled how when her daughter was in elementary school, the students organized a class vote over who was the superior race—Chinese, Taiwanese, Indian, or Caucasians. Her daughter, one of only a handful of white students in the class, did not win. However, because of her daughter's high test scores and grades, Alice said she was often deemed an "honorary Chinese" or "blonde Chinese" and socially accepted by her Asian American peers.

Crude racial stereotypes about students' intelligence reflect not only the way that students perceive of and treat each other inside the classroom, but also how they operate outside the classroom. Several students noted how hanging out with the white kids or hanging out in general can be interpreted as a sign of one's academic failure. Alternatively, hanging out with the Asian kids tends to suggest that you have no social life at all. Alice said that even though her son is in honors classes with mostly Asian students, he does not have many Asian friends. She claimed that this is largely because his classmates usually talk about homework and school projects and he's "not willing to become one of those robots." Sam, who is white, attributed Asian educational values to his son's struggle to live a "normal" teenage life at Mission High:

> The Asian culture does not operate like ours in a social sense. They don't come over to visit [my son] after school . . . I wouldn't say that they're not allowed, but they're not encouraged to go and hang out with—I don't think that they're encouraged to hang out even with other Asian families. There's a lot of studying that takes place. Most of the extracurricular activity is pretty limited to either music or traditional stuff like taekwondo or martial arts or things like that.

While many of the Asian parents and students that I spoke to rebutted Sam's claims, his comments point to the prevalent perceptions of Asian American students and their parents that often drive Asian–white conflicts at Mission.

While stereotypes about academic performance and intelligence shape relations *between* white and Asian students, a sense of academic competition often also affects social relations *among* different racial and ethnic groups. Though both Indian and Chinese students tend to perform well academically, their academic performance is still subject to interracial stereotypes. Mary explained that several Indian parents have made comments to her about Chinese students being more competitive than Indians and have felt threatened by their academic success. Ellie, who is Chinese American, commented that while racial and ethnic stereotypes surround the academic performance of different groups, most people compete with their friends and those in their classes, who are more than likely of the same ethnicity as them. Between white and Chinese, she explained, "We don't really compete with them because we're not like friends with them."

Academic performance also plays a significant role in many Asian American students' identities. Because Maxine, who is of mixed Chinese and Caucasian ancestry, is in honors classes, active in school clubs, and hangs out with mostly other Asian students, she said that she feels more Asian at Mission than white. Whereas Sally, who is Korean American and described herself as nearly failing out of Mission High, said that her poor academic performance made her feel like an outsider at Mission, and led her to hang out primarily among the few African American and Latino kids at school.

African Americans are the ultimate Mission outsiders. In 1990, African American students formed the black student union to promote awareness and combat racism on campus. Four years later, freshman Cameron Jones told *The Smoke Signal*, "When you walk around school, it seem like everybody's afraid of you. They pay real close attention to you in the cafeteria cause they think we're gonna steal something" (Iqbal, 1994). "To be African-American on this campus is to be such a minority," explained Paula, a Mission High teacher, "They really struggle for a sense of place."

The ethno-academic divide at Mission is also furthered by class differences. Though the vast majority of Mission students are from middle and upper-middle class backgrounds, a small number are from working-class or poor families. The only government-subsidized housing project in Mission San Jose is Orysom Village, a 101-unit multifamily and senior complex completed in 2000 amidst intense neighborhood protest. "The Apartments," as students often refer to them, have served as home to many of Mission High's Latino and African American students, as well as to various waves of refugees, including Afghanis and Vietnamese. Mission High youth remain very conscious of their geographies—traditionally differentiating themselves by elementary school enrollment between the elite Weibel Elementary students and the more modest-income Mission Valley or Gomes Elementary students and between the rich "Hill kids" and those from "the flats."

While racial and ethnic segregation would likely exist at any school without the academic pressures, stress, and competition of Mission High, the academic culture at Mission exacerbates some social differences and shapes new ones. The white/Asian social and academic divide is the most prominent of these divisions and helps to shape race and ethnic relationships among all students. But interethnic divisions among Asian students, the lack of representation among Latino and African American students, and social class differences also reinforce the ethno-academic divide.

Everyday Transculturalism

Contentious battles over grades and Mission High's academic culture and educational values have been among the more prominent impacts of Asian immigration in the neighborhood. Much less apparent have been the ways that immigration has fostered positive interethnic relationships and helped to normalize social differences. Particularly striking is how often the very same students, teachers, and administrators who easily acknowledge that race and ethnic differences shape the Mission's social life also say that race does not really matter.

This paradox speaks to the lived experiences of Mission High students, who may hang out in groups formed primarily along racial, ethnic, and cultural lines, but constantly negotiate, mediate, challenge, and push these borders. Although social segregation is a fact of life at Mission, interethnic friendships and dating are fairly common. A *Smoke Signal* survey found that 85 percent of students had a close friend of a different ethnicity and 75 percent never felt excluded due to their ethnicity (Bernstein et al., 2010). Even Leslie, who is white and complained about feeling socially rejected by Asian American students, said that some of her closest friends are Indian and Chinese.

Students reported that despite racial and ethnic divisions, most people are fairly nice to each other and that very little outright social exclusion occurs. Most described Mission High as segregated, but not particularly "cliquish." "It's not like *Mean Girls*," explained Nina, referring to the popular teen flick that mocks social cliques at a predominantly white high school. "You don't have typical troubles between ethnic groups like at other schools," said former Principal Stuart Kew. "So many people are so academically focused, they don't really have time for that" (Marech, 2002).

While high-performing schools reinforce academic divisions among different groups, they also tend to shift the focus away from other social divisions. At Mission High, academic pressures are often greater than social pressures. In contrast to many suburban schools where the popular kids are the cheerleaders and jocks, popular students at Mission are often those that make the best grades and are active in school-wide activities and organizations. "Being smart was being cool," explained Cindy, a recent Mission High grad.

The lack of social pressure to conform or fit in to a stereotypical white ideal has been particularly helpful to Asian American students, who often said that being in an environment where getting good grades, speaking other languages, staying in touch with Asian pop culture, and drinking bubble milk tea are considered normal provided them with a sense of comfort and confidence, and stronger Asian American identities. Anil, a Mission High student whose family moved to Pleasanton, a nearby and predominantly white community, for two years, contrasted his experience of being in the majority and minority:

> At home, at school here, I feel very safe. It's okay to be Indian. It's okay to be Hindu. It's okay to be anything that you want to be. Whereas in Pleasanton, I didn't feel that same way . . . It's the great thing about living here, race doesn't carry as much weight. Because there's so many different races. You know, one card doesn't trump all here. It's, "Hey, I'm Indian," "Hey, what's so special about that? I'm Russian right?" "I'm Chinese" or "I'm from Scotland" or whatever. It's just another thing.

Even white students noted that the environment at Mission has fostered an openness to difference that reduces the pressure on them to conform to traditional stereotypes. Brandy explained:

> [At Mission] you can wear a sweatshirt and put your hair in a ponytail and you'll be normal. But when you go out there [to Pleasanton or Livermore] you have to have like designer jeans, you have to have straight hair, blonde hair, lots of make-up on. Here it's not that big of a deal. So here I feel like you're more okay with who you are. Out there it's more like I need to be like everyone else so I can fit in. Here you're who you are and that's okay.

Mary, a Mission High parent, explained that at Mission there is no pressure to date, do drugs, or "be cool."

Because students' social lives are often limited and ethnically circumscribed, school activities serve as important venues to build cross-cultural ties. Ellie recalled that one year "the white kids" decided to get more involved in planning homecoming, an event she said was traditionally left up to the Asian students. The experience changed her and her other Chinese American friends' perceptions of white students:

> We realized that there's no difference between the whites and the Asians. But I think we always knew that because we have class with them inside but we never hang out with them outside of school . . . And Homecoming, you have to hang out, you have to get to know each other to practice the skits, practice the skits. And that's when we saw them as friends instead of just mere classmates.

Relations built at school also affect a wider web of social relations. Bobby, Jim, and Raj are three second-generation American students, whose parents are from China, Denmark, and India respectively. They have been best friends since kindergarten. While Bobby and Raj both said that their parents still respect many of their cultural traditions, they noted that over the years their friendship has brought their families together. On the issue of discipline, Bobby explained:

> Of the three of us, my [Chinese] parents are the strictest. And then my Danish friend, his parents are the most relaxed. And [Raj's] parents are the most guiding . . . so they're like different parenting styles and how we've all grown up and it kind of just meshes. Our parents have kind of changed too because they know each other. Like over the years, my parents have gotten more relaxed . . . because of the influence of the other two families. And like [Raj's] parents have gotten slightly stricter . . . and then [Jim's] are a bit more caring [about what he does].

Many Mission students and parents said that what has allowed them to bridge these racial and ethnic divides is not their silence about cultural differences, but their willingness to engage them. If you stand in the Mission hallways long enough, you are liable to overhear several racial slurs being thrown about. But inside the classroom you are just as likely to hear serious and thoughtful debate over "the glass ceiling," racial profiling, immigration reform, and other everyday challenges of managing the American color line.

The role of school administrators in addressing the racial and ethnic divide is the subject of some debate. Some claim that the district and school administration have been too quiet about racial and ethnic conflicts and tensions. But others highlight the unique aspects of Mission High culture that support its diversity. For instance, Mission offers an ethnic studies course, started by one of its teachers after 9/11 because of her disappointment in the way that the local community responded to the event. The Ethnic and Race Relations Committee at Mission work to spread cultural awareness and build racial tolerance across the campus and the district. Their students decorate the school for cultural holidays and host educational events, including Multicultural Week, which, next to Homecoming, is Mission's biggest event of the year (Figure 13.3). Though students sometimes complain that Multicultural Week is taken as an excuse for longer lunch periods and joke about the irony of hosting a multicultural event at a school dominated by Asian American students, some say the celebration of cultural foods, dances, and other traditions has made a difference in relieving racial and ethnic tensions, educating students about cultural differences, and building cross-cultural ties. Maxine said that Multicultural Week is one of the best times of the year at Mission because "you get to hang out with people you normally wouldn't."

Other students said that simply being in a diverse environment has made them more aware and tolerant of differences. "The student body at [Mission] inherently taught me a lot about other people and it exposed me to perspectives I would not have otherwise known," said Mike, a 2005 graduate (Staff Writers, 2010). Dawn, a former Mission High student who is white, reported that growing up in the area made her less tolerant of racism. "I'm kinda glad I grew up in an area as a minority so I won't grow up to be some bigot," she said (Radhakrishanan & Zhou, 2004).

Figure 13.3 Students performing at Multicultural Week, one of Mission High's biggest events of the year. Photograph by *The Smoke Signal* staff.

As much as Asian immigration has raised new challenges for race and ethnic relations at high-performing schools like Mission High, it has also raised the hope and possibility of more respectful cross-cultural exchanges. Race and ethnic diversity at Mission has opened up a space for students to push and challenge the boundaries that divide them, while also strengthening their own ethnic identities. Cross-cultural exchanges are aided by school events, classroom discussions, administrative action, and simply the greater number of opportunities afforded by a more diverse student body. And the impacts of students' growing respect for cultural differences extend well beyond the walls of the classroom.

Towards Transcultural Schools and Neighborhoods

> Suburbs offer greater potential for novel issue framing—changing the way people think about important issues rather than talking about different issues entirely—to tap emergent interests, new coalitions among groups, and policy innovations than the engrained image of the city–suburban divide suggests.
>
> (Xavier de Souza Briggs, 2005, p. 320)

The case of Mission High raises important challenges for cities as spaces that help build respect for social and cultural differences. As a center of immigration among highly educated, globally connected, and geographically mobile residents from all over the world, suburbs like Fremont highlight the struggles over space in America's neighborhoods and schools that have attended the vast demographic changes of the past half century, particularly those emerging in high tech regions. On the one hand, it shows increasingly tense race and ethnic relations between white and Asian students and parents because of differences in and stereotypes about academic performance, intelligence, educational values, expectations, and practices. On the other, it shows that the efforts of some to engage their differences, confront intolerance, and make a space for contestation and debate have begun to whittle away at these social and cultural divisions and generate new grounds for cross-cultural engagement.

Mission High also shows that to address the challenges raised by massive immigration and rapidly diversifying metropolitan regions, scholarship and policy must move beyond the black/white, city/suburb dichotomies that so powerfully shaped the dominant thinking about urban social justice, equity, and diversity in the 20th century. Today's metropolitan landscape is far more dispersed, fluid, and messy than at any other time in history. And so too must be our approaches to the possibilities and problems that attend urban diversity in the 21st century.

Finally, Mission High draws attention to the need to think about ethnic diversity outside of traditional public spaces. Instead of directing public policy and planning and design resources to places based on stereotypes and assumptions about where diversity occurs, investigating the lived landscape of diversity—where and how people actually encounter and engage difference in their everyday lives—would result in more substantive and sustainable interventions. Among the central challenges facing cities and suburbs today are their capacity to confront the mundane conflicts that threaten people's ability and desire to live together and to take advantage of the strengths that come from the intermingling of diverse groups within the spaces of their everyday lives.

Note

1 Unless otherwise cited, all quotes are from personal interviews.

References

Aboumrad, A. (2005, February 11). AP statistics at MSJ top in the world. *The Smoke Signal*, 15(6), 1.

Amin, A. (2002). Ethnicity and the multicultural city: Living with diversity. *Environment and Planning A*, 34(6), 959–980.

Bernstein, M., Dhawan, S., Jafarnia, N., & Yuan, S. (2010, June 11). Mission's melting pot: Celebrating MSJ's ethnic diversity. *The Smoke Signal*, 45(9), 1.

Briggs, X. (2005). *The geography of opportunity: Race and housing choice in metropolitan America*. Washington, DC: Brookings Institute.

Brown, P. (2001, May 26). The new melting pot: With an Asian influx, a suburb finds itself transformed. *New York Times*, 1A.

Burayidi, M. A. (2003). The multicultural city as planners' enigma. *Planning Theory & Practice*, 4(3), 259–273.

Chang, S. & Lung-Amam, W. (2010). Born glocal: Youth identity and suburban spaces in the U.S. and Taiwan. *Amerasia Journal*, 36(3), 29–52.
Farr, A. (2002, September 25). Listen up, MSJ: Reverse racism. *The Smoke Signal,* 38(2), 2.
Frey, W. H. (2011, May). *Melting pot cities and suburbs: Racial and ethnic change in Metro America in the 2000s*. Washington, DC: Brookings Institute.
Iqbal, S. (1994, February 3). Mission celebrates Black History Month: Race relations in the past, present, and future. *The Smoke Signal*, 28(4), 1.
Jacobs, J. (1961). *The death and life of great American cities*. New York: The Modern Library.
Kao, J. (2005, March 25). GATE barring academic success. *The Smoke Signal*, 40(9), 5.
Loukaitou-Sideris, A. (1995). Urban form and social context: Cultural differentiation in the uses of urban parks. *Journal of Planning Education & Research,* 14, 89–102.
Low, S., Taplin, D. & Scheld, S. (2005). *Rethinking urban parks: Public space and cultural diversity*. Austin: University of Texas Press.
Marech, R. (2002, May 17). Little Asia: Fremont community largely made up of immigrants with means. *San Francisco Chronicle*.
Radhakrishanan, A. & Zhou, A. (2004, December 16). MSJ on racial perceptions. *The Smoke Signal*, 40(4), 1.
Rishbeth, C. (2001). Ethnic minority groups and design of public open space: An inclusive landscape? *Landscape Research,* 26(4), 351–366.
Rosen, H. (1978, March 9). The typical club is typical of Mission. *The Smoke Signal*, 12(7), 1.
Sanchez, C. (2004, November 3). Children of immigrants seek to define their identity. *National Public Radio*. Morning Edition.
Sandercock, L. (2003). *Cosmopolis II: Mongrel cities in the 21st century.* London: Continuum.
Saxenian, A. (2006). *The new Argonauts: Regional advantage in a global economy*. Cambridge, MA: Harvard University Press.
Shankar, S. (2008). *Desi land: Teen culture, class, and success in Silicon Valley*. Durham, NC: Duke University Press.
Singer, A., Hardwick, S.W., & Brettell, C. B. (Eds) (2008). *Twenty-first century gateways: Immigrant incorporation in suburban America.* Washington, DC: Brookings Institution Press.
Somashekhar, S. (2003, September 6). Mission San Jose kicks off its football resurrection: 2002 varsity season cancelled because of lack of size, experience. *TriValley Herald*.
Staff Writers. (1974, May 16). Mission grads get average test scores. *The Smoke Signal*, 9(13), 1.
Staff Writers. (2010, June 11). When I was your age . . . *The Smoke Signal*, 45(9), 7.
Walter, J. (1987, May 14). Colleges: *The Smoke Signal* looks at the college scene at Mission. *The Smoke Signal*, 21(7), 1.

PART IV

BUILDING COMMUNITIES ACROSS CULTURES

CHAPTER 14

The Road Less Traveled

Transcultural Community Building

Caitlin Cahill

The road to hell is paved with good intentions. This sums up much of historic and contemporary urban planning practice. Rarely are planners from the neighborhoods they work within or are charged with making decisions for, and seldom are issues of positionality, privilege, and relationship engaged with as part of the planning and research processes. This is further exacerbated by a deficit orientation towards lower-income and urban communities of color in particular. Urban scholars Kretzmann and McKnight (1996) argue that the focus on overwhelmingly stereotypical and negative images of urban neighborhoods functions as a kind of "mental map" of the community that is regarded as the whole truth of the neighborhood, rather than part of it, and determines how problems are addressed. There are long-term repercussions to this orientation that reinforces a one-dimensional notion of communities as troubled and broken, while bolstering the expert authority of outsiders. Not only does this deny the long-term wisdom of community members, it also frames neighborhoods and/or community members as "at risk," and serves to justify their disinvestment, dispossession, and dependence (ibid.).

Figure 14.1 "El Teni." Photograph by Georgina Alvarez.

This chapter focuses on the "road less traveled" (ibid, 1996), mapping out the contours of transcultural community building as a theory of change that calls for new ways of thinking about how we work *with* communities. It asks the question: how might research and planning look different when it is done *with* and *by* communities? The case study considers how we work with communities, particularly those that have been silenced, marginalized, or otherwise ignored. Specifically, I will discuss *Easy Targets*, a participatory action research (PAR) project conducted with young people in Salt Lake City, Utah, that focused on the issues of undocumented students' educational rights and access.

On the cusp of dramatic economic, demographic, and cultural transformations, Salt Lake City is an ideal case to investigate the challenges of "achieving" the multicultural city (Thomas, 2008) and the role of transcultural community building in this process. Immigration is one of the most inflammatory issues in Utah (and in the USA as a whole). As Utah's 90 percent white majority has decreased over the last ten years, political and public debate over immigration status, schooling, geographic boundaries, and states' rights have become increasingly antagonistic.

In 2011, Utah passed HB 497, modeled on Arizona's "your papers please" law, authorizing the warrantless arrest of anyone believed to be undocumented, effectively legalizing racial profiling. It is within this hostile context that we do research focused on the experiences of undocumented students and their concerns. My discussion considers the tensions between our participatory research and organizing strategies focused on the educational rights of undocumented students. Our participatory research is informed by the experiences and concerns of undocumented students, holding true to the South African anti-apartheid wisdom that "nothing about us, without us, is for us."[1] But this is complicated in a majority white state, especially if you are undocumented.

Engaging with and excavating power, privilege, and the politics of solidarity and structural responsibility is at the heart of our work. By focusing upon the contradictions we negotiated in our work and the thorny politics of immigration rights activism in Utah, I want to shed light on processes of transcultural community building with the hopes of identifying critical theoretical and practical insights for the field of community builders, planners, and researchers who are committed to the struggle for a more just, democratic, and equitable world.

Transcultural Community Building

"How can 'we' (all of us), in all of our differences be 'at home' in the multicultural and multiethnic cities of the 21st century?" (Sandercock, 2003, p. 1) The subtext of Leonie Sandercock's question is more bluntly articulated: "How can we stroppy strangers live together without doing each other too much violence?" (Donald, 1999, p. 147, cited in Sandercock, 2003) While the grand narrative of urban scholarship celebrates the possibilities of diversity, the romanticization of the "urban encounter," or, even worse, the melting pot, this does not do justice to the realities of city life as a site of contestation and struggle. Renewing critical engagements with race and ethnicity is necessary for urban theory to take seriously the questions of how we might develop meaningful intercultural exchanges and urban spaces (Valentine, 2008; Askins & Pain, 2011; Amin, 2002).

By transcultural, I mean something specific. If a critical multiculturalism involves moving beyond the celebration or tolerance of differences, transculturation involves interrogating these differences. Critical questions are raised such as: Who gets to tolerate? Who has to accommodate whom? Slippery questions of privilege lie just beneath the surface. Linguistic scholar Mary Louise Pratt (1992) conceptualized the term "transculturation" to theorize the agency of oppressed cultures in their relationship with colonialism (replacing reductive concepts of acculturation and assimilation). She describes this in her theory of the contact zone as the "social spaces where cultures meet, clash, and grapple with each other, often in contexts of highly asymmetrical relations of power, such as colonialism, slavery, or their aftermaths as they are lived out in many parts of the world today" (Pratt, 1992, p. 4).

My work follows the lead of María Elena Torre's cutting-edge work on the contact zone and PAR which opens up new ways of thinking about intercultural spaces and community (Torre, 2005, 2009; Torre & Fine, 2008). Critiquing "imagined communities" of consensus, the contact zone is a social space where different cultural groups meet and interact, often in conflict, emphasizing "how subjects are constituted in and by their relations to each other . . . in terms of copresence, interaction, [and] interlocking understandings and practices" (Pratt, 1992, p. 7). Cities immediately come to mind. As spaces of intimate and intense contact across difference, urban contexts exemplify both the transformative possibilities and tensions of transculturalism.

In this chapter, by focusing upon the processes of community building, rather than the city, I want to place emphasis upon on-the-ground, collective, and democratic ways we (all of us) might engage, make change, and work across difference. From a practical and conceptual standpoint, "the city" as a scale of analysis is too big and abstract if you take relationship building seriously. "The City" presumes a distanced mantel from which to understand and act upon urban issues and implies the involvement of an expert authority (urban planner). Even a small city like Salt Lake City is better understood as a "city of neighborhoods." Moving away from the master plan mindset, I want to sketch out transcultural community building as a relational process of social connection, as opposed to something static or fixed. Assuming that community building is an ongoing process we must attend to and be mindful of, critical questions of inclusion being consistently raised: Who is part of this community? Who is missing and why? Whose voices are foregrounded? How do we address dissent? How might we acknowledge our differential relationships to privilege and power? How might we create and sustain participatory democratic practices? (Torre, 2005) Transcultural community building is, first and foremost, an active, ongoing, and dynamic process. A verb rather then a noun, it is a method of engagement across difference (Torre, 2009; Askins & Pain, 2011). This is a radical departure from urban planning as a comprehensive modernist project where the ends justify the means and the messiness of process is erased.

Case Study: Immigration and Education in Salt Lake City, Utah

> I am smart. I am hardworking. I can be your assistant or I can build your house—I can do all these things. But the moment that you shut me out, you cut my arms and legs off, and I cannot move. And there is nothing left, sometimes, but a feeling of . . . desperation. Because then you are reminded that you don't belong here.
>
> (Rafael, undocumented student)

Rafael has lived in Salt Lake City, Utah, since he was a child. In high school Rafael found out he was undocumented. This is a familiar story. Often it is not until they apply for college that undocumented students find out their citizenship status and its serious implications for their future. From this precarious location Rafael articulates a yearning undercut by a deep sense of dispossession that speaks to the profound inequalities of global urban restructuring where a shadow state of exploitation and "permanent temporariness" is quietly tolerated and even encouraged (Yiftachel, 2009, p. 89).

Rafael's experience of disenfranchisement is an example of structural relations unfolding in big and small cities around the world. His story is not new but echoes the experiences of immigrants throughout history who came to the USA to make a better life and struggle to achieve the American Dream in the land of opportunity. What is new is the way that immigrants, in particular young undocumented students, have unwittingly become key players in the political and cultural struggles of globalization (Flores, 2003; Cahill, 2010). Forced to move to find the means of social reproduction, undocumented immigrants find themselves stuck in a cycle of survival; once here in the USA, they find their situation is not so dissimilar from in their countries of origin. Kristin Koptiuch identifies this process as "third-worlding at home," referring to "the effects of a process of exploitative incorporation and hegemonic domination—and its fierce contestation by subjugated peoples—that used to take place at a safe, reassuring distance" (1991, p. 85).

Currently, undocumented immigrants in the USA number about 11.9 million (4 percent of the population) with almost 60 percent coming from Mexico, according to the Pew Hispanic Center (2009). Significantly, nearly half of the Mexicans who are undocumented immigrants have arrived in the USA since 2000. In Utah, the number of undocumented immigrants grew faster than in the country as a whole in recent years (ibid.). The majority of newcomer immigrants have moved to the state's urban areas. In Salt Lake City, for example, more than 40 percent of the West Side residents (the site of our research project) are ethnic minorities, predominantly Latino, and from Mexico in particular. From 2000 to 2004, 75 percent of the enrollment increase in Utah's public school system was students of color (Perlich, 2008). Today, many of Salt Lake City's school districts are minority–majority reflecting the changing demographics.

The theories and practice of community-based PAR are particularly relevant to our understandings of transcultural community building and the potential for intercultural urban exchange. Building upon long-standing traditions of asset-based community development and grassroots activism, PAR reflects a commitment to building capacity and doing research that will be useful to the community. Communities, particularly communities of color, have rarely benefited from the results of research conducted and planning interventions (Breitbart, 2003). To this end, a PAR approach could be understood as a response to the exploitative research practices of "outsiders" (the academy/planners/policy makers, etc.) who have used communities as laboratories.

The *Easy Targets* youth research team reflected the diversity of Salt Lake City's West Side; the most diverse zip code in the state of Utah and home to many immigrant and working-class families. A small group of students (aged 14–18) was recruited and paid a stipend to work on this project as it evolved over the course of two years (2007–2009). Our research team included Latino, White, African American, and biracial young people, faculty and community members. My role as a co-facilitator involved mentoring the youth researchers and working collaboratively on our research. In a community-based PAR process, young people are repositioned as co-researchers as they identify priority issues, frame the questions to be addressed, collect and analyze the data, and develop proposals for change.

Easy Targets was developed out of concerns about friends, family members, and classmates dropping out of high school because they did not see college as a viable

Figure 14.2 *Easy Targets/Los Vulnerables* documentary (DVD case designed by Thomas Nelson). Permission: Caitlin Cahill.

option. As one student put it when faced with filling out a form for college credit: "Mexicans don't go to college." In fact, the high school drop (or push) out rate for Latino students in Utah, documented or not, is unconscionably high. The situation is much worse, of course, for undocumented students. Our research questions focused specifically on the experiences of undocumented students in their pursuit of going to college. Through our research we wanted to identify the questions and challenges faced by undocumented students and ways to address these concerns. We conducted interviews with undocumented students and their family members, administrators, teachers, and legislators. We documented legislative sessions, and community meetings. We decided to produce a documentary for our research in order to create a "safer" space for undocumented students to speak out to a broad public audience.

Significantly and paradoxically, while Utah is one of the most inhospitable states in the country for undocumented communities, it is also one of ten states where undocumented students have the right to in-state tuition (as state residents under House Bill 144, passed 2002). In-state tuition is a big deal. Without it, undocumented students would not be able to go to college because tuition would be at least three times more expensive (in 2012, resident tuition at the University of Utah was approximately $5,200 per year and non-resident tuition was $16,800). In-state tuition makes college possible (although it is still challenging as there is no access to student loans or financial aid for undocumented students), as Isabella, a 19-year-old student we interviewed for our research explains:

> I Love House Bill 144. My parents come home really exhausted everyday. My mom knows that—and to this day—she is 52, she still works in a sweatshop; she works in a factory, and my dad works in construction. So they both actually do. And they've both told me that if I don't get a good education, good skills and a decent job, then I'm going to end up just like them. And they made me promise that I would never work in a sweatshop.

HB 144 opens up the door to college and to a dream of a different future. Not surprisingly, the in-state tuition policy is highly contested. Every year since HB 144 was passed in 2002, legislators have attempted to repeal this right, and there is much public debate in the media. And every year, undocumented students hold their breath and pray that this right will not be taken away. Thankfully, in the 2012 legislative session, undocumented students' right to education was upheld again despite new draconian anti-immigrant racial profiling legislation—although as soon as students graduate they are disenfranchised. These are the contradictions undocumented students negotiate in their everyday lives.

Strategies for Action and Reflection

> So some Latino Students are FINALLY, ACTUALLY doing what is expected of any OTHER US American student in school?? Gee, should they get an honorary Ph.D.

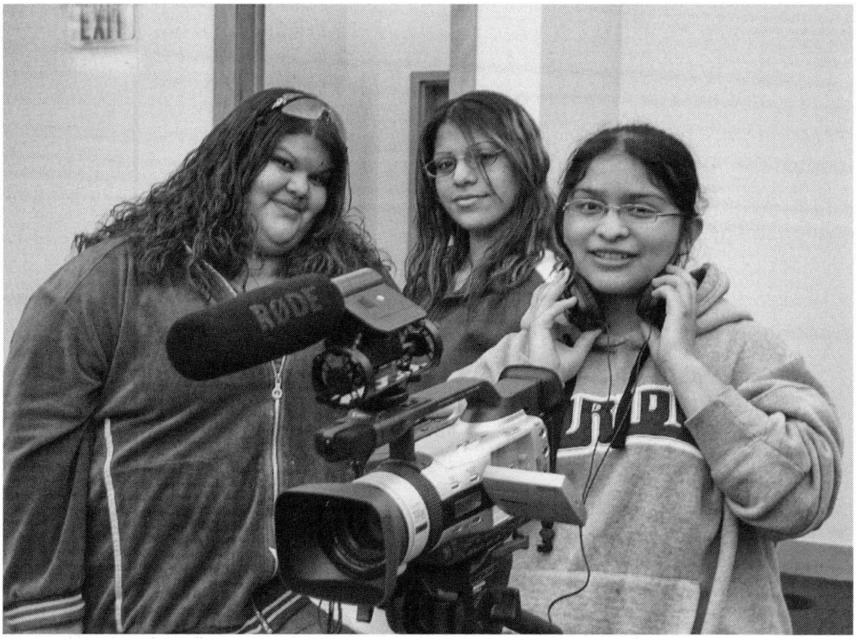

Figure 14.3 Youth researchers film a hearing about House Bill 144 at the State Capital. Photograph by Caitlin Cahill.

> for doing their homework, and not "doing" another gang member? . . . Every hour a teacher spends on an illegal child is an hour taken away from teaching an American student whose parents are LAW-ABIDING citizens.
>
> (Salt Lake Tribune comment board July 2009)

> And by putting you as an illegal person, right, they have stripped you of your humanity . . . and by taking your humanity, you are no longer allowed the same rights, and even the same respect you would give a stranger, and I feel comfortable in saying, an animal. People don't kick dogs, but they feel very comfortable kicking a person who has no humanity or is evil or is undocumented.
>
> (Rafael, undocumented university student)

For obvious reasons, undocumented students do not often speak publicly due to the great risks involved. The xenophobia expressed above in the anonymous online comment board is a constant threatening hum in the background, reinforced by anti-immigrant legislation, that effectively silences and keeps undocumented community members in the shadows. As a result, policy makers, the media, and the general public rarely hear from actual students like Rafael. Instead, the "mainstream" immigration debate is dominated by the fears and concerns of Anglos. Conflating emotional and economic insecurities, the dominant immigration rhetoric assumes a white audience and has become a code word for the dangers of losing "our" jobs, "our" language, "our" culture, and "our way of life" (Sandercock, 2003).

One of the goals of our research project was to intervene and shift how we understand immigration from the perspective of undocumented young immigrants. Tracing the arguments that frame debates over immigration and education, we analyze racist anti-immigrant discourses side by side with narratives of hope articulated by young undocumented students who are working to try to get through high school. Juxtaposing the anonymous comment board post with Rafael's quote above stakes out the polarities of the debate. De-centering whiteness and foregrounding the concerns of those who are "most affected" (while remembering that we are *all* affected, Koopman, 2008), PAR creates space for surfacing counter-stories and poking holes in the dominant discourse, recognizing that "those positioned at the bottom of social hierarchies often have keen insight into the structural arrangements" (Torre, 2009, pp. 111–112). If PAR is to make a meaningful change beyond an "armchair revolution" (Freire, 1997), not only the question of who gets to speak (Spivak, 1988), but also with whom we want to speak, and why, is critical. Engaging questions of audience and purpose also, of course, raises issues of race, privilege, and power.

First, audience. Who in particular do we want to "move" with our research? Undocumented students, the general public, or policy makers? While on the one hand our work focuses on the agency of undocumented students, on the other, we are concerned with what antiracist scholar Mark Warren identifies as "cultivating a sense of 'we'" (Warren, 2010, p. 231). Here we are not speaking about the royal "We," but rather the "we" that has a stake in the welfare of undocumented young people—our "linked fate" (Guinier & Torres, 2002). To this end, we want to reach out to "white people," a shorthand we use to refer to the broad public in a state that is almost 90 percent white. We want to imagine a politics of solidarity that includes raising questions of structural inequities of race and privilege.

Second, what is our purpose? Is it our goal to understand the challenges of undocumented students or is it to change the situation? To expand the pipeline so that more undocumented students are able to go to college? Or are we reframing understandings of immigration to raise public consciousness and inform public policy? Perhaps all of the above. Our work takes place on multiple fronts, engaging diverse stakeholders, and taking different formats "beyond the journal article" (Cahill & Torre, 2007). Here, by examining two strategies used to address undocumented students' educational rights, interest convergence, and critical reframing through PAR, I hope to illuminate the contradictions and contours of transcultural community building practice.

Interest Convergence

Urban scholar Peter Marcuse states, "If the needs of the Included can be linked to the needs of the Excluded, a powerful force for change can be brought into existence" (Marcuse, 2009, p. 252). Following this line of argument, policy makers have made compelling arguments for extending the right to education for undocumented students on the grounds of intergenerational transfer. For example, economist Pamela Perlich's (2008) research on Utah demographic projections suggests that today's and tomorrow's youth, who are increasingly from diverse immigrant backgrounds, will carry a much greater demographic burden than their parents' generation in terms of supporting an increasing share of retirees (2008, p. 11). If adults finance young people's educational expenses, they are making an intergenerational transfer as these youth will eventually pay into Social Security, Medicare, and other expenses for the elderly. Perlich suggests that it is therefore in the interest of the aging population (which represents the white majority of Utah) to ensure that all youth (including immigrant and undocumented students) have sustainable "good" jobs that can pay into the system.

Perlich's scholarship has been used strategically as part of a political platform to reframe the terms of the debate for undocumented students' right to education in Utah. Her argument is an excellent example of interest convergence, the principle that change benefiting people and communities of color only occurs when those interests also benefit whites (Bell, 1980). While interest convergence has been used strategically, educational scholars Alemán and Alemán (2010) suggest caution when this argument is used as a political and activist strategy without foregrounding the centrality of race and racism, as it obscures the overarching goals of social justice.

Figure 14.4 "Monica" still from *Easy Targets* documentary.

Case in point: Monica is still stuck in the shadows, as are thousands of other undocumented students. Interest convergence does nothing to challenge the representation of undocumented young people as illegal, criminal, and not worthy of rights. Recall Rafael's above-stated words "by taking your humanity, you are no longer allowed the same rights, and even the same respect you would give a stranger, and I feel comfortable in saying, an animal."

Still, when we are at the Utah State Capital advocating for the right to in-state tuition policy we may engage the argument of interest convergence. But we don't let this argument stand alone. Transcultural community building involves a conscious and explicit acknowledgment of the structural arrangements that produce inequities; in this case, how racialized stereotypes and discourses work to justify the public assault on undocumented students and produce "illegality" (De Genova, 2002).

Reframing Immigration Politics through PAR

Critical scholars have discussed how the framing of immigration as a "problem" directs attention away from such critical issues as the humanitarian crisis of mass migration and displacement. For example, Lakoff and Ferguson (2006) suggest that immigration might also be understood as a "civil rights problem" or a "cheap labor problem," placing emphasis upon the exploitation of undocumented communities who are forced to live underground and denied basic human rights (Cahill, 2010). One of the key contributions of PAR is the intellectual power of "organic intellectuals" (Gramsci, 1999) whose critical perspectives are developed from an analysis of everyday experiences (Fine et al., 2003). In *Easy Targets* undocumented students speak out, enter the fray, and take part in the national dialogue on educational inequities and immigration rights that too often silences the stories of those who have the most to gain and lose. By participating, students and families put forth a claim to belonging in Utah, motivated by a "desire to participate as 'citizens', that is 'as active contributing members of society who care and think about the world they live in' despite their 'legal' status and marginalization" (Quijada, 2008, p. 77).

Conscious of a potential broad public audience for the documentary, the students we interviewed deliberately spoke back to stereotypes of undocumented immigrants, challenging the dominant discourse. For example, Rafael discussed his desire to work and contribute to Salt Lake because "I don't only belong in this institution, but I also belong in the Valley, and I belong with other people, and I have work to do." Monica, an undocumented student, talked about growing up in Salt Lake City, how it was her home, how she knew no one in Mexico and "would be a lost needle in a haystack" if she was deported. Isabella's mother shared how she had made the decision to cross the border so that her children could have a better education and life, and how her daughter was just a baby and had no choice in the matter. Many of the students strategically mobilized neoliberal discourses of meritocracy, family values, and the "good immigrant." Students discussed their work ethic and their desire to go to college and pursue the "American Dream." The significance of education was articulated not only in terms of personal gain, but also of the well-being of their families and communities.

The documentary serves different purposes for different audiences. While we had wanted to reach out to undocumented students and their families about their educational rights, we also wanted to reach people in power who made decisions that impacted undocumented students' access to higher education. With this in mind, we created different versions of the documentary for multiple audiences, including a bilingual version. With our community partners, we held a press conference where political and business leaders spoke out in support of in-state tuition and gave out a short 5-minute version to all of the Utah State representatives. We screened *Easy Targets* locally in libraries, schools, and aired selections on the radio. Community partners distributed *Easy Targets* DVDs to undocumented families.

As part of our goal was to support undocumented students in going to college, we created other presentations of our research findings from working with community partners. The different research products share different findings of the research process itself. For example, we created a guidebook to higher education for underrepresented students; following this, our community partner Edúcate created a website for undocumented students that tracks public policies impacting undocumented students, and we conducted workshops for undocumented communities on educational issues.[2]

Conclusion: Transcultural Community Building and Political Responsibility

A transcultural community raises critical questions about our collective sense of responsibility, what Guinier and Torres (2002) identify as understanding our "linked fate." Taking seriously the epistemic privilege of the perspective of undocumented students, Guinier and Torres argue the experiences of marginalization reveal problems in our society that need to be addressed: "The first sign of a danger that threatens us all" (Guinier & Torres, 2002, p. 12). As Mark Warren (2010) rightly suggests, "ally" may not be an accurate term as it implies supporting someone else's struggle and not embracing the struggle as one's own. The students' experiences reframe misunderstandings of the immigration "problem" while calling attention to the complex structural processes that connect us all. This is exemplified by the following quote from administrator/scholar Dr. Theresa Martinez from our documentary:

> As far as I'm concerned, anybody in America who benefits from the low cost of goods and services that undocumented workers provide should be on the bandwagon about their children receiving a good education because we are all benefiting from the work of undocumented people. How much more so do we owe their children a chance of a good life? If we don't, think about what that makes us.
>
> That makes us a people who created a serving class . . . we are saying we are going to totally exploit your parent's work but we don't want you to in any way better yourselves.

Drawing clear connections between our relationships to labor exploitation and the critical role of educational access, Dr. Martinez points to our collective responsibility to address structural injustices. This is at the heart of transcultural community building: cultivating a politics of solidarity. But how do we do this? In our work, we

Figure 14.5 "Caution," image created by the Mestizo Arts & Activism Collective 2010.

focused on critically reframing our understandings of immigration politics, reaching out to multiple audiences, and engaging in strategies of interest convergence, while paying attention to relationships of power and privilege.

In *Easy Targets*, our hope was to engage the compassion of the broad public who may be hearing the voices and stories of undocumented students for the first time. Their stories are emotional. We wanted to "prick the psychic amnesia" (Fine et al., 2003), awakening viewers to the injustices undocumented students face in their everyday lives and then motivating those viewers to act upon this newfound knowledge. As Freire puts it, "liberation is thus a childbirth, and a painful one" (Freire, 1997, p. 31). Mary Louise Pratt (1996, p. 15) suggests that "[i]n the process something very powerful happens: white participants experience knowledge not as power but as pain, the way indigenous people have often experienced it." This was often the experience for our white audience members. Pratt suggests, "As we approach this next millennium, that space of pain, sad to say, may be the most constructive space in which the struggle for peace and justice can go forward."

But how do we move from "guilt to solidarity" (Young, 2003)? Moving forward, Iris Marion Young's theory of political responsibility traces our different degrees of influence and relationships to a particular social issue (in this case immigration). For example, how we individually participate in the exploitative conditions of undocumented communities through the services we consume or the privileges we enjoy (cheap food/labor, etc.), as Dr. Martinez points out above. However, an employer or a politician might have a more direct relationship to the exploitative labor conditions of undocumented workers. Or, as our research demonstrates, teachers and school counselors are in key positions of power in terms of informing undocumented students of their educational rights to go to college and pay in-state tuition. Making visible the invisible and often obfuscated lines that connect us, Young's theory of political responsibility demonstrates our complicity in perpetuating the unjust situation of undocumented community members. But at the same time she argues that no one individual is to blame for a structural issue of immigration. Modeling social connection to structural issues, transcultural community building places primacy on our collective relationships to structural and racial injustices and then asks us what can we do about this.

There is much work to do. While undocumented students in Utah still thankfully have the right to in-state tuition (HB 144), their community is under siege. "Attrition through enforcement" legislation such as the "Arizona light" racial profiling bill and other related anti-immigrant policies is making life extremely difficult for undocumented community members to live and work in Utah (popularized by Mitt Romney as the "self-deportation" solution to immigration).

This is the context for our research and its significance. Including excluded perspectives in the development of new knowledge production points to new political possibilities of transcultural community building. Inspired by the insight of Dr. Martin Luther King Jr. "that freeing black people from the injustices that circumscribe their lives, America will be freeing itself as well," Guinier and Torres (2002, p. 293) argue that the experiences of marginalized people of color can be the basis for social transformation. This is the road less traveled. And yet we are following in the footsteps of activists, community members, and all others who have taken great risks to work towards social justice for all of us. Today, inspired by the gay rights movement, brave undocumented students are "coming out" of the shadows to draw attention to the plight of their community, to engage in a national dialogue about immigration that involves us all. Never finished, transcultural community building is an ongoing process of *building* and consistent attention to relationships to race, privilege, and power. Bumpy and slow, it offers planners, researchers, and all of us a promising route towards a more just, inclusive world where "nothing about us, without us, is for us."

Acknowledgments

My work is collaborative and indebted to everyone I have had the privilege to work with on this project over the last five years. Thanks to Jeff Hou for inviting me to participate in the Transcultural Cities Symposium and for his constructive and thoughtful feedback on earlier drafts of this chapter.

This chapter is dedicated to the late, great Matt Bradley, my co-conspirator, friend, and the co-director of the *Easy Targets* project. What a complete joy to work with Matt, who opened up new ways of thinking about and being in the world for all of us. Thanks always to the tireless, fearless, and intrepid *Easy Targets* and Growing Up in Salt Lake City research teams, the change-makers of SLC. I am most grateful to the wonderful research assistants who creatively shaped this project from the ground up including: Sonia Caraveo, Denise Castañeda, Ariana Prazen, and Roberta Targino. Generous funding from University Neighborhood Partners, the Lowell Bennion Center of the University of Utah, and the American Association of Geographers made this project possible. Thanks always to David Quijada, José Hernandez, Alonso Reyna, and everyone from the Mestizo Arts & Activism Collective for their ongoing support.

Notes

1 "Nothing About Us Without Us" is also a slogan of both the disability rights movement and New Orleans residents after Katrina.

2 See: http://educatexcambio.weebly.com/

References

Alemán, E. & Alemán , S. (2010). "Do Latin@ interests always have to 'converge' with White interests?" (Re)claiming racial realism and interest-convergence in Critical Race Theory praxis. *Race Ethnicity and Education*, *13*, 1–21

Amin, A. (2002). Ethnicity and the multicultural city: Living with diversity. *Environment and Planning*, *34*, 959–980.

Askins, K. & Pain, R. (2011). Contact zones: Participation, materiality and the messiness of interaction. *Environment and Planning D: Society and Space*, *29*, 803–821.

Bell, D. A. (1980). Brown v. Board of Education and the interest-convergence dilemma. *Harvard Law Review*, *93*, 518–533.

Breitbart, M. (2003). Participatory research. In N. Clifford and G. Valentine (Eds), *Methods in Geography* (pp. 161–178). London: Sage.

Cahill, C. (2010). "Why do they hate us?" Reframing immigration through participatory action research. *Area*, *42*, 152–161.

Cahill, C. & Torre, M. (2007). Beyond the journal article: Representations, audience, and the presentation of participatory research. In S. Kindon, R. Pain, & M. Kesby (Eds), *Participatory Action Research Approaches and Methods: Connecting People, Participation and Place* (pp. 196–205). New York: Routledge.

De Genova, N. P. (2002). Migrant "illegality" and deportability in everyday life. *Annual Review of Anthropology*, *31*, 419–447.

Donald, J. (1999) *Imagining the Modern City*. Minneapolis, MN: University of Minnesota Press.

Fine, M., Torre, M. E., Boudin, K., Bowen, I., Clark, J., Hylton, D., et al. (2003). Participatory action research: From within and beyond prison bars. In P. Camie, J. E. Rhodes, & L. Yardley (Eds), *Qualitative Research in Psychology: Expanding Perspective in Methodology and Design*. Washington, DC: American Psychology Association.

Flores, L. A. (2003). Constructing rhetorical borders: Peons, illegal aliens, and competing narratives on immigration. *Critical Studies in Media Communication*, *20*, 362–387.

Freire, P. (1997). *Pedagogy of the Oppressed.* Harmondsworth, Middlesex: Penguin Books. First published 1970.

Gramsci, A. (1999). *Selections from the Prison Notebooks.* New York: International Publishers. First published 1971.

Guinier, L. & Torres, G. (2002). *The Miner's Canary: Enlisting Race, Resisting Power, Transforming Democracy*. Cambridge, MA: Harvard University Press.

Koopman, Sarah (2008). Imperialism within: Can the master's tools bring down empire? *ACME: An International E-Journal for Critical Geographies, 7* (2), 283–307.

Koptiuch, K. (1991). Third-worlding at home. *Social Text*, *28*, 87–99.

Kretzmann, J. & McKnight, J. K. (1996). Assets-based community development. *National Civic Review, 85*, 4.

Lakoff, G. & Ferguson, S. (2006). The framing of immigration. buzzflash: http://www.buzzflash.com/contributors/06/05/ con06208.html (Retrieved October 9, 2009).

Marcuse, P. (2009). Beyond the just city to the right city. In P. Marcuse, et al. *Searching for the Just City: Debates in Urban Theory and Practice* (pp. 240–254). New York: Routledge.

Perlich, P. (2008). Utah's demographic transformation: A view into the future. *Utah Economic & Business Review*, *68* (3), 1–3.

Pew Hispanic Center. (2009). *Mexican Immigrants in the United States 2008*. Washington DC.

Pratt, M. L. (1992). *Imperial Eyes: Travel Writing and Transculturation*. New York: Routledge.

Pratt, M. L. (1996) Apocalypse in the Andes: Contact Zones and the Struggle for Interpretative Power. *Encuentros IDB Cultural Center*, vol 15, 1–16.

Quijada, D. A. (2008). Reconciling research, rallies and citizenship: Reflections on youth led diversity workshops and intercultural alliances. *Social Justice, 34* (4), 76–90.

Sandercock, L. (2003). *Cosmopolis II: Mongrel Cities in the 21st Century*. New York: Continuum.

Spivak, G. C. (1988). "Can the subaltern speak?" In C. Nelson & L. Grossberg (Eds), *Marxism and the Interpretation of Culture* (pp. 271–313). Urbana: University of Illinois Press.

Thomas, M. (2008). The paradoxes of personhood: Banal multiculturalism and racial–ethnic identification among Latina and Armenian girls at a Los Angeles high school. *Environment and Planning A*, *40* (12), 2864–2878.

Torre, M. E. (2005). The alchemy of integrated spaces: Youth participation in research collectives of difference. In L. Weis & M. Fine (Eds), *Beyond Silenced Voices* (pp. 251–266). Albany: State University of New York Press.

Torre, M.E. (2009). Participatory action research and critical race theory: Fueling spaces for "nos-otras" to research. *Urban Review, 41* (1), 106–120.

Torre, M. E. & Fine, M. (2008). Participatory action research in the contact zone. In J. Cammarota & M. Fine (Eds), *Revolutionizing Education Youth Participatory Action Research in Motion* (pp. 23–44). New York: Routledge.

Valentine, G. (2008). Living with difference: Reflections on geographies of encounter. *Progress in Human Geography*, *32* (3), 323–337.

Warren, M. (2010) *Fire in the Heart: How White Activists Embrace Racial Justice.* London: Oxford University Press.

Yiftachel, O. (2009). Theoretical notes on "gray cities": The coming of urban apartheid. *Planning Theory*, *8,* 88–100.

Young, I. M. (2003). From guilt to solidarity. *Dissent Magazine*. Spring. http://www.dissentmagazine.org/article/?article=504 (Retrieved July 25, 2008).

CHAPTER 15

Creating Political and Social Spaces for Transcultural Community Integration

Trinh Mai and Kimberly Schmit

Following a global trend, the United States has experienced a significant increase in immigration in the last four decades. Historically a White state, Utah has emerged as one of the new American immigrant gateways; in the past two decades, it has experienced a dramatic demographic shift. From 1990 to 2010 the number of foreign-born in the state nearly quadrupled from 58,600 to 222,638 (Migration Policy Institute, 2012). Latinos led this growth, increasing by 78 percent from 2000 to 2010 according to the 2010 census (US Census Bureau, 2012). In Utah's state capital, Salt Lake City, west side neighborhoods (two zip code areas and fifteen census tracts) comprising 66,701 of the city's 186,440 residents have become home to many of these newcomers. The Latina/o population in west side neighborhoods has seen a 32.4 percent increase in the last ten years, amounting to 46.5 percent of the total area population and 74.5 percent of the wider Salt Lake City population (Dowden, 2012). More than 60 percent of west side residents are minoritized populations. As a preferred site for refugee resettlement, from 2000 to 2009, more than 8,103 refugees arrived in Utah (Department of Workforce Services: Refugee Services Office, 2012). Many of Salt Lake City's refugee-background population reside on Salt Lake City's west side.

This chapter looks at how Utah's flagship university, the University of Utah, has attempted to change its pattern of engagement with new immigrant communities. First, a description of west side neighborhoods and the university's response will set the stage, followed by a literature review on immigrant integration, multiculturalism, and social capital. With this backdrop, we will examine one specific university program to understand its efforts to integrate immigrants, specifically strategies that promote intercultural learning and relationship building and that address power inequity among newcomers and receiving communities. The conclusion will address the strategies discussed, limitations of the study and the work, and implications for future work.

West Salt Lake City

Settling in west side Salt Lake City, newcomers find themselves living within a highly contested space. A long-standing residential area bordered by heavy industrial zones and the Salt Lake City International Airport, many residents refer to a glass wall that separates the west side from the east side of the city, due to historical divisions created by the transcontinental railroad tracks in the mid-1800s, and in the 1990s by the I-15 freeway that parallels the railroad tracks (United Way, 2001). While the neighborhoods enjoy many assets, including a variety of parks, local libraries, increasingly vibrant community centers, a diverse population with a dedication to education and economic upward mobility, historical housing, and active residents and community organizations, the press has historically overlooked this and exacerbates a feeling of isolation with frequent reports of the west side as a "ghetto" with high gang activity and crime (Hunter et al., 2009, p. 295).

The described rapid demographic shifts and historical physical and political oppressors are leaving west side Salt Lake City systems and individuals seriously challenged to experience their neighborhoods and lives in new ways. While many residents do view the area's cultural diversity as an asset, this has historically not been tapped into positively and has been instead allowed to create divisiveness. A high amount of mistrust between immigrant and long-term Anglo residents, as well as between different immigrant groups, continues to exist.

The University of Utah's Response: A Strategic Initiative

In 2001, the University of Utah (U of U), the flagship state university located in the east side of Salt Lake City, responded to this disparity with a strategic initiative. Specifically, the administration was shocked by inequitable graduation and enrollment rates (in 2001 only 25 students from these west side neighborhoods graduated from the U of U) and was keenly aware that the institution was not adequately fulfilling its responsibilities toward all surrounding communities.

Aware of the potential of campus–community partnerships to positively "lead to the mutual restructuring of universities and communities" (Silka, 1999, p. 336), the President of the University appointed a Special Assistant to the President for Campus–Community Partnerships. Her first charge was to conduct nine months of interviews with over 250 west side community residents, leaders, organizations, city officials, and university faculty and administrators. The asset-based community development approach focused on identifying the strengths of the community while placing the highest priority on a needs-assessment done by the community itself, not from above.

This research yielded the following priority areas: (1) *Capacity Building*—building the capacity of west side residents to access and negotiate US systems and to gain successful community integration, while building the capacity of governmental and neighborhood organizations to address issues of health, housing, employment, safety, and environment with newly arriving populations; (2) *Resident Leadership and Empowerment*—promoting the development of resident leadership and increasing the participation of diverse resident leaders in community decision making; (3) *Youth, Education and Success*—increasing access to educational opportunity while building the capacity of educational systems to meet the needs of new arriving and native-born populations; and (4) *Community Building*—building community to address common goals, while addressing issues of mistrust and conflict stemming from differences of income, ethnic groups, religions, race, and political affiliation.

The President and Special Assistant were aware that an institutionalized approach to working in these priority areas was critical to community success. The result was the creation of University Neighborhood Partners (UNP), a university department under the U of U's highest administration—the Office of the President. The Special Assistant and President had also heard directly from residents that in order for UNP's work to be credible and sustainable the university had to come down off its hill in the east side, both figuratively and literally, and must be located in Salt Lake City's west side neighborhoods. In response, UNP leased a residential home, for no cost, from the city. The house, located in a west side neighborhood, opened as an office in April 2003 and serves as a community space and hub for partnership activity occurring throughout seven west side neighborhoods.

Currently in its tenth year, UNP's work continues to focus on building and maintaining long-term collaborations between higher education, west side residents, and community organizations, in order to address the identified priority areas. The mission of UNP is to *"bring together University and west side resources for reciprocal learning, action, and benefit."* The campus–community partnership work is guided by a Board of Advisors composed of ten west side community representatives (non-profits and city programs), ten university representatives, and ten west side residents.

Immigrant Integration

Contemporary policy discourse on immigrant integration is moving away from unilateral assimilationist roots and growing towards models that reflect the rich, complex realities of acculturation (Bourhis et al., 1997; Bourhis et al., 2009; Sandercock, 2003; Yan & Lauer, 2008). Social scientists use the term *acculturation* to describe the process of cultural transformation and have confirmed that the process occurs in both new immigrant communities and receiving communities after a period of sustained engagement (Bourhis et al., 2009). Though both groups have power to influence and impact each other, the receiving communities, especially dominant cultural groups within a country, carry significant advantage with their established institutional power (Bourhis et al., 2009; Penninx, 2003). In order to address this inequity, the integration policies and programs must look beyond newcomers, beyond capacity building at the individual level; instead the work must also target receiving communities and their institutions in order to promote cultural learning and adaptation on this side (Penninx, 2003; Sandercock, 2003).

Sandercock (2003) explores several visions and examples of existing models that work at the individual and institutional level to promote socially just integration of immigrants in *Mongrel Cities*. For this case study, two visions are especially relevant. First is Ash Amin's concept that integration of immigrants has to take place in the "city's micro-publics of banal multicultures" (as cited in Sandercock, 2003, p. 94). These "micro-publics" are spaces such as workplaces, schools, colleges, community centers, and neighborhood houses where people come together and are forced to engage and interact, to create relationships and interdependence. However, these spaces require "organizational and discursive strategies that are designed to build voice, to foster a sense of common benefit, to develop confidence among disempowered groups, and to arbitrate when disputes arise" (cited in Sandercock, 2003, p. 94).

The second guiding vision is Sandercock's own "multicultural perspective" promoting the idea that multiculturalism is both a political and social practice. Political practice means upholding the right to difference and the right to city. The *right to difference* can be defined as the right to maintain and express unique cultural identities, and the *right to city* can be understood as the right to occupy and participate in public spaces as an equal. A "multicultural political structure" is required at society's base to protect from discrimination and promote inclusion. On the social level, according to Sandercock (2003), multiculturalism requires daily negotiations and cultural evolutions made possible through intercultural interactions at micro-public spaces discussed prior. Furthermore, in addition to addressing the material needs or threats of both receiving and new communities in resettlement, attending to emotional needs is a priority, especially validating and addressing fear and intolerance in both new and receiving communities (Sandercock, 2003).

Neighborhood or community centers are one of these "banal micro-publics" that can promote intercultural dialogue and address power inequities. Sandercock (2003) highlighted a neighborhood house, stating that it succeeds in four key areas: differentiated benefits (broad level of benefits for many diverse groups), inclusive participation (personal relationships in all communities to promote participation in all areas including decision making), varied discourse (cultural dialogues and resolution of conflicts due to difference, leadership capacity building), and inclusive definitions (sense of common identity and belonging).

Yan and Lauer (2008) found that these centers may play a key role in building the social capital of ethno-culturally diverse immigrants. One definition of social capital is "the ability of actors to secure benefits by virtue of membership in social networks or other social structures" (Portes, 1998, p. 6). Lin's (2000) research found that disadvantaged social groups such as women or minoritized populations are disadvantaged in their social capital, having less access to social networks that could lead to better opportunities in education, employment, etc., and so implying that these groups must build "bridging" capital with more advantaged, dominant groups in order to progress. Much of the social capital literature, however, focuses only on the disadvantaged group, such as immigrants, gaining in these "bridging" relationships with dominant receiving communities, implying that the benefit is not reciprocal.

Working from and building on this literature, the following case study will examine one of UNP's programs, the Hartland Partnership Center, a community center bearing much resemblance to the neighborhood houses described above. We will focus on the strategies that foster shifting power structures and relationship building. We will use our and our students' personal experiences (secondary data in the form of student evaluations) as participants/actors in these spaces to examine transcultural processes and to argue that these exchanges result in mutual benefit for all stakeholders and communities.[1]

Case Study: UNP/Hartland

Figure 15.1 UNP Hartland Partnership Center runs community and educational programs out of a donated three-bedroom apartment. Photograph by Kimberly Schmit.

UNP's Hartland Partnership Center opened in September 2004, in a three-bedroom apartment at the Seasons at Pebble Creek Apartments (formerly known as Hartland Apartments) (Figure 15.1). Due to initial success with engaging the community, youth programming expanded into a much-needed second two-bedroom apartment in 2006 and a Head Start Preschool was added by a local community organization in February 2007.

The 300-unit apartment complex is an affordable first home for first generation Latino immigrants, as well as for many families who have come to the USA with refugee status. At the time of the Center's inception, over 70 percent of the residents were non-native English speakers and 90 percent lived at or below the poverty threshold (Demographic Statistics, 2005). This community continues to represent a denser concentration of immigrants and people living in poverty than in its surrounding west side neighborhoods where 60 percent of residents are minoritized populations (but not necessarily new immigrants) and approximately 28 to 48.6 percent of residents earn $25,000 or less (May, 2011). Early community assessment also revealed widespread fear and mistrust of Hartland residents by neighbors due to lack of communication and cultural understanding between them (Smith & Munro, 2008). On the other hand, cultural pride and strong social networks supported the newcomer families as they faced the challenges of living in a new community and navigating new systems. Community leaders identified this apartment complex both as a space of opportunity because of its rich diversity but also saw it as a threatening space because of reports of higher incidence of crime and conflict.

Since its inception, the Hartland Center has functioned as a *community–capacity building* partnership that brings together Hartland residents, university faculty, students, and community agencies to co-design and implement programs with the families living at the complex and the surrounding communities. Involved partners work together to build upon one another's strengths in an effort to facilitate programming in the areas of resident leadership, language acquisition, youth programming, information, referral and capacity building, citizenship, health, family financial literacy, early childhood education, and life skills. Undergraduate and graduate students, under faculty supervision, work at the Hartland Center for practicums, research experience, and service learning credit. Most of the students and faculty as well as community organization representatives are from Salt Lake City's dominant, White, middle-class communities.

When the Center opened in 2004, the initial work of myself, as the On-site Coordinator, and another UNP staff member, who was a local male Somali leader, was to spend time meeting the residents of the complex. Going door-to-door, we began to build relationships with many local leaders who went on to become our first Resident Committee leaders. While this work resulted in a diverse initial Committee, most of the other programs at the Center were primarily accessed by African-born women and children. In fact, an early conversation revealed that many in the community thought that the Center was only for African women. Further, due to the clipboards that partners carried as we went door-to-door, leaders informed staff that many in their community also thought the Center was operated by the government and might not be safe for everyone. Upon learning this, strategic and variant efforts were put in place to support the Center being a space for more of the residents of the complex. For example, emphasis was further moved from traditional and formal modes of building relationships (i.e., clipboards) to more informal modes (i.e., conversation and relationship building) and under the guidance of resident leaders, programs that were gender- and ethnic-specific were offered. While this proved successful, resident engagement in the Centers has highly varied over time, depending on ethnicity, gender, age, and amount of time living at the complex and/or in the United States. Further, the community is highly mobile and requires an ongoing effort to ensure that the Centers are represented by and working with the multitudes of communities who live in the complex.

In contrast with assimilation integration initiatives, UNP/Hartland's focus continues to be both on the skills and tools of individuals *and* on changing the systems and institutions they interact with to accommodate new voices and shared power around the table. As new arriving populations develop the skills and social capital that they need to be successful in a new country, these "New Americans" also share their knowledge and diverse cultural perspectives with schools, social systems, and native-born families, enabling dominant systems and communities to grow and change in ways that expand and reflect each new generation of immigrants. In order to promote this mutuality and reciprocity in acculturation, UNP/Hartland grounds all of its work in two areas: socially, to build relationships, and politically, to redistribute power. Most programs may have their primary focus in one area; however, they often overlap in both areas of practice.

A Social Practice: Relationship Building

> If I am going to write about a successful story, I want to do it in regard to the ways the Hartland Center is fostering in neighbors a sense of belonging, creating the opportunity to talk to each other, play, and look forward to a brighter future, in which the multicultural, multiracial matters become assets to enrich our lives, and recreate a strong society.
>
> (UNP Student Evaluation, 2010)

Upon entering the Hartland Partnership Center, one is immediately immersed into a dynamic multicultural space. The words, "Welcome! Come in. How are you? How is your family?" serve as the initial greeting upon entering the living room. Multiple languages at many levels of fluency fill the air; residents and university students of all ages sit side by side on the couch, discussing a program, a bill, an upcoming wedding. The phone rings, is answered in English by a Spanish speaker, and is immediately handed off to a Somali staff member. Resident Committee leaders leave one room as staff from a local English language school enter another. University faculty and students carry out "student supervision" while eating in the kitchen. The living room walls surround, filled with art from various cultures, a University of Utah flag, community information in many languages, awards received by the Center, and photographs celebrating parties, events, and relationships. Children run in, home from school, backpack set aside and ready to use the computer.

Socially, the Hartland living room is a "micro-public" space where the various stakeholders come together to interact, to create relationships and interdependence. To begin, students and staff, who are traditionally seen as service providers, are asked to spend time sitting in the living room as individuals, socializing with residents. Student interns are instructed to practice their professional roles and be mindful of their professional boundaries. However, they are also encouraged to occupy this space as people and as representatives of their many personal communities. Students come to learn about people's cultures, but they are also asked to share and teach about their own personal beliefs and values. Topics of discussion in the living room range from questions about access to higher education, discussions on marriage, sex, religion, and food. As people sit, they share their experiences and find commonality in their mutual interest in learning about the other. These daily conversations normalize the "other" and enable sharing of people's daily successes and struggles. Trust is built, and issues such as power and privilege can be discussed on a more personal, caring level. A White social work student (2010) said in her final evaluation, "I have a more realistic view of refugees [rather than a romantic one]. They're like me with good and bad days." Such a statement is telling in that regular intercultural dialogue is necessary to experience connection and build community among diverse peoples.

Figure 15.2
The Hartland Resident Committee organizes an annual community party where diverse communities of residents and stakeholders share food, music, and fun. Photograph by Kimberly Schmit.

Figure 15.3 End of Summer Community Celebration, 2009: building community as a social practice. Photograph by Kimberly Schmit.

Another strategy is creating spaces for people to have fun together. Recreational activities such as game nights, picnics in the park, parties, and field trips are mechanisms for cross-cultural learning (Figures 15.2 and 15.3). These events reduce people's barriers and defenses and build community through shared laughter and enjoyment, qualities that transcend language and social barriers. These activities have helped students and residents alike to build shared memories, connections, and community. Furthermore, events and field trips situate newcomers in mainstream, public spaces, increasing their visibility and reducing their marginalization (Figure 15.4). Student evaluations repeatedly have highlighted the impact of recreational activities (2010):

> Through the Hartland Center both families learned about the Head Start Center at Hartland and different members of these families came to the game night to chat and play board games with us, intern social work students, and other neighbors. I was happy to walk in the Hartland Center, and see these families in the vans that the program provided to go to PIP [community picnics], sitting next to other families who did not speak Spanish, or have a similar skin color, but smiling because they knew they were neighbors.

Finally, the biggest success for me is the relationships that these trips have built. First, the relationships between students and residents have grown because of time

Figure 15.4 Hartland residents and university students on a Splore (not-for profit that provides accessible outdoor adventures) field trip in Millcreek Canyon, Salt Lake City. Photograph by Ellen Goldberg.

spent together, not working on paperwork or solving a problem, but time spent relaxing and laughing and being regular human beings.

A Political Practice: Redistribution of Power

UNP seeks to support newcomers to occupy more public spaces with equal power, especially institutional spaces such as universities. One of UNP's goals is to create structures and positions that facilitate and recognize the power of newcomers in these institutional spaces. Central to all aspects of the Hartland Center is the work of the Resident Committee. The Committee is made up of six to eleven residents who represent the ethnic diversity of the Hartland community. Each member is paid for a commitment to a one-year term. The primary roles of the Committee are as liaison and advocate for bringing the Center and the community into partnership. Members guide appropriate programming choices for the Center and help build relationships between residents and other Center stakeholders.

In a letter to the Hartland community, the Resident Community defined their goals and methods:

> Our main goal as the Resident Committee is to work with people who live at Hartland to make our community a better place to live . . . The Resident

> Committee is made up of six residents who represent you, the residents of Hartland. We work closely with Center staff to address community challenges and successes and encourage residents to get involved with Center activities.
>
> (Personal communication, 2007)

The work of the Committee serves as an example of a strategic structure that situates immigrants in a position of power. Results of this work include: increased resident ownership of the Center, the creation and implementation of culturally appropriate programming that is more effective, and other stakeholders being moved into the role of *learner*.

One faculty in nursing sums it up well when she describes what she learned when she approached the Committee to tell them she wanted to teach a sex education class for the youth at Hartland. "My time in the meetings has taught me to ask the question rather than provide the answer." The result of her idea, a surprise to this faculty, was that the leaders did not want this class taught. They were adamant that this was the role of the parent. However, as the conversation continued, the faculty and Committee decided to work together. One, the faculty would provide resources to the parents, and two, they would work together to co-create and implement healthy relationship classes.

As a UNP partnership manager and first facilitator of the Committee, I had the privilege to learn, first hand, the power of working with the leaders. Initially, I very much depended on the leaders to guide the programming, to connect the Center to the community and to teach me the pros and cons of multicultural meetings. However, I did not engage my own opinions and culture into the dialogues. While the leaders shared very personal aspects of their identities with me, mine remained invisible.

However, as time went on and our relationships became more personal, there came a pivotal moment when the leaders asked me to share. During a discussion regarding the role of religion in financial planning, a leader turned to me and asked, "What is your religion?" I remember stopping, losing my breath and then responding to a room full of what I perceived to be devoutly religious people, whom I had grown to respect greatly, that I did not practice religion and that I did not believe in God. The Committee members listened to my hesitant answer and immediately delved into asking many questions that explored my beliefs.

At that moment things shifted tremendously. I realized that in this community building endeavor, it was imperative for me to share my identity, especially if I was going to ask others to share theirs. By choosing to be vulnerable and give up my protected position of power I was entering the dialogue in a new way: interdependency was being created. Awareness was born in me that in order for the neighborhood/community building endeavor to be successful, all partners had to be as willing as residents to participate as individuals, not just system representatives. Through this power-shifting endeavor we would be more apt to form relationships that lead to positive personal and system growth. Ultimately, this event led to on-site policy that supports all partners in learning about how their own personal stories and identities are integral to the creation of more socially just, shared space.

The Resident Committee, serving as an accessible, organized, critical advising body, has also facilitated active community participation in academic research and teaching. In research, Committee members have acted as informal gatekeepers as well as cultural consultants (paid and unpaid) and co-researchers. Their participation has connected research with underrepresented communities. They have worked with researchers and communities to build trust and heal from past exploitation. Because of trust and relationships formed over a sustained period of time, honest critiques and feedback from community members emerge: "Why are only refugees studied and not immigrants?" or "I want to study White people and their beliefs and practices" [because they're always studying us] (personal communication, 2009, 2010). Academics can then address these questions and express their own questions and assumptions. Long-term research relationships between community and university partners bring these many truths to the table, making it possible for intercultural exchange as well as opportunities for more voices to bear on the traditionally elitist work of academic research. This model leads to a power shift, resulting in co-ownership of the knowledge creation process, so residents or community organizations also see themselves and are formally recognized as knowledge creators.

As a social work professor, I have had the opportunity to support a fellow social work faculty in building a research partnership with the Hartland Resident Committee and witnessed first hand the intercultural process that has unfolded in the past four years, impacting all partners involved. Interested in conducting research with youth of refugee background, particularly focusing on the correlation between attachment or relationships and mental health, when my colleague first approached the Resident Committee, they responded with strong opposition, expressing concerns that, once again, their community will be stigmatized with mental illness; in addition, they questioned why only the youth were being recruited and not the entire family. Additional conversations, however, led to uncovering some common ground that both the faculty and the Committee were able to build upon. The faculty was open to receiving feedback and relinquishing some of her "expertise" as researcher. The Committee reviewed and suggested changes on standardized questionnaires; however, it did not help to recruit participants for the study. The researcher eventually hired cultural liaisons to join her team, understanding the benefit of this cultural investment. She has returned at different points to share and discuss results of the data with the Committee and other community stakeholders, each time building trust and relationship between her and the Committee, but also between the university and the community in the context of research. Notably, the research results have shown a different, more positive picture of mental health for these communities—that attachment or family relationships of youth of refugee backgrounds are protective factors, positively mediating the traumas related to refugee experiences.

Besides research, UNP also seeks to bring community leaders to university spaces as teachers. Some examples have been positioning Resident Committee members as teacher's assistants in university classes, having Committee members conduct presentations in classrooms and supporting them to write ethnographies to serve as required class readings. In the course "Immigration & Resettlement: Community & Interdisciplinary Perspectives," UNP employs an immigrant community leader as a teaching assistant and seeks to center community knowledge in a university classroom. Our teaching assistant acts as the bridge between his communities and the classroom. He humanizes theories and policies about immigration and integration and shares his direct lived experience and expertise. His stories have power to silence a full room and engage students on a deeper level. He bridges "us" and "them," *us* being the studiers and *them* being the people studied. He assists us in connecting students with community members and in coordinating community projects.

As a social work professor and co-facilitator of this course, I have learned first hand that time and effort must be invested in working to address real challenges and benefit from difference; business as usual must be disrupted at the academy. For example, our teaching assistant fulfills a different role; he does not complete much of the responsibilities of a typical teaching assistant. His style of facilitation reflects more what I perceive to be traditional, directive styles of teaching, and his presentations and lectures are circular versus linear, which sometimes conflict with my style and academic culture. These differences pose challenges, and in order to fully benefit from these differences, I must be conscientious in not using my authority to make efficient decisions, but to invest the time and effort in having many conversations about how best to channel our differences. In addition, our teaching assistant prefers face-to-face meetings, and the institution and I often use email. Furthermore, he does not have reliable transportation and depends on students and colleagues, including myself, for support in this area. Making space for "presence" and "difference" (Sandercock, 2003) and prioritizing intercultural education in the classroom require reprioritizing existing values. It has meant for me slowing down and learning new ways of working and living. Sharing power in the classroom with a community leader has led to personal growth and integration of additional cultural values and practices for me and my students.

Another strategy at Hartland that leads to intercultural learning and power redistribution is that decisions are made by conversations with many stakeholders rather than by predetermined procedures or guidelines set by a small group of authorities. For example, when students came to the Center in short shorts and revealing dresses, my social work faculty's instincts were to respond by implementing a dress code, assuming that this dress is unprofessional and would offend immigrants from more traditional cultures. However, conversations with many stakeholders surprised me and contradicted my assumptions; people prioritized inclusion and representation of many cultural practices over standardization and a dress code. A Muslim community representative (2010) stated that students respect her culture and what she wears, and so she respects their cultures and what they wear. The following student quote describes this shared decision making process and her perception of its benefit (UNP Student, 2008):

> Sharing a common attitude of respecting and valuing each team member creates an atmosphere where we can learn about each other's strengths, gain new information about how to approach a situation, and often produce better outcomes.

Conclusions: Limitations and Implications for Future Work

UNP Hartland Partnership Center has taught us that transcultural community integration is a daily social and political practice. Facilitating long-term relationship building and addressing inequities in power and privilege (between dominant, receiving communities and immigrant communities) have been two significant endeavors in promoting community integration. Both efforts are interrelated and interdependent. Formal organizational values explicitly promote relationship building, reciprocity, multiple truths, and equal voice; these values lay the foundation for daily social and political practice. They facilitate people showing up and engaging personally as individuals, exploring and crossing boundaries such as professional roles, ethnicity, and class. Moreover, structures and programs actively, formally and informally promote relationship building. The Center is an informal space decorated, designed, and operated by a diverse team of community representatives, with the intention of creating a safe, inclusive and relevant space for as many as possible. Leaving an imprint on many are programs that bring people of diverse backgrounds together to laugh, play, eat, and relax—sharing their common humanity and creating memories together. Politically, UNP Hartland has created structures and positions to recognize and compensate individuals from immigrant communities for their knowledge and expertise, repositioning them as resident leaders, teachers, and researchers. This strategy promotes social justice and builds on the capacity of individuals from different stakeholder groups to cross boundaries and promote transcultural integration. Finally, UNP Hartland prioritizes dialogue and process in decision making; thus, policies and programs adapt to changing communities and stakeholder groups.

A look at limitations of the case study reveals critical potential for future work. To begin, the above study explores the social and political practices from our, the authors, perspectives and experiences. A more strategic research project to explore the effectiveness of these efforts from various stakeholders' lenses and voices would help to reveal a more complex picture of *if* and *how* integration *is* or *is not* occurring. Centering the newcomer in this analysis is imperative.

Further, our findings are a result of analyzing what is happening with residents who are already occupying the space. Current demographic data that look at who lives at Hartland in relation to data that show who is accessing the Center would undoubtedly lead to questions of why the Center is a choice for some and not for others. For example, according to a 2008 assessment, the majority of the residents at the complex are Latino/a and the majority of the residents who come to the Center are from African backgrounds. Is this because three Somali residents staff the Center? Does the Resident Committee membership need to be more proportionate to the demographics of who lives at the complex? Is there a relationship between the university and community organizational partners being majority White and American-born with the programming attracting a disproportionate amount of people with refugee backgrounds versus first generation, Latino/a immigrants? Does the current political climate affect Latino/as accessing the Center? Is the Center seen as a safe space?

Lastly, a more rigorous look at how power and location intersect could reveal the tension created when the receiving community travels into the geographic community of the newcomer. In fact, whose community is it? Does the newcomer have the same access to travel into the receiving communities' neighborhoods? Why or why not?

Note

1 Analysis of this case study is conducted primarily through the lenses of a social work professor, a UNP academic partner, who is a first generation, middle-class Asian American immigrant, and a UNP partnership manager who is a fourth generation, middle-class White American. As university employees, we both represent a dominant privileged institution of the receiving community.

References

Bourhis, R. Y., Moïse, L. C., Perreault, S., & Senécal, S. (1997). Towards an interactive acculturation model: A social psychological approach. *International Journal of Psychology*, 32(6), 369–386.

Bourhis, R. Y., Barrette, G., El-Geledi, S., & Schmidt, Sr. (2009). Acculturation orientations and social relations between immigrant and host community members in California. *Journal of Cross-Cultural Psychology*, 30(3), 443–467.

Demographic Statistics (September 14, 2005). Evergreene Management Group: Hartland Apartments.

Department of Workforce Services: Refugee Services Office (2012). Statistics and Reports. Retrieved March 30, 2012, from: http://refugee.utah.gov/home/index.html

Dowden, J. (2012). University Neighborhood Partners Area Maps. Bureau of Economic and Business Research, University of Utah.

Hunter, R., Munro, S., Dunn, L., & Olson, K. (2009). Bridging university and community: The power of collaborative partnerships for social change. In K. Mohran, J. Shi, S .E. Feinblatt, & K. W. Chow (Eds), *Public University and Regional Development* (pp. 289–311). Sichuan, China: Sichuan University Press.

Lin, N. (2000). Inequality in social capital. *Contemporary Sociology*, 29(6), 785–795.

May, Heather. (2011, January 4). The west side: Where an ill wind blows. *Salt Lake Tribune*. Retrieved from: http://www.sltrib.com/sltrib/home/50786628-76/health-west-residents-lake.html.csp?page=1

Migration Policy Institute (MPI) (2012). MPI Data Hub. Utah. Retrieved March 30, 2012, from: http://www.migrationinformation.org/DataHub/state.cfm?ID=UT#1

Penninx, R. (2003). Integration: The role of communities, institutions, and the state. Migration Fundamentals. Migration Policy Institute. Retrieved November 15, 2010, from: http://www.migrationinformation.org/Feature/display.cfm?ID=168

Portes, A. (1998). Social capital: Its origins and applications in modern sociology. *Annual Review of Sociology*, 24, 1–24.

Sandercock, L. (2003). *Cosmopolosis II: Mongrel cities of the 21st century*. New York: Continuum.

Silka, L. (1999). Paradoxes of partnerships: Reflections on university–community collaborations. *Research in Politics and Society*, 7, 335–339.

Smith, Y. & Munro, S. (2008). Anthropology & occupational therapy in community-based practice. *Practicing Anthropology*, 30(3), 20–23.

United Way of the Great Salt Lake Area. (2001). Community needs assessment. Unpublished document.

University Neighborhood Partners. (2008). Student Evaluations. Unpublished document.

University Neighborhood Partners. (2010). Student Evaluations. Unpublished document.
US Census Bureau. (2012). Census of population. Retrieved March 30, 2012, from: http://2010.census.gov/2010census/popmap/
Yan, Miu Chung & Lauer, S. (2008). Social capital and ethno-cultural diverse immigrants: A Canadian study on settlement house and social integration. *Journal of Ethnic and Cultural Diversity in Social Work*, 17(3), 229–250.

CHAPTER 16

Transcultural Participation

Designing with Immigrant Communities in Seattle's International District

Jeffrey Hou

Emerged from the citizen revolts against urban renewal projects in the 1950s and 1960s, public participation has become a common practice in many North American cities today. Ballots, public hearings, and citizen committees are now routine ways for soliciting public input and consent. Design workshops, games, and surveys provide additional ways for involving citizens and communities in the making of place (Hou, 2011). In neighborhoods and university campuses from coast to coast, community design centers have emerged as a way for professionals and educators to assist distressed communities in need of outside help (Curry, 2004). These practices have built upon and resulted in a growing literature on participatory planning and design that deals with specific techniques and methods to engage the public (see Sanoff, 1979, 2000; Hester, 1984, 1990). Recently, growing interest in humanitarian design and public interest design further embodies the principles of participation and social justice.

As a democratic practice, participatory design provides opportunities for people to be involved in decision making that affects their lives. At its best, it also provides opportunities for creating places that embody the diverse cultures and social values of a community. However, in spite of the promises of access and empowerment, the actual practice of citizen participation often falls short of the moral rhetoric and the public's expectations. Over the years, scholars have questioned the real purpose and effectiveness of participation in redistributing power in the planning and design process. The availability of opportunity for people to participate does not necessarily mean that their participation is effective and meaningful (Francis, 1999; Hester, 1987, 1999; Hou & Rios, 2003; Hou & Kinoshita, 2007). The opportunity to participate often does not translate into actual power in the decision-making process. At worst, participation could be used to co-opt the constituents and placate the real issues through tokenism, manipulation, and therapy (Arnstein, 1969).

For populations such as recent immigrants, low-income residents, and other marginalized social groups, the common participatory practice poses additional problems. Methods and techniques that work for mainstream communities may not have the same efficacy for engaging social groups that do not share the same values or protocols of interaction (Hou & Kinoshita, 2007). For recent immigrants and refugees in particular, cultural and linguistic barriers, as well as the lack of legal and technical knowledge, present a major challenge for participation. Cultural differences and social conflicts further add to the complexity of participatory design practice in the context of diverse communities (Umemoto, 2001). In the face of increasingly diverse urban populations, the established repertoire of techniques and methods is no longer adequate to address the complexity of a multiple public.

Given the desire for diverse cultures to be reflected and supported in the built environment of the contemporary city, how can marginalized social groups, such as immigrants, youths, elders, and low-income populations, participate more effectively and meaningfully in the making of place? What methods and approaches can result in a more democratic and culturally inclusive process that reflects the changing demographics and identity of a community? How can participatory design facilitate cross-cultural learning and transcultural placemaking? Using the community design process in Seattle's International District as a case study, this chapter examines specific methods for engaging the predominantly Asian American populations there, including youths and elders. Based on interviews with participating stakeholders and community organizers, it assesses outcomes and contributions of the participatory processes in transcultural placemaking.

Planning and Design in a Multicultural Immigrant Community

Located at the edge of downtown Seattle, the International District is a historical neighborhood with a predominantly Asian American population. Since the late 1800s, the area has been home to successive waves of immigrants from China, Japan, the Philippines, and more recently Southeast Asia, resulting in the area's diverse history, cultures, and identities. As a hub of Asian immigration, the district has long served as the center of the city's Asian American communities (Chin, 2001; Abramson et al., 2006). Today it continues to support a significant population of low-income, non-English speaking, and elderly immigrants after a long period of struggles.

In the 1940s, the district suffered dramatic decline in population and economy with the internment of Japanese American citizens during World War II. This was followed by an exodus of middle-class residents to the suburbs through the 1960s and 1970s. The construction of Interstate freeway 5 through the neighborhood further eroded its declining residential base. Following the new fire codes in the 1970s, the high cost of renovating the historic buildings discouraged many property owners from upgrading their apartment buildings and single-residency hotels, leaving a large section of the district's building stock in severe disrepair.

The recent efforts in neighborhood design and planning in the district have been organized to address both persistent problems and new challenges. Despite the efforts of many local community organizations to renovate the buildings, many buildings and storefronts remain vacant. On the other hand, the expansion of downtown development during the dot com boom of the 1990s and the recent effort by the city to increase the housing density in the city center presented recurring threats to the district in terms of displacement and loss of cultural identity (Abramson et al., 2006). In 1998, a Neighborhood Strategic Plan was produced by a citizen task force as a part of the city's broader neighborhood planning process. In 2003, to implement many of the recommendations from the Neighborhood

Strategic Plan, a district-wide urban design master plan was developed through a joint effort by several community organizations with funding from the South Downtown Foundation, an entity created to mitigate the impact of a nearby sports stadium.

However, despite these largely community-based efforts, the ways local residents could participate in these processes were mainly limited to community meetings conducted in English and indirect representation through the various community-based organizations. The limited participation by the district's mainly non-English speaking, immigrant residents in particular raises questions on whether the needs and desires of the diverse residents and users of the district have been adequately reflected and represented in these plans.

Engaging a culturally diverse community in the planning and design process has been a focus of several service-learning design studios and community outreach projects in the Department of Landscape Architecture at the University of Washington (UW). Since 2002, through the studios and other outreach efforts, a number of methods and techniques have been experimented with to engage the diverse stakeholders in the International District in planning and design projects to improve the district's environment and livability. The projects include designs for streetscape improvement, installations for a night market event to activate a neighborhood open space, and renovation of a children's park in the neighborhood. This chapter examines the processes and outcomes from two of the projects in which specific methods of engagement have been applied.

"Photovoice" and the Maynard Avenue Green Street

Located on a steep hill overlooking South Downtown, the South Maynard Avenue links two distinct neighborhoods within the International District—Chinatown and Japantown. Designated by the city as a Green Street,[1] Maynard Avenue connects two important open spaces in the district—the "Danny Woo International Community Garden" on top of the hill and the "Hing Hay Park" in the heart of Chinatown. However, despite its strategic location, the northern end of the street was poorly used partly because of the steep slope and lack of businesses and amenities on the street. The recent development of a large vacant property, facing the street, into an affordable housing complex provided the catalyst for a proposed streetscape improvement. The new design was expected to better serve the community by enhancing walkability and urban sustainability in the area through widened sidewalks, increased vegetation, and natural drainage. As a small-scale, hands-on project, the proposed streetscape improvement also provided an opportunity to engage the local community in an actual design process—a process that was not available in the previous planning efforts in the district.[2]

To engage the community within a limited time frame, the participatory process targeted the primary users of the street—the non-English speaking, elderly immigrants who lived in the apartment buildings nearby. Based on past experience, we anticipated cultural and language barriers in communicating with the elders and their discomfort in expressing their opinions to outside professionals. To overcome these barriers and possible discomfort, we selected the "photovoice" technique as a way for the elders to register and communicate their observation of the built environment in the neighborhood, which could in turn inform the design.

Photovoice is a developed technique used in a variety of contexts, from participatory evaluation in public health to needs assessment in the design of urban open space (Hou, 2005). The method enables researchers and practitioners to gain "the possibility of perceiving the world from the viewpoint of the people who lead lives that are different from those traditionally in control of the means for imaging the world" (Ruby, 1991, p. 50, cited in Wang & Burris, 1997). In the Maynard Avenue Green Street project, the photovoice exercise involved recruiting a group of elderly residents, including immigrants from mainland China, Hong Kong, South Korea, and Taiwan, providing them with disposable film cameras, developing the photographs, and conducting follow-up interviews with the elders through bilingual designers who collectively speak English, Ho'lo Taiwanese, Japanese, and Mandarin Chinese. To keep the exercise simple and accessible, the elders were asked to photograph what they liked and disliked about the neighborhood environment.[3] The developed photographs then served as the subject of conversation between the elders and the design team in follow-up interviews. Through the interviews, the design team was able to probe deeper into the underlying concerns and preferences embedded in the photographs.

Because most of the interview participants were also gardeners at the nearby Danny Woo Garden, many of the photographs were taken from the garden. Since the 1970s, the garden has been an important place for elderly immigrants in the neighborhood to grow food, exercise, and socialize. It is also one of the few open spaces in the area that are actively used by the residents. The photographs from the garden gave the design team abundant clues on how open space and outdoor environment were perceived and utilized by immigrant residents in the community (Figure 16.1). They also revealed specific meanings that were given to various structural and physical features of the neighborhood environment by the residents. For example, in response to the Asian-style structures in the garden that were photographed repeatedly, the elders talked about how the familiar forms help make a new environment feel like home. The culturally familiar forms are not limited to physical structures and architectural forms. Vegetation, in terms of street planting and plants grown in private front yards, also carries cultural meanings and significance. Other photographs include shots of paths and retaining walls in the garden, revealing certain aesthetic and cultural preferences that were important to the gardeners. Another key finding from the interviews was the importance of open space in supporting social interactions and networking, a critical aspect of social life for the elderly immigrants. In a series of photographs, instead of documenting the built environment, an elderly man took pictures of friends he has met in the garden. Through these photographs, he talked about how he befriended many of his neighbors through gardening as well as taking lessons in a nearby community cultural center. He mentioned that talking with friends was important to the lives of elderly people.

While the repeated images may confirm a similar pattern of uses and reactions, the interviews also show that the same element may carry different meanings for different residents. For example, one elderly woman took a picture of the streets near

Figure 16.1
Photographs taken by elderly residents from the photovoice exercise.

the site to show how the steep slope presented a challenge for her, given her heart condition. On the other hand, a younger, middle-age woman took pictures of the same streets to show how she used the sidewalks everyday as a place to exercise by walking. In terms of aspects of the neighborhood that were disliked, one woman photographed trash on the street, while another photographed places where

neighborhood beyond the park site. Using the photovoice technique similar to the Green Street project, the selected participants were given disposable cameras to photograph their activities on a typical day in the neighborhood as a way for them to describe how they utilize the physical environment in the neighborhood. The photographs from the exercise, along with the park designs they produced, provided the medium for conversation. The interviews yielded some important insights about the use of open space in the district. One such finding was that while many elders try to engage in outdoor activities to stay healthy, they tend to get their exercise by simply walking around the district instead of visiting a park. An improved park will thus provide more motivation and incentives for the elders to engage in more outdoor activities. It can also serve as a social space for the elders.

The results from the participatory activities described above have informed and strongly influenced the subsequent design discussion. The collaborative design alternatives produced by the WILD youths and the UW students were unveiled in a community open house attended by many neighborhood residents, members of various community organizations, and city staffs. Ballots were collected through which the majority of open house participants expressed remarkably similar preferences to those of the preschool kids who voted on the proposed designs at a separate event. The results from the process enabled the project to receive significant city funding for implementation. A professional design team was then selected to develop the final design based on community inputs. Construction was completed in March 2012, and the initial user responses have been overwhelmingly positive (Figure 16.5).

Figure 16.5 A view of the completed International Children's Park with expanded play area and features for multigenerational uses. Photograph by Jeffrey Hou.

Transcultural Participatory Design

Do the above-described methods succeed in getting the immigrant residents to participate more effectively and meaningfully in the process of participatory place-making? Do they enable and encourage cross-cultural learning and communication between community stakeholders and designers? Based on the responses from the participants, including youths, schoolteachers, elders, and members of community organizations, the answer seems positive. More specifically, their responses suggest the following critical dimensions of successful transcultural participatory design in immigrant communities:

Social. In the two cases above, much of the transcultural processes have taken place through social interactions that came as part of the participatory design activities. The activities enable elderly residents and youths to become more active participants in the public forums in which they have been able to interact formally and informally with people inside and outside the community than they would have from passively receiving information. Both the photovoice method and the intergenerational design workshops enabled a level of interactions beyond the typical protocols of community meetings. Through interpreters and bilingual professionals, the elders clearly enjoyed the chance to share their stories and perspectives in their native tongue and with the younger designers. Such opportunities were often not available in a typical public process. The social interactions helped break down the traditional, structural barriers in the community and also between community stakeholders and the professionals. These interactions echo what Amin (2002) describes as "banal transgression" between different cultural groups in the everyday setting, as well as what has been described elsewhere as "informal processes" that bridge community differences (Hou & Kinoshita, 2007). Transcultural participatory design therefore needs to create opportunities for social interactions.

Representational. The "photovoice" and "design buffet" methods have served as vehicles for cross-cultural communication that produces far greater cultural insights than typical public processes. Specifically, the photovoice activities in both projects helped overcome cultural and linguistic barriers and provided the designers and students with a broader understanding of the life experience of residents in the neighborhood beyond the immediate scope of the projects. The photographs taken by the elders and the follow-up interviews for the Green Street project allowed for interpretations and conversation about the cultural meanings of features in the built environment. It allowed the professionals to literally see things through the eyes of the elders. They revealed how the seemingly mundane details and nuances in the physical environment may hold significant cultural meanings for the elderly immigrants. Both sets of interviews also allowed for a better cross-cultural understanding of how local, especially elderly, residents view and utilize the streetscapes for their daily activities. In the design buffet, rather than passively responding to questions, participants used models to convey what they would like to see in the park. Together, the variety of tools has broadened the ability of the participants to register and express their thoughts. They have served as the medium for cultural representation and cross-cultural communication.

Mediated. While the activities and methods described above clearly provided the opportunities for cross-cultural learning and interactions, the extent to which these needs and preferences could be interpreted and revealed depended greatly on the ability of the designers and organizers as transcultural agents. In both projects, language ability and prior knowledge of the community's diverse cultural traditions have been important for discerning and interpreting the results of photovoice exercises, visual preference feedbacks, and the designs created by the participants. How the activities were designed and to the extent that they were effective also depended on knowledge of what might or might not work for the particular audience, which requires an understanding of the culture of the community. The facilitators' ability also determined how cultural differences could be expressed and mediated, and how such differences could play a positive role in the community process rather than a distraction or dividing force. The results from the intergenerational workshop further demonstrate the importance of representing and mediating differences in the community to arrive at a greater understanding of diverse needs in the community. Our experience in the International District suggests the critical importance of cultural and transcultural competency as a necessary skill for planning and design professionals to work effectively in diverse communities.

Collaborative. As communities become more diverse and their challenges become more complex, it is increasingly difficult for design professionals to completely handle the public process on their own, especially in communities where their knowledge and experience is limited. Working with community partners, and sometimes other professionals with additional skillsets, is desirable if not necessary. The collaboration between the community partners and the design studio at UW has been key to the effectiveness of the participatory activities in the two projects. In particular, the community partners played an important role in outreach and mobilizing. They helped the UW studio connect with the residents and audience in the community and navigate the community process. The students and the faculty at UW in turn provided the labor and expertise in design and participation. The community partners helped develop and craft the participatory activities, going over seemingly mundane details such as how long or short the activities need to be and the level of complexity that required understanding and sensitivity to the cultural norms of the community. Furthermore, developing a successful collaboration is often a transcultural act as well, especially between institutional and community partners. For example, the education goals of a university curriculum don't always align with the social and political goals of a community-based youth program. An effective and rewarding cross-cultural collaboration requires the partners to understand and respect each other's needs and priorities, as well as manage possible conflicts.

Engaging, Empowering, and Learning from Immigrant Communities

As urban communities in North America and elsewhere become increasingly diverse, there is a growing need for public process and participatory practices to respond to emerging cultural differences, conflicts, and complexity. Tools in transcultural participatory design need to be further developed to meet the increasing demand for design and planning professionals to work in multicultural contexts and navigate economic, social, and cultural changes, in order to produce culturally responsive designs that serve diverse stakeholders. The transcultural participatory process as described in this chapter shows some possibilities in this direction. In the Green Street project in particular, the process allowed designers to see beyond the general goals of walkability, beautification, and sustainability and negotiate the culturally specific needs and preferences of the users. It helped prevent a cookie-cutter solution by creating a project that reflected and addressed the diversity in the community. In the International Children's Park project, the interactive participatory activities provided not only an opportunity to better understand the needs and preferences of the different user groups, but also enabled the designers as well as the participants themselves to understand how different uses and preferences may coexist and mutually enhance the intergenerational use of the park. The fact that the participants from different age groups all favored the same design suggested the design's multiple cultural responsiveness.

Yet, producing culturally responsive design and enabling professionals to learn from the perspectives of the communities should not be the only goals of transcultural participatory process, nor should the process stop at the completion of a design. In the face of institutional, political, and cultural barriers, transcultural participatory design is also a means of empowering immigrants and other marginalized individuals to become active members of the society across multiple barriers. In the case of the International District, by providing community participants with opportunities for learning and playing a more active role in the design process, the transcultural participatory methods have enhanced participants' ability to engage in future public process. As one community organizer reflected upon the change in the community, "the Children's Park studio transformed stakeholders' perspective, expectation and experience about a project through community engagement." Another organizer also described how the local businesses and residents now have a better understanding of how things work, particularly how projects come together and how public process works, and that once the residents became more comfortable with public process, they would ask "what happened to what we said at the last meeting?"

The outcomes of engagement, empowerment, and learning suggest that participatory design has come full circle in terms of addressing its intended goals as developed in the early decades. They address a fundamental social justice issue in terms of unequal access and power, in this case as the result of cultural, economic, and institutional barriers facing immigrant communities. In supporting learning and understanding across cultural barriers, transcultural participatory design can help build capacity for individuals and groups to navigate and traverse the shifting cultural terrains of today's urban communities. In facilitating and creating social interactions, transcultural processes help build connections and alliance across social and cultural boundaries. Through collaboration, transcultural process can lead to greater capacity within and between the communities to address new and continuing challenges in today's changing cities.

Notes

1 Enacted as a Seattle land use code initiative in 1993, the Green Streets Program designates certain urban corridors as Green Streets for the purpose of improving the pedestrian environment by narrowing the roadway and increasing the amount of "open space" for sidewalks and landscaping (Gangnes, 2003).
2 This project was initiated by the Inter*Im Community Development Association with the author and two graduate student interns leading the conceptual design process.
3 Altogether, six residents participated in the photovoice exercise.
4 About twenty high school age youths and thirty adults participated in the workshop, which was developed as part of an adult ESL (English as a Second Language) class. All materials were presented in Traditional Chinese and English.
5 About twenty preschool children participated in the workshop, with assistance from three bilingual preschool teachers. The culturally diverse children speak a wide range of languages, including Spanish, Chinese, Vietnamese, and Tagalog.

References

Abramson, D., Manzo, L., & Hou, J. (2006) From Ethnic Enclave to Multi-ethnic Translocal Community: Constructed Identities and Community Planning in Seattle's "Chinatown-International District." *Journal of Architecture and Planning Research,* 23(4): 341–359.

Amin, A. (2002) Ethnicity and the Multicultural City: Living with Diversity. *Environment and Planning A*, 34: 959–980.

Arnstein, S. R. (1969) A Ladder of Citizen Participation. *Journal of the American Institute of Planners*, 35(4): 216–224.

Chin, D. (2001) *Seattle's International District: The Making of a Pan-Asian American Community*. Seattle, WA: International Examiner Press.

Curry, R. (2004) Community Design Centers. In Bell, B. (Ed.), *Good Deeds, Good Design: Community Service through Architecture*, pp. 61–70. New York: Princeton Architectural Press.

Francis, M. (1999) Proactive Practice: Visionary Thought and Participatory Action in Environmental Design. *Places,* 12: 2.

Gangnes, D. A. (2003) Seattle's Green Streets Ripe for Modernization. *Daily Journal of Commerce.* October 2. Online, available: http://www.djc.com/news/co/11149502.html (accessed February 20, 2005).

Hester, R. T. (1984) *Planning Neighborhood Space with People*. New York: Van Nostrand Reinhold.

Hester, R. T. (1987) Participatory Design and Environmental Justice: Pas De Deux or Time to Change Partners? *Journal of Architectural and Planning Research,* 4(4): 289–299.

Hester, R. T. (1990) *Community Design Primer*. Mendocino, CA: Ridge Times Press.

Hester, R. T. (1999) Place of Participatory Design: An American View. In Hester, R. & Kweskin, C. (Eds), *Democratic Design in the Pacific Rim: Japan, Taiwan, and the United States*. Mendocino, CA: Ridge Times Press.

Hou, J. (2005) Speaking Images: A Case of Photovoice Application in Community Design. Paper presented at Visualizing Change: Association for Community Design Annual Conference. New York, March 30–April 1.

Hou, J. (2011) Citizen Design: Participation and Beyond. In Banerjee, T. & Loukaitou-Sideris, A. (Eds), *Companion to Urban Design.* London: Routledge.

Hou, J. & Kinoshita, I. (2007) Bridging Community Differences through Informal Processes: Reexamining Participatory Planning in Seattle and Matsudo. *Journal of Planning Education and Research,* 26(3): 301–313.

Hou, J. & Rios, M. (2003) Community-driven Placemaking: The Social Practice of Participatory Design in the Making of Union Point Park. *Journal of Architectural Education,* 57(1): 19–27.

Ruby, C. (1991) Speaking for, Speaking about, Speaking with, or Speaking alongside: An Anthropological and Documentary Dilemma. *Visual Anthropology Review*, 7(2): 50–67.

Sanoff, H. (1979) *Design Games*. Los Altos, CA: W. Kaufmann.

Sanoff, H. (2000) *Community Participation Methods in Design and Planning*. New York: Wiley.

Umemoto, K. (2001) Walking in Another's Shoes: Epistemological Challenges in Participatory Planning. *Journal of Planning Education and Research,* 21: 17–31.

Wang, C. & Burris, M. A. (1997) Photovoice: Concept, Methodology, and Use for Participatory Needs Assessment. *Health Education and Behavior*, 24(3): 369–387.

CHAPTER 17

Urban Agriculture as "Agricultural" Producer

Adam Prince

I would like to begin this chapter by considering a report commissioned by the Fair Food Foundation and carried out by Policy Link and Michigan State University, entitled "Healthy Food For All." In it the authors rightfully observe that "around the [United States], there are growing movements underway to transform the present food system into one that promotes healthy people, healthy places, and healthy economies" (Treuhaft et al., 2009, p. 5). The specific organizational forms of these so-called movements vary, but usually include some combination of "farmers' markets, urban-based community-supported agriculture (CSA), produce delivery services, 'good neighbor' programs where liquor and convenience stores owners are asked to supply fresh fruits and vegetables, garden projects, and environmental and nutritional education programs" (Guthman, 2008a, p. 432). In Oakland, California, the substance of this movement has gained traction within a growing community of urban food producers and advocates who are working in a variety of ways to address food insecurity within the city's most historically under-served communities. Even the most superficial survey of the scene reveals that nearly all of Oakland's diverse enclaves exhibit some form or another of community-based food production, from modest windowsill herb patches to more ambitious roof-top projects to distributed farming networks that make use of the city's numerous vacant or underutilized lots. Further, a constellation of affiliate organizations provides a framework of much needed institutional support.

For a great many of these organizations, community building and empowerment figures prominently among their stated objectives. And while I do not deny the efficacy of food as a galvanizing matter of concern, a survey of available literature reveals that more work should be done with respect to understanding how the site-specific interactions and negotiations engendered by collective food production actually strengthen and empower communities. Recent theories in urban studies that consider culture in relation to place and "placemaking" might prove to be useful in illuminating these questions insofar as they draw meaningful connections between situated practices, community building, and the production of various forms of political agency.

In the space that follows, I direct my attention to the site of Keller Plaza community garden, located in North Oakland, California. First, a collection of anecdotal observations will set the stage. Then, I talk about the ways in which, rather than producing empowered communities "in general," community-based urban agriculture is actually producing novel cultural forms, which I refer to as "agricultures." Finally, I conclude with several observations concerning the rather popular, albeit somewhat misguided, idea that "alternative" food practices alone can and will effectively ameliorate urban food insecurity.

Methodological and Theoretical Orientation

The privileged gaze of the researcher has long bedeviled place-based methodologies. Indeed, few scholars working in an ethnographic mode would deny the significance of both situating one's research and accounting for one's "unique" subject position. But a simple recognition of the problematic relationship between researching subjects and the subjects of research is not enough. Methodologies dedicated to taking cultural processes seriously must now be produced and performed *with* subjects and, by extension, respond to all the vicissitudes of space and place. And so it seems that beginning in the thick of it, so to speak, in order to get a very palpable sense of who and what is *already there*, might begin to carve a path through this historically fraught enterprise and into a better understanding of the prefix "trans" we researchers are, with increasing frequency, placing in front of the *verb* "cultures."

Akin to the complex stories that follow, my worldly inhabitation is informed by a dense and overlapping web of theoretical commitments and experiential knowledge. My fieldwork was conducted in and around North Oakland, California, with the bulk of it taking place from August 2010 to January 2011—although my interest in community-based food initiatives goes quite a bit further back. My analysis draws on a range of work within critical social theory, especially in the realms of anti-oppression and social justice, and work on new cultural geographies. Although examples of the former are too numerous to list, Paulo Freire immediately comes to mind. In the case of the latter, I am thinking of figures like Anna Lowenhaupt Tsing, James Clifford, and Homi Bhabha, among others. Further insight has been gained from a growing body of scholarship that attempts to make sense of alternative food practices in the context of place-based and/or culturally inflected practices (Hou et al., 2009; Guthman, 2008a, 2008b; Mares & Peña, 2010). Finally, the present work is brought to life by the many generous contributions of Planting Justice as well as other contributors to this book.

The Locations of (Trans)Cultures

Cultural engagement is a leading actor in the production of urban geographies, such as they are in history. At the same time, cities respond with an equally salient productive force as urban planning and development, the flows and distribution of capital, zoning, policy, and so on, proliferate and police claims made upon the city. Often times, the most productive spaces of engagement are located at the locus of competing cultural, spatial, economic—which is all to say, political—claims. "Concrete" material forms and "fluid" cultural formations are endlessly made and unmade moment-to-moment. Anthropologist Anna Tsing has elsewhere used the metaphor of "friction" to talk about these sorts of "awkward, unequal, unstable, and creative qualities of interconnection across difference" (2005, p. 4).

Further, in the full light of latter-day globalization, the demarcations and imbrications of cultures are articulated and capitulated at various scales, from the individual to the collective and from the local to the global. Indeed, mobility seems to be the name of the game. Against the backdrop of complex transnational flows, cultures always find a way to surreptitiously transgress "tidy" territorial borders. And it is precisely within this constant shuttling back and forth between scales micro and scales macro that we might locate the tacit promise for place-based struggles to have far-reaching influences and effects. How, for instance, might a community of urban food producers in Havana, Cuba, provide a useful model for organizing residents in, say, Buenos Aires (or vice versa)?

But culture does not exist at a level of generality that enables us to address it in a monolithic sort of way. Our senses of various cultural norms and forms—and the ways in which they affect one another in history—are diverse, often contradictory, and their force is grasped as much in that from which they are distinguished as in what they capture. For all of the ways in which cultures might appear to us as whole or stable objects, there are countless other ways in which they are instead spontaneously constituted and maintained through provisional webs of heterogeneous forms. Cultural designations and significations, whether through consent or compulsion, are characterized by multiple layers and (site) specificity, animated *in place* by multiple and overlapping histories and contexts. As my earlier reference to culture-as-verb alluded, all of these observations are nested within culture as an interminable process of becoming. For these reasons and more, anytime we speak of cultures, we are always already speaking of transcultures.

Before proceeding, I would like to take a moment to acknowledge that limited space has to an extent circumscribed the depth and orientation of my analysis. Regrettably, a sustained look at the racial politics of urban agriculture has been left for another day, as has greater attention to the cultural appropriateness of many alternative food initiatives.[1] There is also much more to be said about how transcultural spaces and places might be evaluated in regards to distributions of power and privilege. While asking forgiveness for these omissions, I hope that the following close reading might nevertheless bring its readership into a better understanding of alternative food projects vis-à-vis community building and empowerment.

History and Context: Oakland, CA

The pervasiveness of "food insecurity" in Oakland, California, is by now a well-documented and widely accepted fact. A great many of the residents living in its flatlands neighborhoods have few options for healthy, culturally appropriate, or affordable food. This already difficult set of circumstances is further intensified by other factors such as purchasing power and access to transportation (McClintock & Cooper, 2009). Geographer Nathan McClintock (2011, p. 89), in his discussion of the political economy of neighborhood development in Oakland, explains that:

> The socioeconomic terrain demarcating poverty and affluence in this Bay Area city of 423,000 (2010 estimate) roughly follows the contours of its physical geography of flatlands and hills. Census data reveals that the vast majority of Oakland's people of color live in the flatlands. Between a quarter and a third of people in the flatlands live below the poverty line; median income is 25 percent lower than the citywide average . . . It is precisely in these flatlands neighborhoods that the city's food deserts can also be found. And it is here that food justice movements have taken root.

While this brief observation falls a few brush strokes short of painting a comprehensive picture of the heterogeneous histories and contexts underwriting Oakland's present food crisis, it does begin to give us a sense of the reality many of its residents face on a daily basis. Furthermore, it delineates the dire set of circumstances out of which many of Oakland's community-based alternative food projects *necessarily* emerged.

In addition to the myriad of informal projects scattered about the city, recent years have seen organizations such as People's Grocery, City Slicker Farms, Mo' Better Food, Phat Beets Produce, and Planting Justice convert vacant lots and underutilized park land in West and North Oakland into productive landscapes in order to provide flatlands residents with fresh produce either via community-supported agriculture (CSA), cooperatively owned grocery stores, sliding-scale farm stands, or farmers' markets (McClintock, 2011). These efforts can be situated within a long history of struggles for food justice in the Bay Area that flows back through the Black Panther Party's survival programs to the destruction of traditional Ohlone foodways by the Spanish missionary complex and the subsequent annexation of California to the United States.

Keller Plaza is built atop these thick histories. Hemmed in by State Route 24 at the corner of 53rd and Telegraph Avenue, it is located just south of the Oakland/Berkeley boundary in Oakland's Temescal neighborhood. Not unlike the majority of Oakland's flatlands neighborhoods, unequal access to healthy, affordable food was brought upon the Temescal through protracted structural violence, the bulk of which has disproportionately impacted racial and ethnic minorities. A brief history of the development of the Temescal provides meaning in understanding the shifts of certain population's relationship to their food sources (Zandi, 2010).

The Temescal neighborhood initially developed during the latter part of the nineteenth century as a transit space connecting downtown Oakland and the University of California, Berkeley, by way of today's Telegraph Avenue. Like a great many other "in between" spaces, it was constituted through difference: not here, not there. In other words, that which was believed to be "unfit" for other neighborhoods was thus projected onto it. Because the sale of liquor was prohibited within a one-mile radius of the University campus and liquor licenses were heavily regulated in downtown Oakland, Temescal experienced a proliferation of drinking establishments, with as many as fifteen bars by the 1880s. This was the preamble to subsequent structural developments, racist redlining practices, and neoliberal institutional and organizational arrangements which colluded to produce a set of circumstances in which liquor became more accessible than affordable nutritious food (Zandi, 2010).

Diasporic flows play a significant role in shaping the cultural and physical geography of place. Different figurations of place, and the different conferrals of authorities and identities upon places, are shaped and transformed by complex and overlapping comings and goings. Neighborhood "character" can be here one day, gone the next. For the time being, the Temescal neighborhood is home to Oakland's highest concentration of Ethiopian and Eritrean immigrants. It is not uncommon to hear the area along the 51st block of Telegraph Avenue referred to as "Little Ethiopia" and a unique assemblage of restaurants, retail spaces, cultural centers, and places of worship serve as visual registers of the community's enduring presence in the neighborhood. Many of the estimated 20,000 Ethiopians and Eritreans living in the Bay Area immigrated in the early 1980s after the Refugee Act of 1980 was ratified to "assist [refugees] to achieve economic self-sufficiency as quickly as possible after arrival in the United States" (US Department of Health and Human Services, 1980). Violence associated with the Eritrean War of Independence, exacerbated by several periods of significant drought and famine, precipitated the migration of thousands. These stories are significant insofar as the majority of the residents presently living at Keller Plaza are Eritrean or Ethiopian. The remaining Keller Plaza community is comprised largely of African Americans and people of Latino or mixed descent.

Growing Agricultures/Agricultures Growing

The origins of the garden at Keller Plaza can be traced to a time of crisis for the community. On November 23, 2006, a Thanksgiving Day gathering at the apartment complex turned deadly. The ensuing triple-murder trial divided the Eritrean community, with the defendants' supporters sitting on the opposite side of the courtroom from the victims' relatives (Berton & Lee, 2011). Responding to a deep sense of fear and distrust within Keller Plaza, Christian Church Homes of Northern California, the private non-profit corporation responsible for managing the apartment complex, hired Planting Justice to organize the production of an abundant, nutritious, and culturally relevant community garden (Zandi, personal communication).

Planting Justice is a North Oakland-based non-profit organization dedicated to "empowering disenfranchised urban residents with the skills, resources, and inspiration to maximize food production, economic opportunities, and environmental beauty in their neighborhoods" (http://www.plantingjustice.org). It grew out of the work that Haleh Zandi and Gavin Raders accomplished as co-founders of the Backyard Food Project, a for-profit business they founded together in the summer of 2008. Similar to Planting Justice, Backyard Food Project sought to address a major disconnect and shortcoming of the alternative food movement and edible landscaping in general, which tend to include only those who can pay a high premium for these services, leaving most low-income and people of color behind (Zandi, 2010).

According to Haleh, Planting Justice is the first organization of its kind to combine ecological training and urban food production with a grassroots door-to-door organizing model. This allows Planting Justice to vastly increase their educational community outreach, helping to recruit volunteers, decentralize fundraising sources, and provide local jobs training for young community organizers. In Haleh's own words, "Community garden participants are a part of the tangible and replicable solution to the interconnected economic, ecological, and social crises facing our region and nation; crises that are most acutely felt by communities facing long term unemployment, violence, and incarceration" (personal communication, March 22, 2012).

In order to maximize community involvement at Keller Plaza, Haleh and Gavin went door-to-door throughout the apartment complex to hand out fliers and speak with residents about their feelings toward the garden. To the extent that it was possible, they also engaged residents in the design and layout of the garden as well as the various crops to be cultivated. Since breaking ground on February 1, 2010, the scope and scale of the project has grown considerably. At the time of this writing, the garden boasts thirteen raised beds, a rather eclectic herb spiral, a covered picnic area, a rainwater catchment system, composting, and an ADA (Americans with Disabilities Act) compliant path system for differently enabled residents. Further, weekly "culinary arts lessons," which feature simple recipes that can be made from the garden's harvests, extend the reach of the garden into the community's domestic experiences.

A brief inventory of the garden's abundant harvest provides a wonderful parallel for the diversity of individuals and influences collaborating in its production and maintenance: Kale and collards, inter-planted with onions and garlic, appear alongside tomatoes surrounded by basil. Tena Adam, a medicinal herb commonly used in traditional Ethiopian coffee ceremonies, cohabitates with chamomile, sage, basil, and lemon verbena. A rather dynamic actor in its own right, Tena Adam, with the scientific name *Ruta chalepensis*, is native to North Africa. Its arrival to the Americas can be traced back to Spanish conquistadores (McGown, 2006). Given the history of Spanish colonialism I referred to earlier, could Tena Adam be a repeat visitor to the Temescal? I am tempted here to track the multilayered histories brought together through this seemingly innocuous specimen, as I am sure they

Figure 17.1 The centrally located garden area: herb spiral, several raised beds, covered gathering spaces, and rainwater catchment barrels. Photograph courtesy of Planting Justice.

would lead to a better understanding of the ways in which humans and non-humans collaborate to weave complex multi-species histories. But, for now at least, I will keep with the task at hand.

In an email exchange, Haleh provided me with an anecdotal account of the day-to-day experiences of the garden:

> Older Ethiopian and Eritrean ladies will walk through the garden paths with their heads and hands to the sky saying what we think to be prayers. A young Latino boy eagerly asks us how long until the jalapeño peppers will be ready to harvest. Another older African–American woman sits and watches while we plant vegetables and herbs, often joining us when her body feels strong enough. The staff always expresses how well the garden looks and how exciting it is for the residents to participate in this sacred practice of growing food for the community. As participants learn about bio-intensive, organic gardening techniques, they are developing a collective responsibility for the health and happiness of their neighbors.

This story characterizes the garden as a place of performativity, produced and maintained through various activities that may or may not have much to do with the cultivation of food. It also shows how ethnically and generationally inflected prac-

Figure 17.2
Gavin Raders of Planting Justice working alongside a Keller Plaza resident. Photograph courtesy of Planting Justice.

tices play key roles in these performative acts. While the youth tend to take on the lion's share of the labor, elderly residents frequently stop in to say hello and exchange information about the plants in season. "At times," Haleh recalled, "it has proven difficult to engage both constituents in the garden at the same time." For a while, she dealt with this issue by having one workshop in the afternoon when youth could participate after school and one workshop in the morning so the elderly could participate without being disturbed by the youths' activities. Although due to time restrictions and resource constraints the workshops are now back down to one per week, they seem to be working out OK so far.

The garden also functions as a discursive space for participants to testify to their aspirations, their fears, their anxieties, and their grievances, peer to peer.[2] On one occasion, an elderly Habesha (of Ethiopian or Eritrean ethnicity) woman planted a herb that was particularly special to her. In time, it was found by the other gardeners—old and young alike—to be of no "practical use" to the community and, perhaps more notably, it proved to be a nuisance because its growth pattern took over the herb spiral to the detriment of its other inhabitants. "Whenever I pruned back this herb," Haleh said "she would become very upset." In the end, the herb had to be removed, although a portion was potted and given to the woman.

Now, my intention here is not simply to be yet another finely tuned voice in the growing chorus of community garden success stories. Rather, I endeavor to gather together a sufficient collection of details in order to support my advance into an exploration of transcultural exchange. To this end, it is important to emphasize that the aforementioned accounts, whether antagonistic or affiliative, are animated by reciprocity with regards to cultural differences and sustained by a commitment to equity in diversity. Whatever their outcome, these transcultural interactions and

Figure 17.3 Gavin (third from right) works with residents and volunteers to plant starts. Photograph courtesy of Planting Justice.

negotiations perfomatively produce the garden as a site of individual and collective meaning as relationships are forged between residents who might not otherwise be in conversation, collaboration, or contestation.

As individuals and communities collectively work to establish and maintain various urban agricultural projects, they learn organizing skills and develop social networks that have the potential to contribute to the production of a wider set of contexts. Locally situated identities thus enter zones of liminality or, more to the point, *already* liminal identities multiply and proliferate. Herein lies the possibility for Keller Plaza residents to transgress the existing material and conceptual boundaries circumscribing their community. Through their engagement in these spatial and relational practices, they create associations to a broader network of peers with whom the shared practice of food cultivation is the connective tissue.

The Production of Agricultural Subjects and Spaces

As a site of cultural engagement, Keller Plaza community garden contains the potential to engender entirely new transcultural forms, what I call *agricultures*. In contradistinction to the rather limited (and limiting) concepts of acculturation and assimilation, with tendencies toward the leveling of multicultural contours, the term *agricultures* designates emergent transcultural forms that are allied through systems of food-related values, knowledges, and practices. Further, *agricultures* are characterized by the development and sharing of skills, the establishment of alternative social networks, and the participation in a day-to-day politics of shared creative expressivity.

Sociologist John Brown Childs writes that "[the] interaction[s], developed through interpersonal relations of people engaged in common tasks, [are] producing working groups of activists whose roots are in communities and organizations, but who also form bridges among diverse peoples as they address substantial, albeit often varied corrosive dilemmas" (2003, p. 10). This begins to get at the heart of the emancipatory potential of Keller Plaza community garden. More than just a place to cultivate food, the garden is an intermediary space for productive engagement out of which might emerge a whole spectrum of politics of collective self-determination intrinsic to the development of a more empowered and politically potent food citizenship. Through their participation in the collective process of food production, the residents of Keller Plaza are not so much freed from their attachments to the dominant food system—though a normative analysis might have us believe this to be the case—as invited into a new choreography of partners: humans, vegetables, systems, artifacts, and so on; some of whom are not likely yet known but who may prove to be nonetheless indispensable to the constitution of the communities and cultures we might yet inhabit.

There is always a radical openness inhering in present cultural diversity, and the multitude of forms that might emerge as well as the effects they might have on one another can never be fully predicted in advance.[3] Individuals and communities endlessly modify existing urban patterns in meaningful, interesting, and, perhaps most notably, unexpected ways. Seen through this lens, the logic of normative planning and development models are profoundly destabilized and the notion of

Figure 17.4 Residents and volunteers enjoy a meal in the gathering space adjacent to the garden. Photograph courtesy of Planting Justice.

designing transcultural space is rendered strangely paradoxical. This is precisely because transcultural space occurs *as a result of* the negotiations between and among various cultures within the built environment. Transcultural space is activated through the overlapping of heterogeneous cultures. Or, to put it another way: transcultural spaces are produced between and among an ineradicable plurality of actors representing a multitude of diverse and oftentimes conflicting subject-positions, negotiating their differences and acting in concert. A mouthful, to be sure! But such an extravagant sentence seems necessary to call attention to the complexity of the creative and ongoing narratives of transcultural exchange.

Urban Agriculture as *The* Solution?

I began this chapter by invoking the struggle for food justice in the United States. In the space that followed, I attempted to illustrate, through the example of the Keller Plaza community garden, the ways in which politically empowered *agricultures* emerge out of locally situated spatial and relational politics of the everyday. Additionally, I have tried to forge alliances between community-based urban agriculture and transcultural exchange. But before going further, a last minute detour is necessary in order to call attention to several pernicious tendencies within the more mainstream branches of the alternative food movement. The reason for this digression is simple: I want to insist that the alternative food movement's ability to produce substantive change is indispensably linked to the extent to which it mobilizes against systemic injustices.

Although alternative food practices are widely understood to be inherently counter-hegemonic, Julie Guthman (2008b) has effectively argued that, in many cases, food politics has been at the leading edge of neoliberal regulatory transformations, given their emphasis on consumer choice, localism, entrepreneurialism, and self-improvement. From this vantage point, emphatic cries for "farming the city" should be met with a fair amount of skepticism insofar as they often put forward urban agriculture as the singular solution to urban food insecurity. So too should approaches that are ostensibly "bringing good food to others."[4] This is because the proliferation of food insecurities and injustices must be understood in relation to what Eric Holt-Giménez refers to as a "corporate food regime" that wreaks untold destruction on diverse cultural and ecological lifeways (Holt-Giménez & Wang 2011)—and that always disproportionately denies certain communities access to fresh, affordable, and healthy food (all the while inflicting catastrophic ecological and environmental impacts on all communities). More simply put, the current urban food crisis is a symptom of deeply entrenched structural inequalities which, try as they might, no amount of "fork voting," or urban farm stands, or slow(er) food practices are equipped to properly resist.

While the prevailing focus on food consumption is useful to the extent that it galvanizes a wide range of actors around a common cause (as well as satisfies actual existing needs), it often crowds out more radical analyses oriented toward dismantling structural inequalities. Many of the assessments offered up by the alternative food movement's most visible and outspoken figures, however initially appealing and sensible they may seem, tend to produce recommendations that more closely resemble hackneyed "pull yourselves up by the bootstraps" formulations than an emancipatory *agricultural* politics, properly so-called. To be clear, my point here is not to devalue the indispensable contributions that many urban food projects and practices have made in the struggle for food justice. Nor am I suggesting that urban residents should cease organizing, collaborating, agitating, and advocating—between and among cultures—for the continued growth and development of grassroots alternatives to the corporate food regime. Rather, I am insisting that struggles for food justice must, above all, be clear about the differences between superficial change and structural change (Holt-Giménez and Wang, 2011).

This seems to suggest that a different responsiveness and responsibility must emerge within the design and planning professions. One that will require reimagining these practices—which have historically and conventionally been figured as apolitical—in order to collectively ensure that struggles for food justice are being facilitated and supported from "the top" as well as from "the ground." Where community-led efforts to increase food access and build stronger food systems exist, agencies should use policy instruments to implement their recommendations (Treuhaft et al., 2009). Further, the myriad of organizations and initiatives that make up the grassroots must come to grips with the fact that they might be complicit—albeit unintentionally so—in the interpellation of a particular cultural lifeway, at the expense of transcultural *lifeways*. This is where the notion of acting as an ally takes on a special significance, informed by a spirit of reflexivity and a commitment to working *with* communities rather than for them. To the extent that these sorts of alliances are possible, what begins to emerge is an alternative transcultural food system—a food system for and by a diversity of *agricultures*.

Notes

1 Please refer to Julie Guthman's essay entitled "Bringing good food to others: investigating the subjects of alternative food practice" (2008a).
2 Dale Carrico has written elsewhere that "a peer is emphatically not an equal, nor an intimate, but one who registers in their alterity both their equity in respect to us and their diversity from us" (http://amormundi.blogspot.com/2009/05/more-on-freedom.html).
3 I should mention, however, that the *openness* that inheres in present diversity is always set against various forms of power, coercion, control, oppression, and so on. Therefore, to simply assume that, because of this openness, anything at all times is possible would be to ignore the material conditions circumscribing it.
4 This expression is borrowed from Julie Guthman's (2008a) essay of the same name.

References

Berton, J. & Lee, H. K. (2011, June 1). Oakland brothers guilty of triple murder. *San Francisco Chronicle*, p. C1.

Bhabha, Homi (2004). *The Location of Culture.* New York: Routledge.

Childs, John Brown (2003). *Transcommunality: From the Politics of Conversion to the Ethics of Respect*. Philadelphia, PA: Temple University Press.

Guthman, Julie (2008a). Bringing good food to others: Investigating the subjects of alternative food practice. *Cultural Geographies* 15: 431–448.

Guthman, Julie (2008b). Neoliberalism and the making of food politics in California. *Geoforum* 39 (3): 261–264.

Holt-Giménez, E. & Wang, Y. (2011). Reform or transformation? The pivotal role of food justice in the U.S. food movement. *Race/Ethnicity: Multidisciplinary Global Contexts* 5 (1): 83–102.

Hou, J., Johnson, J. M., & Lawson, L. J. (2009). *Greening Cities, Growing Communities: Learning from Seattle's Urban Community Gardens*. Seattle: University of Washington Press.

McClintock, Nathan (2011). From industrial garden to food desert: Demarcated devalution in the Flatlands of Oakland, California. In Alison Alkon and Julian Agyeman (Eds), *Cultivating Food Justice: Race, Class, and Sustainability*. Cambridge: MIT Press.

McClintock, N. & Cooper, J. (2009). *Cultivating the Commons: An Assessment of the Potential for Urban Agriculture on Oakland's Public Lands.* Oakland, CA: HOPE Collaborative/City Slicker Farms/Food First.

McGown, Jay (2006). *Out of Africa: Mysteries of Access and Benefit Sharing.* Edmunds, WA: The Edmonds Institute (in cooperation with African Centre for Biosafety).

Mares, T. M. & Peña, D. G. (2010). Urban agriculture in the making of insurgent spaces in Los Angeles and Seattle. In Jeffrey Hou (Ed.), *Insurgent Public Space: Guerrilla Urbanism and the Remaking of Contemporary Cities.* New York: Routledge.

Treuhaft, A., Hamm, M. J., & Litjens, C. (2009). *Healthy Food For All: Building Equitable and Sustainable Food Systems in Detroit and Oakland*. Retrieved November 1, 2010, from: http://www.policylink.org/atf/cf/%7B97C6D565-BB43-406D-A6D5-ECA3BBF35AF0%7D/Healthy%20Food%20For%20All-8-19-09-FINAL.pdf.

Tsing, Anna Lowenhaupt (2005). *Friction: An Ethnography of Global Connection*. Princeton, NJ: Princeton University Press.

US Department of Health and Human Services. (1980). Retrieved from the DHHS website in the Administration for Children & Families section: http://www.acf.hhs.gov/programs/orr/policy/refact1.htm

Zandi, Haleh (2010). Creating Food Security and Practicing Food Sovereignty: Links between Ecology, Cultural Diversity, and Human Rights in the Food Justice Movement of the East Bay, California. Unpublished manuscript.

PART V

STRUGGLES FOR TRANSCULTURAL CITIES

CHAPTER 18

What's Parks Got To Do With It?

Latino Children, Physical Activity, and the Parks System in Lancaster, Pennsylvania

Mallika Bose and Kirk Dimond

Public parks and open spaces are an integral part of modern urban life: from supporting recreational and psychological needs of urban residents to playing a role in youth development, workforce development, community capacity building, and public health (Walker, 2004). The role of urban parks is also significant in light of the increasing incidence of adult and childhood obesity in the United States.[1] Obese children suffer from a host of health problems including high blood pressure, high cholesterol, and Type 2 diabetes (Freedman et al., 2007) and childhood obesity is a predictor of adult obesity (Serdula et al., 1993). Obesity is concentrated in minority and disadvantaged communities (Kumanyika & Grier, 2006; Schoenborn et al., 2002), especially African–American and Latino communities (Day, 2006). At the same time different ethno-cultural groups exhibit different motivations and preference for leisure and recreational activities (Floyd, 2001; Floyd et al., 1994). The fact that minority low-income inner city residents live in a built environment characterized by well-connected networks of streets with sidewalks, higher densities, and access to public transportation—all elements that are supposed to promote active living—and yet suffer from high rates of obesity and obesity-related health problems points to our incomplete understanding of how the built environment is linked to active living of minority groups living in urban contexts.

The relatively few studies that have examined the impact of parks on urban minority communities have focused on large multicultural cities like Chicago (Gobster, 2002) or Los Angeles (Loukaitou-Sideris & Stieglitz, 2002; Wolch et al., 2005). Meanwhile, there are minority populations in many smaller cities located across the United States—a trend that is likely to continue, if not grow—that have fewer resources and capacity to deal with the needs of their immigrant and minority communities relative to larger urban centers. To address this gap, this exploratory study uses Lancaster, Pennsylvania, as a case study to examine how parks are responding to the physical activity needs of minority groups in small towns in the United States. In doing so, this study utilizes the framework of transcultural place-making which explicitly recognizes the dynamic nature of ethno-cultural affiliations (Hollinger, 1995), the legitimate rights of different groups in society, and their right to difference (Sandercock, 2003).

Lancaster, a city of approximately 59,000 residents (US Census Bureau, 2010), is located between Harrisburg and Philadelphia. About 40 percent of its residents are Latino, with the majority of them being of Puerto Rican origin (74.3 percent). The earlier Latinos in Lancaster were mostly Puerto Ricans and they remain the largest Latino group. However in the past fifteen years, the Latino community in Lancaster has diversified to include Dominicans, Mexicans, Cubans, and Peruvians, among others. The early Puerto Ricans living in the city were mainly employed in poultry processing plants, tobacco warehouses, or foundry work (Escobar-Haskins & Haskins, 2007). Even in the 1980s when the Latino population constituted a significant proportion of the residents of Lancaster, they were not adequately represented in the fire department, police department, or the school district, nor did they enjoy political clout. Since the 1990s the Latino population, especially Puerto Rican, has begun serving on governing bodies like the city council and the school district. Even though the Latino community has made strides in gaining visibility, they still face challenges related to education, housing, and under-employment. Like many other cities, Lancaster has suffered from the problem of "White flight" from the city to the surrounding suburban areas, leading to a concentration of Latinos in the city (Escobar-Haskins & Haskins, 2007).

This chapter continues by situating the study in the context of the evolution of parks in the USA and the relationship between parks and physical activity. Next, the findings are presented through three themes: (1) access to parks by the Latino population in Lancaster, (2) role of parks in meeting the physical activity needs of Latino children in Lancaster, and (3) the parks planning process in Lancaster. The chapter concludes with a discussion of the role of parks in meeting the needs of Latino children in the context of equity, environmental justice, and transcultural placemaking in Lancaster.

The Role of Parks in Society

Parks, in their many forms, serve a multitude of purposes in American culture. In the United States, parks became an integral part of society starting in the mid-nineteenth century with New York's Central Park as the first major project that successfully addressed many social issues of the times. Different time periods brought new goals and styles to parks, reflecting the social power, worldviews, and economics of each era (Cranz, 1982). *Pleasure grounds* (1850–1900) were built to accommodate strolling, carriage racing, bike riding, picnics, rowing, classical music, and non-didactic education. Despite being intended for use by all city residents, due to their location at city edges they primarily served the upper middle class. The *reform park* (1900–1930) emerged in response to the inequality of access to the pleasure grounds. These small, rectilinear parks emerged on city blocks, serving children, immigrants, and the working class for activities such as supervised play, gymnastics, crafts, Americanization classes, dancing, plays and pageants. The *recreational facility* (1930–1965) was built in both urban and suburban locations to allow for more active recreation with the suburban family being the main beneficiary. The *open space* system (1965–1982) dismantled the idea of standardized controlled locations for park activities and proposed that recreation could take place anywhere. Park design became site specific and conceptualized as part of a system of parks and open spaces that benefited residents, workers, poor urban youth, and the middle class. Recently, Cranz and Boland (2003, 2004) proposed a new observed park type: the *sustainable* or *ecological park*, designed to benefit residents, wildlife, cities, and the planet, that includes a variety of spaces, ranging in size and order, consisting of native plants, permeable surfaces, ecological restoration, green infrastructure, and exhibiting resource self-sufficiency.

With the emergence of the ecological/sustainable park type, Cranz and Boland note a change in park design motives suggesting an "overt concern for ecological fitness" as a new phenomenon in park purpose (2003, p. 45). While they describe the potential for the new park type to develop into a more socially inclusive park than previous types, social sustainability is not its focus. For Latino cultures, parks are seen as social spaces (Low et al., 2005; Loukaitou-Sideris, 1995), as sites for celebrating events like birthdays, reunions, anniversaries, and picnics involving large groups of people with hammocks, card-tables, grills, flags, boom boxes, and other items brought from home to "claim the territory, mark the turf, and in some way privatize the public space" (Loukaitou-Sideris, 1995, p. 94). Such activities are not associated with ecological objectives of protecting wildlife, neither are they considered to be health promoting.

We argue that resources for ecological parks are disproportionately allocated for solitary relaxation and individualistic physical activities like jogging, trail walking and biking, activities desired primarily by white middle-class users, rather than accommodating the needs of minority groups like Latinos in urban areas (Low et al., 2005). This may result in conflict between park patrons and managers as minority users attempt to modify/adapt their park spaces to suit their needs, especially when such uses are delegitimized. To overcome this problem, Low et al. (2005) suggest a potential sixth park type, a "*culturally diverse park*," that combines features of the ecological park with the needs of different groups in the transcultural city.

Parks and Physical Activity

Research indicates that the manner in which park facilities are used differs across race and ethnicity (Floyd, 2001; Wolch et al., 2005). Even though picnicking and group interactions are popular park activities in the Latino community (Baas et al., 1993; Loukaitou-Sideris, 1995),[2] physical exercise remains an important reason for park visitation for ethnic groups (Sasidharan et al., 2005). Since Latino youth and adults report lower levels of leisure-time physical activity than their white counterparts (Day, 2006), and park visitation is linked to higher rates of physical activity, it is important to examine the role of parks in meeting the physical activity needs of Latino youth.

A recent review of qualitative studies (McCormack et al., 2010) examining urban parks and physical activity indicates that a variety of recreational opportunities (both passive and active), age-appropriate play equipment, ball fields, facilities for picnicking, natural areas, and amenities like water fountains and clean restrooms are factors that promote park use. Minorities are more likely to use urban community parks relative to regional parks that are usually located in the outskirts of cities and metropolitan areas (Loukaitou-Sideris & Stieglitz, 2002). These urban parks are typically used more intensively while suburban parks are used primarily on the weekend and for organized sports activities (Loukaitou-Sideris & Stieglitz, 2002). At the same time ease of access to parks as well as fees for services pose barriers to park use by African–Americans and Latinos (Byrne & Wolch, 2009).

In a time when increasing rates of obesity in adults and children, especially in minority communities, is a major public health concern (having significant cost implications), the use of parks for physical activity becomes a key social and economic matter. This study examines how parks can respond to the health and physical activity needs of the Latino community in Lancaster by examining access to parks, patterns of park use for physical activity by Latino youth, and the parks planning process in Lancaster.

Examining Access to Parks in Lancaster

The parks system in Lancaster City consists of a regional park, 2 community parks,[3] 11 neighborhood parks, 10 mini parks, and 6 other public open spaces (Figure 18.1). Twenty school sites also play some role in providing important open spaces serving the neighborhoods (Thomas Comitta Associates & Arro Consulting, 2009).

Is there any difference in access to parks in Lancaster City (with a high proportion of Latinos) relative to the surrounding suburban areas (with a much lower Latino population)? What is the relationship between access to parks and minority status of residents? To answer these questions we compare the park area per capita of Lancaster City with the surrounding suburbanized area comprising of the twelve

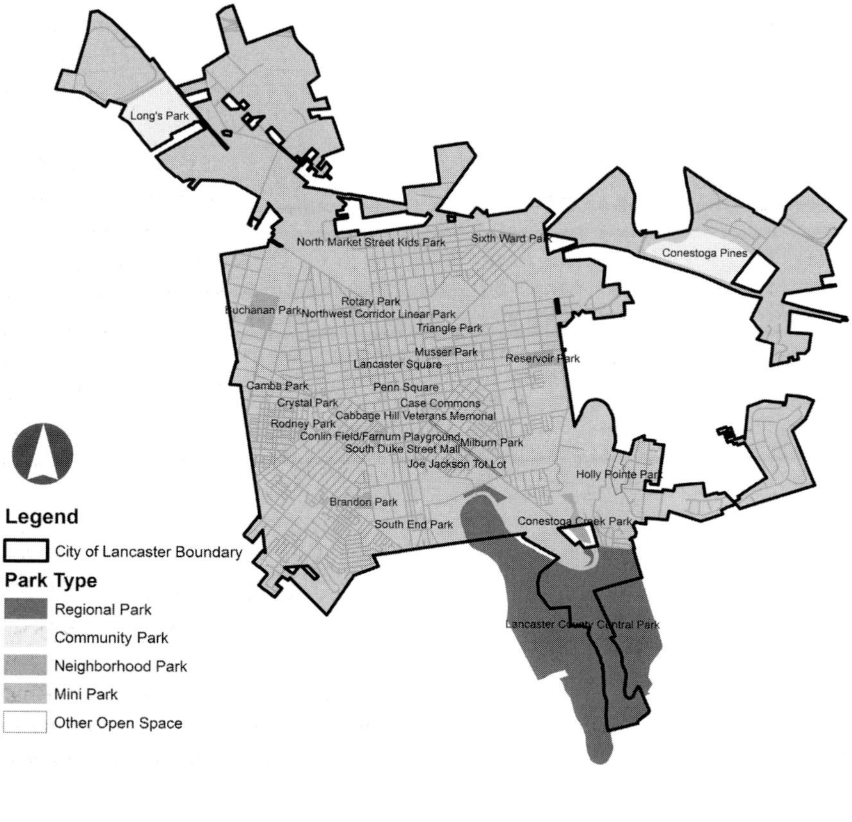

Figure 18.1 Parks in the City of Lancaster.

municipalities of the Lancaster Inter Municipal Committee or LIMC.[4] Table 18.1 reveals the discrepancy between the park acreage available to the residents of the city and of the outlying suburban areas: 2.20 acres/1000 people in the city versus 14.43 acres/1000 people in the LIMC area. Thus Lancaster City fails to meet the park standards recommended by the National Recreation and Park Association (NRPA) by a large margin while the LIMC municipalities enjoy adequate park area.[5] When you exclude regional parks and other open spaces (that usually consist of undeveloped land) the park area available to residents of the suburban zone is still within the recommended park standard (9.08 acres/1000 people) while that of the city is even lower (1.14 acres/1000 people). When we examine the availability of parks by census tracts for the LIMC area (Figure 18.2), we see that there is some variation by census tracts within townships/boroughs of the suburban zone, with the majority of census tracts enjoying sufficient access to parks. In contrast, within the city, all tracts have insufficient park acreage.

At the next level of analysis we utilize the lens of ethnicity to examine the distribution of parks within the City of Lancaster. We group census tracts in the central city based on the percentage of Latino population (Group I < 33 percent, Group II (33–66) percent, and Group III > 66 percent) and then examine park acreage and number of park facilities available within them (Figure 18.3 and Table 18.2).[6] Table 18.2 indicates that in terms of mini and neighborhood parks, the park acreage in Groups III (0.52 acres/1000 people) and II (0.99 acres/1000 people) census tracts is well below the recommended park standard of (1.25–2.50) acres per 1000 persons, while that for Group I (1.43 acres/1000 people) lies within the recommended value. Thus census tracts with a higher proportion of Latinos have poor access to parks in comparison to census tracts with a lower proportion of Latinos.[7]

Table 18.1 Comparison of park acreage in Lancaster City and surrounding suburban (LIMC) area

	Population	Park area in acres	Population/ park area in acres	Park acres/ 1000 population	NRPA standards
LIMC tracts without Lancaster City	178925	2581.73 (1625.53)*	69.30 (110.07)*	14.43 (9.08)*	6.25–10.5 acres per 1000 persons for mini + neighborhood + community parks
City of Lancaster tracts[1]	56080	123.58 (63.69)*	453.80 (880.51)*	2.20 (1.14)*	
LIMC total	235005	2705.31 (1689.22)*	86.87 (139.12)*	11.51 (7.19)*	

Source: Based on 2010 Census data.

Notes

1 In computing the parks area for the City of Lancaster we use the census tracts (1–12, 14, and 147) that are part of the original City of Lancaster and do not include the outlying annexed areas which are more representative of the surrounding suburban areas in terms of settlement patterns and density.

* The numbers in parentheses refer to only mini, neighborhood, and community parks, while the first number includes regional parks and other open spaces in addition to the three park types.

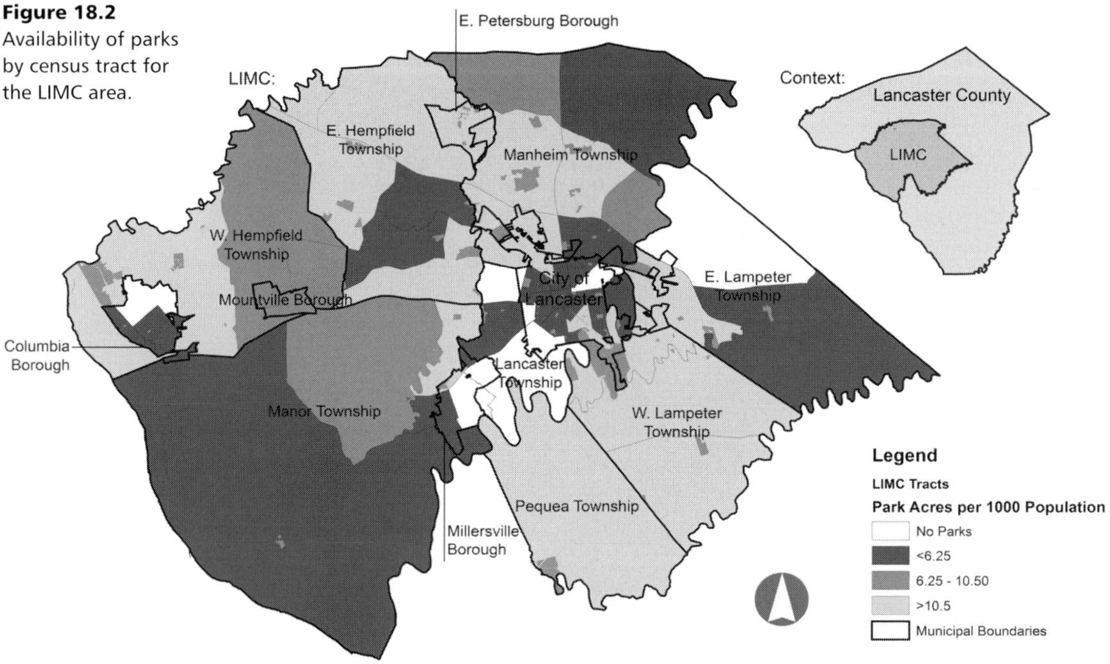

Figure 18.2 Availability of parks by census tract for the LIMC area.

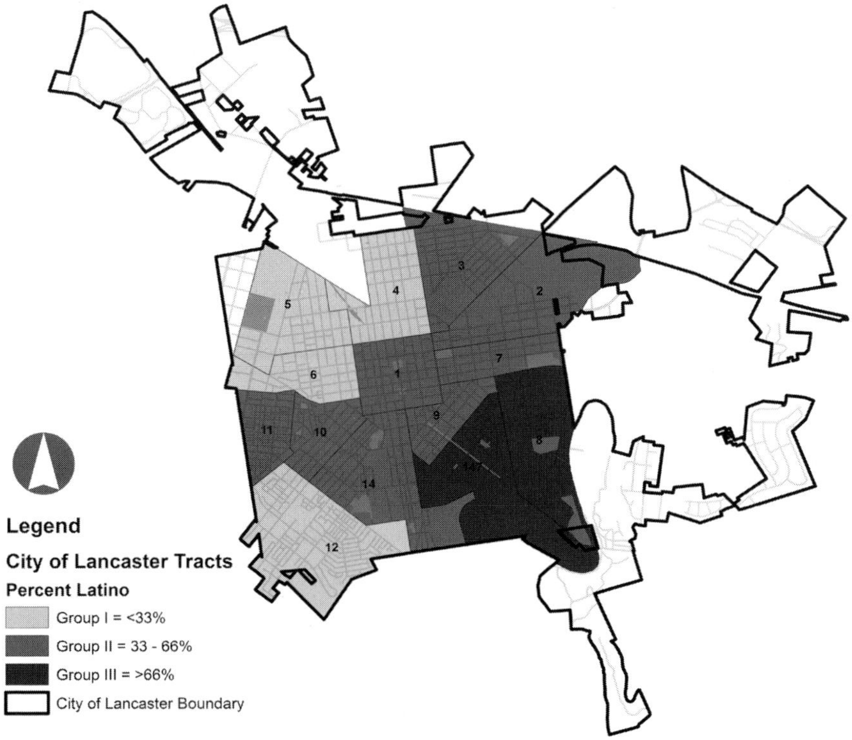

Figure 18.3 Census tracts in Lancaster City grouped by Latino population.

Table 18.2 Comparison of park area and facilities by selected census tracts

Census Tract Groups	Mini Parks – count/acres	Neighborhood Parks – count/acres	Community Parks –count/acres	Other Open Spaces – count/acres	Regional Parks – count/acres	Total Parks – count/acres	Number of Facilities	Total Population	Community Park (acres/ 1000 population)	Mini + Neighborhood Park (acres/1000 population)	Population/Facility
Group III – Tracts with > 66% Latino (Tracts 8, 147)	2/1.35	1/2.9*	0	1/2.63	1/46.15	5/53.03	14	8240	0	0.52	588.57
Group II – Tracts with 33–66% Latino (Tracts 1, 2, 3, 7, 9, 10, 11, 14)	6/2.94	7/27.78	0	1/1.25	0	14/31.97	64.5	30916	0	0.99	479.32
Group I – Tracts with < 33% Latino (Tracts 4, 5, 6, 12)	3/0.32	2/23.92	0	1/0.09	0	6/24.33	24	16924	0	1.43	705.17

Source: Based on 2010 Census data.

Notes
* Since Conestoga Creek Park located in census tract 8 is an undeveloped wooded site it has not been included in the computation of park area.
NRPA Park standards: Mini + Neighborhood Parks = 1.25–2.50 acres/1000 persons; Community Parks: 5–8 acres/1000 people.

When we look at the number of facilities, we find that the population per facility is highest for Group I census tracts (with the lowest proportion of Latinos), followed by Group III and then Group II. This is related to the type of parks that are located within each of the census tracts. Generally speaking, community parks have the greatest number of facilities, followed by neighborhood parks. Since there are no community parks in the census tracts that comprise the central city, it is the location of neighborhood parks per capita that is associated with the per capita availability of facilities: 2 in Group I, 7 in Group II, and 1 in Group III. Thus even though Group III census tracts (with the highest proportion of Latino population) have the lowest park area per capita (0.52 acres/1000 population), they have access to more facilities per capita than Group I census tracts (with lowest proportion of Latinos).

This analysis of park accessibility by area and facilities indicates that within the city, census tracts with a higher Latino population do seem to have lower than desirable park area and facilities.[8] It is also interesting to note that both community parks (Long's Park and Conestoga Pines Park) are located in census tracts in recently annexed areas of the city on the periphery—problematizing access by minority communities concentrated in the southeast section and central area of the city and likely to be dependent on public transportation, or walking/biking. When we move our focus from park area to the facilities in the parks we find that census tracts with a higher number of neighborhood parks per capita have greater access to park facilities. As a result, residents of Group I census tracts, in spite of enjoying the highest per capita park acreage, have lower number of park facilities per capita relative to Groups II and III census tract residents.

Exploring the Context of Physical Activity of Latino Children

The method of photo-voice (Strack et al., 2004; Wang & Burris, 1997) was used to collect information regarding the spatial context of physical activity from the perspective of Latino children. Fourteen Latino children (ages 6–10) were recruited from two Boys and Girls Clubs in the City of Lancaster. Each child was given a disposable camera and asked to take photographs of places that they associated with their physical activity. No attempt was made to define physical activity. The photographs were then developed/printed and used for a de-briefing session with the participating children to understand how physical activity was conceptualized by the participants, and where such activities occurred and with whom. The focus of the interviews was on understanding the context of physical activity from the participating children's point of view.

The photo-voice method resulted in 184 photographs, 144 of which represented outside environments (78.26 percent) while the remaining 40 (21.74 percent) denoted indoor environments (Table 18.3). Parks represented the lion's share of the outdoor photographs at 45.14 percent (Figure 18.4) followed by amenities at or near home (25 percent) and infrastructure elements (24.31 percent) like streets, alleys, and parking lots (Figure 18.5). The Boys and Girls Club featured prominently in the indoor photographs (57.5 percent); followed by playing with toys at home (12.5 percent) and referents (12.5 percent) like bikes, skateboards, and scooters located indoors but referring to outdoor activities. Thus physical activity in the lives

Table 18.3 Details of photographs associated with physical activity

	Average (%)	Min (%)	Max (%)	Comments
Outside	78.26%	21.43	100	
Infrastructure	24.31%	0%	50%	Parking lots (near home, church and parks visited), streets, alleys
Amenities in and around home	25%	0%	71%	Stoop, backyard, patio, front yard, neighbor's yard, friend's yard, ice-cream shop
Parks[1]	45.14%	0%	90.91%	Central Lancaster Park, Conlin Field/Farnum Park, Musser Park, Brandon Park, NW Corridor Linear Park, Crystal Park
Other	5.55%	0%	15.00%	Blue sky, trees, referents like skateboard, bike, and basketball
Inside	21.74	0%	78.57	
Boys & Girls Club	57.5%	0%	100%	Basketball, foosball, pool, air hockey, board games, reading books, running around, kickball, freeze tag, art projects, computer games
Church	7.5%	0%	100%	Playing in recreation area of church
Referent	12.5%	0%	50%	Bike, skateboard, scooter
Playing with toys	12.5%	0%	42.86%	At home
TV/Video games	7.5%	0%	50%	Sedentary activities at home
Weights	2.5%	0%	50%	At home

Note
1 The two participants who did not have photographs of parks mentioned that they visited parks, but could not take photographs because the days were getting shorter and they could not get around to the park during daylight hours.

of Latino children of this study overwhelmingly occurs in outdoor places, with parks playing a prominent role. However in addition to parks, many other everyday environments like parking lots, alleys, and streets/sidewalks also support the physical activity of Latino children. We call these places interstitial places and hybrid places, in the sense that they are developed to accommodate the transportation infrastructure but adapted to fulfill the physical activity needs of the Latino children of Lancaster City.

All except one of the parks represented in the photographs were neighborhood parks. In the case of the one child who frequented the Regional Park (Central Lancaster Park), it was his closest park, and connected to his home by a trail. About 80 percent of the participating children frequented parks that were closest to their homes; the remaining 20 percent frequented parks that were near the Boys and Girls Club that they visited for after school activities. Ten of the fourteen children (or 71 percent) that participated in this study frequented parks within a half-mile network distance from their homes. The remaining four children frequented parks that were located just outside the half-mile network distance from their homes. In all instances, the children walked to parks—by themselves, or accompanied by

Figure 18.4 Parks frequented by study participants.

Figure 18.5 Interstitial spaces used for physical activity by study participants.

siblings, relatives, parents, or friends. The preponderance of neighborhood parks photographs can be construed to indicate that neighborhood parks are an important setting for physical activity for Latino children. The fact that almost all the children used the closest park, and walking was the mode of transportation, points to the importance of parks within easy walking distance.

Analysis of the photographs and interviews reveal that physical activity is interpreted as a variety of sports-related and other activities, the majority of which are located outdoors. Basketball, baseball, football, and soccer were popular sports activities that were recognized as physical activity. However, not a single participant engaged in playing any of these games through an organized league; rather, they played informally with friends and family at the park, in the streets, and in their front yards/backyards. Running around in the backyard, playing catch and tag in the backyard and streets, riding a bike or scooter on a hilly bank or parking area were also popular activities. Only one participant frequented a skate park, even though it was outside the half-mile walking zone around his home. However, a trail connected the skate park to his home. Climbing trees, digging holes, running in fields, and using play equipment at play areas (slides, see-saws, swings, monkey bars) featured prominently in the photographs. In describing these activities a common refrain was that it was fun and exciting.

Understandably, indoor physical activities centered around the Boys and Girls Club—basketball in the indoor gym, foosball, pool, and board games. A few kids took photographs of sedentary activities like watching TV and playing video games, even though the directions asked for photographs related to physical activity. Some children played in the recreation area in their church before and after service. Interestingly, trees and the blue sky were also represented in the photographs. When queried about it, the participants responded that they liked to look at trees and the sky. They enjoyed the wide expanse of grassy fields and looking for clouds of different shapes in the sky. In most cases the participants engaged in physical activity with others: friends, siblings, or even other kids that they met at the park. Over and over again the participants characterized their physical activity as fun. Thus the notion of physical activity was associated with nature, play, excitement, and socialization.

Park Planning in the City of Lancaster

Parks and recreation planning in Lancaster is a multi-level, multi-agency endeavor. City agencies playing an important role in parks and recreation planning include the City's Department of Public Works, Administrative Services, Public Safety, Mayor's Office of Special Events (MOOSE), Economic Development and Neighborhood Revitalization, and the Lancaster Recreation Commission.[9] The Lancaster County Department of Parks and Recreation coordinates park and recreation facilities on the county level while the Lancaster School District serves local students. The Spanish American Civic Association (SACA) is a prominent agency representing the Latino community including their park and recreation needs. The city also supports two Boys and Girls Clubhouses that serve indoor recreation needs of youth in the community.

Several planning documents at the scale of the city, LIMC, and county tackle issues related to parks and recreation planning (see Table 18.4). The city level documents (Thomas Comitta Associates & Arro Consulting, 2007a, 2007b, 2009) focus on assessment of and improvements to parks in the city (Urban Park, Recreation and Open Space Plan, 2009) and specific neighborhoods (Northeast Revitalization Initiative for the City of Lancaster, Pennsylvania, 2007 and Southwest Revitalization Initiative for the City of Lancaster, Pennsylvania, 2007), while the central county (or LIMC) and county level planning documents deal with comprehensive planning which includes parks and recreation planning. The only exception is the Greenscapes report (2009) which focuses on green infrastructure planning at the county level.

Review of the documents reveals that parks and recreation planning is conceptualized as a rational process (Sandercock, 2003), with the focus on the development of parks in the likeness of the open space system (Cranz, 1982) centered on improvement of existing parks and creation of a network to facilitate accessibility and use. Discrepancies were also observed between the reports. For example, according to the Growing Together report (ACP-Visioning & Planning, LLC, 2006, page 10.4), the City of Lancaster has adequate community park area (30.0 acres), a surplus (88.9 acres) of neighborhood parks, and a deficit (4.5 acres) of mini parks using population estimates for 2030. On the other hand, the more recent Urban Park, Recreation and Open Space Plan (2009) states that the city lacks adequate community park acreage but has sufficient mini and neighborhood park area.[10] It is important to note that the 2009 report does point out the existence of under-served areas within the city. This strategy of improvement, in spite of improving the overall park system, will not alter the existing inequities of park availability and accessibility, primarily experienced by the minority population living in Lancaster City.

According to the reports, a variety of public participation methods have been used in the parks and recreation planning process in Lancaster (see Table 18.4). The methods employed differed from project to project and as a general trend the more recent projects involved some effort to reach out to minority groups, including Lancaster's sizable Latino population. However there is no discussion of specific methods used to reach out and encourage participation from Latinos in Lancaster City. Lancaster's Urban Park, Recreation and Open Space Plan process included a survey that was provided in both English and Spanish, but few Spanish language surveys were completed, and no reasons offered for the poor participation by the Latino community. Similarly the Growing Together (ACP-Visioning & Planning, LLC, 2006) plan acknowledges the rapid growth of the Latino population, and incorporated focus group sessions with Latino and other minority groups. The report, however, does not discuss the extent of participation by the Latino community, the specific ideas generated from these focus groups, or how these ideas were incorporated into the parks and recreation planning process. Such an approach suggests that the needs of the Latino community are not considered to be different from the majority (white) community. This is evidence of parks planning and management practices conceptualized in terms of universal guidelines and principles that are ultimately based on the needs of the majority group in society.

Table 18.4 Public participation methods used for parks and recreation planning efforts in Lancaster, PA

Document/Report	Public participation methods used	Comments
Lancaster County Comprehensive Plan: Policy Plan. 1999.	• Newspaper survey • Regional meetings • Community forum • Citizen questionnaire • United Way Focus Group meetings	*Focus*: vision and goals of Lancaster County Comprehensive Plan, including key focus areas *Scale*: Lancaster County
Balance: The Growth Management Element. 2006	• Taskforce of 60 people • Public meeting series; community forum • Media coverage • Surveys (telephone) • Lancaster County Planning Commission website/newsletter • Special forums with municipalities and municipal associations	*Focus*: growth management *Scale*: Lancaster County
Greenscapes: The Green Infrastructure Element. 2009.	• Taskforce consisting of representatives of County departments, other organizations, businesses, citizen groups with an interest in green infrastructure • Public workshop • Focus group with key green infrastructure stakeholders (rural/urban green infrastructure, parks and recreation, taskforce members)	*Focus*: green infrastructure *Scale*: Lancaster County
Growing Together: A Comprehensive Plan for Central Lancaster County, Pennsylvania. 2006.[a]	• Phase I: – Kickoff meeting – Stakeholder workshop – Public meetings; special focus groups with High School/College students, African–American and Latino individuals – Community visioning summit consisting of presentations, small group work, and surveys • Phase II: 3 public forums	*Focus*: comprehensive planning *Scale*: Lancaster Inter-Municipal Committee (LIMC) comprising of 11 municipalities in Central Lancaster County.

Table 18.4 Continued

Document/Report	Public participation methods used	Comments
Urban Park, Recreation and Open Space Plan for the City of Lancaster. 2009.[b]	• General opinion surveys • Public meetings • Site-specific opinion surveys • Plan reviews through websites (City, School District, Lancaster Recreation Commission and others), newsletter	*Focus*: parks and recreation planning *Scale*: Lancaster City
Northeast Revitalization Initiative for the City of Lancaster, Pennsylvania (NERI). 2007.[c] *Southwest Revitalization Initiative for the City of Lancaster, Pennsylvania* (SWRI). 2007.[d]	• Revitalization advisory committee • Hiring of neighborhood resident as planner • Neighborhood workshops (2 in NERI and 4 in SWRI)	*Focus*: physical and strategic vision plan; neighborhood revitalization plan *Scale*: specific neighborhoods; NERI: Musser Park, East Side, Ross portions of Lancaster City; SWRI: The West End, Cabbage Hill, Prospect Heights

Notes

a ACP-Visioning & Planning, LLC (2006).

b Thomas Comitta Associates & Arro Consulting (2009).

c Thomas Comitta Associates & Arro Consulting (2007a).

d Thomas Comitta Associates & Arro Consulting (2007b).

Physical Activity and Parks Systems Planning in the Transcultural City

Analysis of the location and distribution of parks in the LIMC area indicates that park area per capita is much lower in the city relative to the surrounding suburban zone. Since the majority of the Latino population is concentrated in the city (Latino population in the city is 39.3 percent versus 7.65 percent in the surrounding twelve townships/boroughs), and the city supports a larger proportion of low-income families (21.7 percent of families below the poverty level in Lancaster City versus 5.9 percent in the surrounding twelve townships/boroughs)—this represents an environmental justice issue that warrants closer scrutiny. Within the city itself, there is evidence to suggest that access to parks is poor in neighborhoods with a higher concentration of Latinos. From an equity standpoint the per capita park area within the city should be higher than that in the surrounding suburban areas since low-income families are likely to experience restricted mobility, suffer higher risk of obesity-related diseases, are less likely to use fee-based private facilities, and live in denser urban areas without large front yards/backyards that provide adequate play spaces.

Analysis of the photo-voice material reveals the "lived spaces" (Lefebvre, 1991) of physical activity and play of Latino children and attests to the important role of outdoor environments, especially neighborhood parks, in meeting their physical activity needs. Furthermore, the children utilize a variety of ordinary places/landscapes (Jackson, 1994) to engage in play and physical activity. Streets, parking lots, and alleys, all part of the transportation infrastructure of the city, are adopted/adapted for use as play areas and sites for physical activity. This appropriation and adaptation of ordinary landscapes/places for supporting play and physical activity of Latino youth in the city (Rios, 2009) can be characterized as an example of hybrid placemaking (AlSayyad, 2001). This study, by examining *how* and *where* Latino children engage in play and physical activity, provides a window into the spatial practices of Latino children. However, it is unable to inform us as to *why* Latino children appropriate/adapt interstitial spaces for their play. Future work needs to explore this avenue of research.

Review of parks and recreation planning documents indicates that (parks) planning in Lancaster is conceived of as a rational utopian enterprise (Holston, 1999) with the intent of maximizing the benefits of the park system for the average citizen of Lancaster. Unfortunately, this average citizen is modeled on the normative characteristics of a resident from the dominant (majority) group in society. A parks planning process that responds to the needs of such a generic citizen is unlikely to meet the needs of the Latino community. Instead, what is needed is an *appreciative* model of planning (Ameyaw, 2000) that brings professionals and citizens together to cooperate and develop their vision of the city despite their cultural differences. Such an approach would include:

- A conceptualization of the planner as a mediator between different groups in society (versus an expert/technocrat) with a focus on deliberation and negotiation between different groups in the transcultural city (Ameyaw, 2000; Burayidi, 2000; Sandercock, 2003).

- Recognition of the "right to the city," such that different ethno-cultural groups in the city are empowered to rally for their vision of the city without having to bracket their needs/vision to that of dominant groups in society (Lefebvre, 1991). In the Lefebvreian (1991) sense, the term refers to the radical restructuring of the power relations implicated in the production of urban space, such that control shifts from capital and/or the state to those that live in urban areas/cities (Purcell, 2002). In other words, urban inhabitants by virtue of living in the city (not by virtue of national citizenship) are endowed with rights, including the right to participation and appropriation (Purcell, 2002). Participation refers to the direct participation of urban inhabitants in the processes that lead to the production of urban space in their city, while appropriation is the right of inhabitants to physically occupy urban spaces and actively work towards the development of urban spaces that meet their needs.
- Recognition of the legitimacy of cognitive *and* affective modes of knowledge acquisition, leading to the use of alternative modes of information gathering (like storytelling, use of talking circles, informal conversations) in addition to surveys, meetings, and focus groups during the (parks) planning process (Burayidi, 2000; Sandercock, 2003).
- Recognition of the right to difference and acknowledgment of the needs and legitimate rights of minority groups in society (Sandercock, 2003). This can only work with an empowered citizenry and in a climate where minorities and other disadvantaged groups feel that they are respected and that there is a political commitment to addressing their needs.

In the context of parks planning and management in Lancaster, such a process would demand meaningful participation of the Latino community in all matters related to the planning, development, and management of parks and parks programs. It would not seek to create a shared sense of place between the Latino and Anglo community; rather it would recognize and respect the rights of both groups and engage in a process of negotiation to accommodate their diverse needs through an engaged parks planning process.

Acknowledgement

The authors would like to thank Sharon Taverno for collecting the photo-voice information.

Notes

1 CDC website: http://www.cdc.gov/obesity/childhood/ (accessed on February 20, 2011).

2 We recognize that there are differences in sub-groups within Hispanic communities. For example a study (Gobster & Delgado, 1993) found that Puerto Ricans didn't play soccer and picnicked less, while engaging in basketball and swimming more often than other Hispanics.

3 Both community parks are located in the periphery in the recently annexed areas of the City.

4 We used GIS maps obtained from Lancaster County to compute the area of parks within selected census tracts in the city and the surrounding LIMC municipalities. We then used Census 2010 data to compute park area per 1000 population for Lancaster City and the twelve surrounding municipalities. In computing these figures we only included park areas located within the city boundaries, so for example only those parts of Lancaster Central Park (a regional park) within the City boundary are included and not the entire 544 acres.

5 The commonly used parks standards are derived from the NRPA Guidelines authored by Fogg (1981, 1990), though there has been a major rethinking of the philosophy of parks guidelines (Mertes & Hall, 1995) since then.

6 A small section of the city at the northwest corner is excluded from this analysis since it forms a small portion of census tract 11805, and it has a density and settlement pattern that is more akin to the suburban areas surrounding the city of Lancaster.

7 This is not to suggest that all census tracts within each of the three groups are homogeneous in terms of access to parks. Within each of the three groups, there are census tracts that do not have any parks at all: census tracts 6 and 12 in Group I, census tracts 2 and 11 in Group II, and census tract 8 in Group III.

8 The issue of access is mediated by factors like available transportation choices and ability to use fee-based recreational facilities. A more nuanced analysis of access to parks should take into account such factors.

9 The Lancaster Recreation Commission is a not-for-profit organization. It is a community partnership of the City of Lancaster, School District of Lancaster, and Lancaster Township. The Commission provides over 400 neighborhood-based recreation and learning opportunities for all ages—with an emphasis on affordable programs for children.

10 Also the calculations of park area within the original city boundaries (that is not taking into account the recently annexed areas on the outskirts of the city) reported earlier in the chapter contradict the report findings, especially for the residents in the center city. However since the 2006 report does not clearly indicate the method of computing parks area, and we did not talk to the authors of the report, we can only speculate about the factors contributing to this discrepancy.

References

ACP-Visioning & Planning, LLC. (2006). *Growing Together: A comprehensive plan for Central Lancaster County, Pennsylvania*. Lancaster, PA: Lancaster County.

AlSayyad, N. (2001). Hybrid culture/hybrid urbanism: Pandora's box of the "Third Place." In N. AlSayyad (Ed.), *Hybrid Urbanism: On the Identity of Discourse and the Built Environment*. Westport, CT: Praeger.

Ameyaw, S. (2000). Appreciative planning: An approach to planning with diverse ethnic and cultural groups. In Burayidi, M. A. (Ed.), *Urban Planning in a Multicultural Society*. Westport, CT: Praeger, pp. 101–114.

Baas, J., Ewert, A., & Chavez, D. (1993). Influence of ethnicity on recreation and natural environment use patterns: Managing recreation sites for ethnic and racial diversity. *Environmental Management*, 17, 523–529.

Balance: The Growth Management Element. (2006). Lancaster, PA: Lancaster County Planning Commission.

Burayidi, M. A. (2000). Urban planning as a multicultural canon. In Burayidi, M. A. (Ed.), *Urban Planning in a Multicultural Society*. Westport, CT: Praeger, pp. 1–14.

Byrne, J. & Wolch, J. (2009). Nature, race, and parks: Past research and future directions for geographic research. *Progress in Human Geography*, 33(6): 743–765.

Cranz, G. (1982). *The Politics of Park Design: A History of Urban Parks in America*. London: Massachusetts Institute of Technology Press.

Cranz, G. & Boland, M. (2003). The ecological park as an emerging type. *Places*, 15, 44–47.

Cranz, G. & Boland, M. (2004). Defining the sustainable park: A fifth model for urban parks. *Landscape Journal*, 23(2), 102–122.

Day, K. (2006). Active living and social justice: Planning for physical activity in low-income, black and Latino communities. *Journal of the American Planning Association*, 72, 88–99.

Escobar-Haskins, L. & Haskins, G. F. (2007). *Latinos in Lancaster County: Voices, Perspectives, Myths and Realities*. Lancaster, PA: Alegre Research and Demographics.

Floyd, M. (2001). Managing parks in a multicultural society: Searching for common ground. *Managing Recreation Use*, 18, 41–51.

Floyd, M., Shinew, K., McGuire, F., & Noe, F. (1994). Race, class and leisure activity preferences: Marginality and ethnicity revisited. *Journal of Leisure Research*, 26, 158–173.

Fogg, G. E. (1981). *Park Planning Guidelines Revised*. Arlington, VA: National Society for Park Resources.

Fogg, G. E. (1990). *Park Planning Guidelines*. Arlington, VA: National Recreation and Park Association.

Freedman, D., Mei, Z., Srinivasan, S., Berenson, G., & Dietz, W. (2007). Cardiovascular risk factors and excess adiposity among overweight children and adolescents: The Bogalusa Heart Study. *Journal of Pediatrics*, 150(1), 12–17.e2.

Gobster, P. (2002). Managing urban parks for racially and ethnically diverse clientele. *Leisure Sciences*, 24, 143–159.

Gobster, P. & Delgado, A. (1993). Ethnicity and recreation use in Chicago's Lincoln Park. In P. Gobster (Ed.), *Managing Urban and High-Use Recreation Settings, General Technical Report NC-163* (pp.75–81). St. Paul, MN: USDA Forest Service North Central Forest Experiment Station.

Greenscapes: *The Green Infrastructure Element*. (2009). Lancaster, PA: Lancaster County Planning Commission.

Hollinger, D. A. (1995). *Postethnic America: Beyond Multiculturalism*. New York: Basic Books.

Holston, J. (1999). Spaces of insurgent citizenship. In J. Holston (Ed.), *Cities and Citizenship* (pp. 155–173). Durham, NC: Duke University Press.

Jackson, J. B. (1994). *A Sense of Place, A Sense of Time*. New Haven, CT: Yale University Press.

Kumanyika, S. & Grier, S. (2006). Targeting interventions for ethnic minority and low-income populations. *Childhood Obesity*, 16(1), 187–207.

Lancaster County Comprehensive Plan: Policy Plan. (1999). Lancaster, PA: Lancaster County Planning Commission

Lefebvre, H. (1991). *The Production of Space*. (D. Nicholson-Smith, Trans.). Malden, MA: Blackwell. (Original work published 1974.)

Loukaitou-Sideris, A. (1995). Urban form and social context: Cultural differentiation in the uses of urban parks. *Journal of Planning Education and Research*, 14, 89–102.

Loukaitou-Sideris, A. & Stieglitz, O. (2002). Children in Los Angeles' parks: A study of equity, quality and children's satisfaction with neighborhood parks. *Town Planning Review*, 73, 467–488.

Low, S., Taplin, D. & Scheld, S. (2005). *Rethinking Urban Parks: Public Space and Cultural Diversity*. Austin: University of Texas Press.

McCormack, G., Rock, M., Toohey, A., & Hignell, D. (2010). Characteristics of urban parks associated with park use and physical activity: A review of qualitative research. *Health and Place*, 16(4), 712–726.

Mertes, J. D. & Hall, J. R. (1995). *Park, Recreation, Open Space and Greenway Guidelines*. Arlington, VA: National Recreation and Park Association.

Purcell, M. (2002). Excavating Lefebvre: The right to the city and its urban politics of the inhabitant. *GeoJournal*, 58, 99–108.

Rios, M. (2009). Public space praxis: Cultural capacity and political efficacy in Latino placemaking. *Berkeley Planning Journal*, 22, 92–112.

Sandercock, L. (2003). *Cosmopolis II: Mongrel Cities of the 21st Century*. New York: Continuum.

Sasidharan, V., Willits, F., & Godbey, G. (2005). Cultural differences in urban recreation patterns: An examination of park usage and activity participation across six population subgroups. *Managing Leisure*, 10, 19–38.

Schoenborn, C., Adams, P., & Barnes, P. (2002). Body weight status of adults: United States, 1997–1998. *Advance data from vital and health statistics of the National Center for Health Statistics*, 330, 1–15.

Serdula, M., Ivery, D., Coates, R., Freedman, D., Williamson, D., & Byers, T. (1993). Do obese children become obese adults? A review of the literature. *American Journal of Preventive Medicine*, 22, 167–177.

Strack, R.W., Magill, C., & McDonagh, K. (2004). Engaging youth through photovoice. *Health Promotion Practice,* 5(1), 49–58.

Thomas Comitta Associates & Arro Consulting. (2007a). *Northeast Revitalization Initiative for the City of Lancaster, Pennsylvania*. Lancaster, PA: Lancaster City Parks Committee.

Thomas Comitta Associates & Arro Consulting. (2007b). *Southwest Revitalization Initiative for the City of Lancaster, Pennsylvania*. Lancaster, PA: Lancaster City Parks Committee

Thomas Comitta Associates & Arro Consulting. (2009). *Urban Park, Recreation and Open Space Plan for the City of Lancaster, Pennsylvania*. Lancaster, PA: Lancaster City Parks Committee.

US Census Bureau. (2010). 2010 Census Summary File 1, Table PCT 11. Retrieved July 25, 2011, from: http://factfinder2.census.gov/faces/tableservices/jsf/pages/productview.xhtml?pid=DEC_10_DP_DPDP1andprodType=table

Walker, C. (2004). *The Public Value of Urban Parks*. Washington, DC: Urban Institute.

Wang, C. & Burris, M. A. (1997). Photovoice: Concept, methodology, and use for participatory needs assessment. *Health, Education and Behavior*, 24(30), 369–387.

Wolch, J., Wilson, J., & Feherenbach, J. (2005). Park and park funding in Los Angeles: An equity-mapping analysis. *Urban Geography*, 26, 4–35.

CHAPTER 19

Placemaking in Between Urban Redevelopment

Little Indonesia in Taipei

Hung-Ying Chen

> On June 15th, 2005, Jamaican athlete Asafa Powell with the speed of 9'77" broke the world record of 9'78" set by U.S. athlete Tim Montgomery in the 100-meter race at the Athens Olympic Game. The distance from Little Indonesia to Taipei Main Station is not even 100 meters. However, although thousands of people come through the Taipei Main Station each day, it is rare to see anyone visiting Little Indonesia except the Indonesian workers. It is such a pity to ignore the nearest shortcut to the paradise of Indonesian cuisine. You may regret missing such fantastic experience of multiculturalism. Do not miss the great opportunity . . . due to the mere distance of 9'77"!
>
> (POT Supplement No. 359)

Hidden within old, slum-like urban blocks, Little Indonesia is located to the southeast of Taipei Main Station, the main transportation hub in Taipei (Figure 19.1). The area known for its bustling traffic and home to several mega urban projects has also been a hot spot for urban redevelopment in Taipei. Sited in the heart of the sprawling city, it seems as if Little Indonesia has successfully held out against the threat of urban redevelopment for the past decade, which prompts the questions: how has "Little Indonesia" competed with developers who have long eyed this geographically strategic area in the city? What are the competing forces of placemaking that has de-territorialized and re-territorialized Little Indonesia in the geography of Taipei? These questions arouse my curiosity toward this tiny, seemingly messy, but energetic street block. The existence of Little Indonesia also questions the predominant view of urban development that does not account for the rights of transient migrant workers like the Indonesians, who despite the lack of official citizenship have developed their networks and places in their adopted city.[1]

The fieldwork for this study was conducted from April 2007 to May 2008. At the beginning, I learned that as a Taiwanese and an outsider to Little Indonesia, it was very awkward for me to intrude into the life of the Indonesian "locals." To overcome such a barrier, I decided to work as a waitress on weekends in one of the larger Indonesian restaurants. I also volunteered for a local NGO—Migrant Workers Concern Desk (MWCD). In the process of learning from a place which seemed to have an impenetrable boundary built around itself, I was surprised to learn that the boundary is not purely drawn along the lines of race, class, and gender; on the contrary, the barriers around Little Indonesia and its invisibility from the outside world were more a result of the disapproval of its presence by the mainstream society. Its obscure border reveals instead a dynamic process of boundary making.

Figure 19.1 The big outdoor advertisement signifies the location of Little Indonesia in Taipei. Photograph by Hung-Ying Chen.

Placemaking by Transnational Migrants

The populations of Indonesian migrants in Taipei are composed of migrant workers, marriage migrants, and Oversea Chinese Indonesians who arrived here around the time of the 1998 Chinese Massacre as well as much earlier. Since the early 1980s, Indonesian domestic workers have become the largest foreign workforce in Taiwan.[2] The stability of migrant workers is highly influenced by the government policy toward migrant workers and trends in the labor market. With the support of recruitment policy, the number of Indonesian migrant workers in Taiwan doubled every year during 1998 to 2002. By 2001, the number of Indonesian workers for the first time exceeded that of the Filipinos, reaching 91,132. The large number of Indonesian workers also led to the establishment of niches of "hometown business" in the Taiwanese urban fabric. It was in this context that Little Indonesia in Taipei was born, as a social and spatial byproduct of the national labor recruitment policy.

Unlike experiences in the USA and Europe, migrant workers in Taiwan cannot formally become entrepreneurs unless they acquire citizenship by marrying Taiwanese citizens. Since 1992, under the "Employment Service Act" migrant workers in Taiwan must work in certain occupations, and they are only guaranteed one-year work contracts. In 2003, the Council of Labor Affairs in Taiwan established a quota system for the number of migrant workers from specific countries and determined the maximum length of legal stay at six years (later extended to nine years through new legislation in July 2007). With no legal right to become an entrepreneur, migrant workers are not allowed to rent a place, much less operate a stable business. However, some marriage migrants have been able to get assistance from their husbands of local Taiwanese origin to run businesses before officially acquiring citizenship. The legal system in effect created a structural inequality among ethnic communities and businesses in Taiwan.

Some scholars have investigated the lives of migrant workers and some of their autonomous placemaking in Taiwan. Lan (2003) observes how Filipino workers transform streets and sidewalks into a weekend enclave in the Chungshan District, which became known as Little Manila. She applies the concept from Erving Goffman's Dramaturgical Analysis to point out that the working place of domestic workers functions as their "front stage." On the other hand, in the limited free time they could enjoy the public place, the city turns out to be their "back stage." Such a metaphor presents an opposite notion to the perception of the city space as the front stage. Wu (2002) conducted his research on recreational activities of both Filipino and Thai workers in Taiwan. He points out that Filipino and Thai workers have different routines for their holiday activities and thus create different levels of territorialization. Needless to say, there is little privacy for foreign domestic workers in Taiwan since they are required to reside with their employers due to the nature of their work. Paradoxically, their sense of privacy is achieved by being in a public space with their friends during their days off. Unfavorable living and working conditions have compelled migrant workers to create alternative spaces for themselves in the public spaces of Taipei.

Given the challenges faced by migrant workers, how do places such as Little Indonesia struggle for their existence and sustain themselves in the City of Taipei? In this chapter, I discuss different forces that have contributed to the precarious existence of Little Indonesia: (1) dislocations and relocations of ethnic entrepreneurs in Taipei Main Train Station and the making of a neoliberal landscape in Taipei; (2) everyday life of migrants in Little Indonesia; (3) the contradicting views of Indonesian workers through constant policing and stereotypical media coverage in contrast with the City's promotion of so-called "weekend carnivals" in Little Indonesia.

Caught in (between) Urban Redevelopment

The incremental formation of the current Little Indonesia in Taipei is a result of three periods of expansion and relocation. The first period, the origin of Little Indonesia, occurred from 1998 to 2000,[3] as some ethnic businesses were established inside the Jin-Hua Department Store on the second level of Taipei Main Station. At that time, as many as 100 stores crowded an 11,900-square meter area inside the station that included many ethnic restaurants and shops.[4] Food and goods from the Philippines, Thailand, Vietnam, and Indonesia transformed the generic commercial space inside the station into a distinct gathering place for many foreign migrants. Later, some successful storeowners started branches in New Taipei City (formerly Taipei County) and Taoyuan County. Shortly after this, from 2001 to 2003, a vibrant ethnic marketplace had taken hold at the station, and the area was nicknamed the "United Nation businesses." At that time, stores not only crowded inside the Taipei Main Station but also expanded into the area south of the station.

However, this diverse and vibrant marketplace did not last too long. Multiple changes at the Main Station, including increasing regulations on businesses, gradually pushed the "undesirable" ethnic businesses out of the station building. By 2003, three Indonesian restaurants, two Filipino restaurants, and one Thai restaurant were lost. More relocation occurred in 2005 and 2007, first following the forced closure of Jin-Hua Department Store due to a breach of contract by the turnkey company. At that time, most stores were forced to move to the underground shopping street that became known as "Little Indonesia" and "Little Manila."

The second major relocation was caused by the transferring of operating rights for the commercial space inside the station to a private company, the Breeze Corporation. Breeze attempted to rebrand the property for a more affluent and mainstream clientele and was intent on pushing away foreign workers.[5] The transformation of the commercial operation inside the Taipei Main Station reflects a broad neoliberalization of the city through privatization of public space and resources. The twelve-year lease to the Breeze Corporation (with option to extend for six more years) brings over NT$50 million a year to the Taipei Railways Administration, a state-owned company. After nine months of preparation, the Breeze Center opened with great fanfare in the media.[6] The space now features a "Cosmopolitan Food Plaza" that the CEO, Mr. Liao, claimed as inspired by his experiences traveling abroad (Figure 19.2).

During the transaction, however, the Taiwan Railways Administration expelled the Southeast Asian ethnic businesses from inside the building. Instead of individual tenants, the turnkey company now has full control of the commercial space inside

Figure 19.2 The lobby of the Taipei Main Station with newly opened Breeze Center on the second level. Photograph by Hung-Ying Chen.

the station and selected brand-name shops to operate inside the space.[7] This process resulted in the gentrification of the station and privatization of public space (as the Taiwan Railways Administration is state-owned). With the right to manage the space, the turnkey company also decides who is or is not welcome into the space—an invisible exercise of social exclusion. As an example, during the period of business solicitation at the Breeze Center, the owner of Sari Restaurant in Little Indonesia applied to rent a space inside the Food Plaza. However, her application was rejected after the Breeze Center management found out about the location of Sari restaurant. The management argued that the style of the Sari restaurant did not fit the image of the Breeze Center. But, underlying the rejection was clearly the fear toward certain ethnic consumers and emphasis on the institutionalized process of gentrification. Here, Robertson's "what will fly globally and what will not" (1995) comes to mind.

Another important force that contributed to the emergence of Little Indonesia in its current form was the process of urban redevelopment around the Main Station, formerly the core of Taipei metropolitan area. In the early 1970s, under the initiative to rebuild the area around the Main Station, former Mayor Lee Teng-Hui invited KMG Architects and Engineers to draft a plan. The "Taipei Main Station Regional Renewal Project" was submitted by KMG and, thereafter, the area was planned as the "Taipei Main Station Special District," which was approved by the Ministry of Interior in 1980. Later, a master plan was approved in the 1990s. In 2004, a new government initiative to double the number of foreign visitors identified the area as an important "national gateway," and an international design competition was launched by the National Government's Council for Economic Planning and Development. The result was an urban mega project consisting of six sub-developmental projects. However, fierce competition for urban redevelopment projects and the complex political terrain (Chatterjee, 2006) have stalled the redevelopment of the area.

While the proposed projects to make Taipei Main Station a major transportation hub seem to have provided a prospective future for this area, the development has not been smooth. Successive construction for subway lines, high-speed railway, an airport express train, etc. has created endless disturbances in the area for over twenty years. In the current location of Little Indonesia, most of the landowners could not stand the deteriorating conditions in the neighborhood and have since moved out. For many property owners, it is more practical to rent their properties to ethnic business rather than participate in a redevelopment project that carries high risk. However, although the foreign workers were stakeholders, they had no right to participate in the decision making process of the redevelopment projects. What they did was to take advantage of the cheap rent resulting from the stalling redevelopment process.

Over the years, the community of Indonesian restaurants and shops and the gathering places of Indonesian guest workers have managed to survive under the transformations in the center of Taipei. So far, they have been able to find a niche in the process of urban transformation in the lag time in the midst of the massive and complicated urban redevelopment process.

Life in Little Indonesia

Sunday is the most vibrant day of the week in Little Indonesia as the Indonesian workers have their day off and gather there in large numbers. A "normal" Sunday schedule here goes like this: at 5 a.m., employees at the local restaurants begin preparations for the long day ahead. Around 8 a.m., some people who have already arrived start singing karaoke but the activity doesn't peak until the afternoon and early evening. The "ghost dance club" during weekdays now has a long queue outside its premise on the weekend. Most of the restaurants are open until 8 p.m. For Indonesin migrant workers, Little Indonesia has become their weekend home. During the week, however, Little Indonesia is unremarkable, occupied with routine preparation for the weekend (Figure 19.3). For the few Taiwanese locals who frequently visit the area—cab drivers waiting for passengers, second generation Taiwanese–Indonesian couples, and some foreign backpackers staying in the nearby hostels—Little Indonesia allows them to imagine a foreign land at home or a foreign land within another foreign land.

Businesses in Little Indonesia are generally multifunctional—restaurants combined with karaoke, remittance services, money exchange, and disco. They also sell snacks along with electronic products, magazines, books, and hijabs. In some cases, one can find seven to eight types of activities squeezed into one tiny store with roughly 36 square meters of space. In addition, spaces outside the stores, including building arcades and the stairwells of vacant buildings, are all utilized by street vendors. The most common types of business in Little Indonesia are the 3-in-1

Figure 19.3 Some stores in Little Indonesia choose to close on weekdays since most of their main customers—the migrant workers—have to work. Others take the time to prepare for the weekend festivities. Photograph by Hui-yu Huang.

restaurants (combining restaurant, grocery store, and karaoke), express parcel services, and catalogue shops that send furniture and electronic goods to family members in Indonesia.

The variety of businesses in Little Indonesia serves the needs of Indonesian workers trying to survive in an alien environment while maintaining ties to their distant home. In particular, food in Little Indonesia alleviates their nostalgia for authentic Indonesian cuisine. Here, one can find different varieties of Indonesian cuisine prepared by cooks from different regions of Indonesia. The tastes can vary from store to store, and not all can be easily found elsewhere in Taiwan. To compete with one another, each restaurant claims to be more "authentic" than the next. In addition, some restaurant owners have been creative in inventing new delicacies by combining with local Taiwanese favorites. For example, at two of the biggest restaurants in Little Indonesia, the owners combine *ikan goring* (Indonesian-style fried fish) and *tempat* (fried soybean) with *hsiensuzhi* (salted crispy chicken)—a popular Taiwanese fried food found in many night markets. My former boss even found locally grown Indonesian banana *pisang rastali* that he has been using to make authentic *pisang goreng* without having to import them from Indonesia. The dish became the specialty of the restaurant that I once worked for.

Moving from the former location inside Taipei Main Station to the adjacent area outside of the Main Station, the location change brought a significant shift in the pattern of ethnic business. Previously, a large commercial space would be subdivided into multiple units based on private deals made between the renters and sub-renters. After moving to the current location outside the station, small business owners have operated as street vendors, setting up shop alongside the storefronts in a parasitic manner. These vendors pack their equipment and merchandise in portable cases in order to move quickly and evade police inspection. Some migrant workers and Chinese Indonesian students take up part-time jobs as vendors to make extra income. For example, Jay, a factory worker known for his diligence, repairs discarded battery chargers and sells them to other Indonesian workers. Yasbi, a college student, always attracts lots of customers every weekend. He started his business by using his laptop to install the latest pop songs into customers' mobile phones. Sometimes he also sells second-hand cameras and MP3 players. While Yasbi seemed successful, he said to me, "you don't understand, every Sunday you guys have fun here and there, eating and chatting, I try hard to pretend to be calm, however, I am actually very anxious that I might get caught one day." Such high insecurity and uncertainty requires Yasbi to keep a low profile and hence he does not advertise his business.

In Little Indonesia, "hope" and "help" are two important sources of commodity because the Indonesian customers often feel hopeless and despondent. The unequal relationship between the workers and the employers, brokers (or mediators), and the government leaves the Indonesian workers in a vacuum, unable to obtain help except in Little Indonesia. This unenviable setup makes the various services in this neighborhood profitable. As customers in Little Indonesia enjoy food, music, and the company of others, many illegal brokers also hang around looking for their prey. They search for workers whose faces look downtrodden and offer to help them return to Indonesia by claiming connections with the Immigration Bureau, Police, or other official entities. Such middlemen usually advertise their connections with an official bureau to show their ability in functioning as an "alternative travel agency," capable of forging documents and obtaining a flight back to Indonesia for the customer. Even though the worker eventually realizes he or she has been cheated, there is no recourse as he or she might be arrested before the middleman is caught. As the local organization Migrant Workers' Concern Desk (MWCD) found out, many runaway workers have no way out but to gamble away their only hope and trust these dubious "mediators."

On a brighter side, Little Indonesia does provide important opportunities for social networking, especially mutual support among female guest workers. One day I met Father Suwendi in a barbershop run by an Indonesian marriage migrant where I realized the existence of such networks. Father Suwendi works at MWCD, a service-oriented organization for migrant workers of different nationalities in Taiwan. During the week, he and several Filipina and Vietnamese sisters visit many detention centers to see workers who have been jailed and are now waiting for deportation. Members of MWCD document the cases for the workers, call their employers and help claim their unpaid salaries, while asking support to pay for airplane tickets. Every weekend, another migrant group IPIT (Ikatan Pekerja Indonesia Taiwan), founded by Indonesian sisters, occupies the same office space as MWCD and offers language classes and advocacy work as well as a gathering space to sing and dance. Some of the Taiwanese employers initially tried to block their employees from participating in these activities because: (1) they were concerned that fraternizing with other migrant workers would teach their staff tips for running away; (2) they were concerned that their staff would become harder to manage or control after they gained more self-confidence through these classes. However, after observing the classes and activities, most employers became very supportive of their employees' participation.

Father Suwendi had planned to invite some famous Indonesian pop stars to hold concerts in Taiwan. I had heard about similar practices in Little Manila in Taipei and some Nepali communities in Seoul but was initially ignorant of the power these performances would have. When I saw these activities with my own eyes, I was surprised by how such popular entertainment can help migrant workers overcome cultural and geographical barriers. I was also impressed by the willingness and flexibility of a religious cadre to weave together a transnational social space for migrant workers.

Weekend Carnivals and the Hidden Attacks

> Relations of domination are, at the same time, relations of resistance.
>
> (Scott, 1990, p. 45)

> Importing foreign labors [to Taiwan] may actually become de facto immigration. Can [the society] tolerate more foreign immigrants at all? Can we live together with them? I think, *if we conduct a survey on whether our citizens are willing to have a Thai-village, a Filipino village, or even a Malay village in Taiwan, to share*

> *our employment opportunities with foreigners, I am afraid that people here will hardly agree.*
>
> (Zhao, 1992, p. 145; emphasis by the author)[8]

Every year the Taipei City Government holds the Thai Songkran Water Festival in the Red House Theater Plaza in the popular Ximen area to acknowledge the contribution of Thai workers to the city. Limousines dispatched from different companies unload Thai workers one after another at the plaza. After all the workers are gathered, the show starts with a traditional Thai dance, followed by people splashing water and flour at each other. In a critical look at the event funded by the City Government, one may find it resembles some sort of Southeast Asian cultural circus, or a temporary and superficial patch to treat workers' homesickness. While such events might be designed out of good intentions, the tolerance toward the collective existence of migrant workers at the festival sharply contrasted with the thinly veiled opinion behind the words of Zhao, the former minister of the Council of Labor Affairs. The contrast is further highlighted in the forced eviction of ethnic businesses from the Taipei Main Station to make way for the shiny and modernized "Cosmopolitan Food Plaza" created by the Breeze Corporation.

Little Indonesia shares a very similar carnivalesque atmosphere every weekend. In contrast to repetitive and bland factory or domestic work, the excitement in an ephemeral, informal urban landscape provides a way for the guest workers to release their stress and anxiety. However, stigma created by unfavorable media coverage accompanied with rude inspections by the police and immigration officers often disturb the peace of this place. As a result of frequent harassment, visitors and business people in Little Indonesia have become very sensitive and defensive towards Taiwanese outsiders. In the following section, I use some examples to illustrate the institutional attacks and their impact on the local business. These examples provide clues for understanding how spatial governmentality (Gordon, 1991; Ferguson & Gupta, 2002) is employed in Little Indonesia through the surveillance of media and police, and how the processes of territorialization and de-territorialization often go together in Little Indonesia.

We begin with police inspection and surveillance. The relocation from the Taipei Main Station to the area outside the station not only changed the spatial pattern of Little Indonesia, but also resulted in new methods of spatial surveillance. Previously within the Taipei Main Station, police officers rarely conducted inspections inside of Jin-Hua Department Store on the second floor of the building even though a police branch office was on the first floor. As described by Mr. D:

> In the past when we were still in Jin-Hua, [police] could come but rarely did, except for special cases of robbery and theft . . . This was because they were afraid to get into trouble in someone's department store.[9] Now it is different. To claim credits for capturing runaway workers, they enjoy visiting here every weekend.

In the current location, police officers can patrol the streets and alleys of Little Indonesia without hindrance. Private stores are the only exception. Prior to the enactment of Article 67 of the Immigration Act that authorizes police officers to enter private spaces in the name of law enforcement, the police were more careful in keeping better relationships with the Indonesian store operators because they could go to court for illegal entry; but now they can do as they like. However, not all police officers are abusive, and some older police officers from the local branch have worked with local entrepreneurs to build long-term, positive relationships.

But overall, Little Indonesia has become a hot spot for police (from various branches) and immigration officers to capture runaway workers. As a popular place for many kinds of migrants, it is a convenient location for these officers to fulfill their quotas. In one recent instance, twelve police and immigration officers converged in one single restaurant. They included uniformed as well as plainclothes police officers from different branches in the city. According to them, their joint presence was a mere coincidence.[10]

Persistent tension among the shops, restaurants, visitors, police, and local residents in the neighborhood are reinforced by negative media coverage and xenophobia in Taiwan. For Little Indonesia, the biased media coverage has presented the neighborhood as a ghetto that justifies police presence and enforcement. The media coverage of Little Indonesia can be classified into two types: (1) the exotic cuisine and bustling street life; (2) illicit activities such as sex trade, human trafficking, and illegal financial transactions. The negative reporting perpetuates an image of social problems concentrated in one location. Furthermore, it contributed to the general hostility towards guest workers. For Indonesian migrant workers, they must struggle both in the space of production (factories and employers' homes) and the larger urban space of reproduction and recreation. Little Indonesia is burdened with a commonly accepted narrative that depicts the gathering spaces for ethnic minorities and aliens as exotic and different, and as locations for the sex trade and crime.

Daily life in Little Indonesia is clearly affected by the combination of police surveillance and negative media images (Figure 19.4). In Taipei, media coverage about Little Indonesia provides justification for further surveillance and enforcement of unfavorable laws against Indonesians. Such collective social surveillance shows the operation of synopticon, through keeping the public eye on a specific group via TV, newspapers and magazines, such social consent provided by a general "public," to guide, but not control (Bauman, 1989). It reaches an effective goal to rectify the behavior of another specific ethnic minority group first; later the media discourse introduces pressure for the government to act to protect the safety of local residents. This invisible pressure eventually seeps into daily encounters between the ethnic gathering and the governmental apparatus. In practice, however, the only thing that the surveillance can confirm is the ethnic identity of those under surveillance and a landscape of ethnic spectacles under the gaze of untrained eyes. In other words, one can argue that the sense of enclosed territoriality in Little Indonesia, accompanied by the discourses of isolation and ghetto pathology, is actually fictitious because the constant surveillance has already made the place highly permeable (Chiou, 2007; Ferguson & Gupta, 2002).

In addition to police surveillance, maintaining a peaceful relationship with the surrounding neighborhood is also a constant struggle for Little Indonesia. Residents

Figure 19.4 A scene from a series of massive crackdowns on underground businesses by police in March 2008. Photograph by Hung-Ying Chen.

in nearby neighborhoods often file complaints to the City's Environmental Protection Bureau regarding the noise from karaoke and playing of Indonesian music. In response, Indonesian business owners quickly lower the volume when enforcement officers approach. But sometimes they are still caught. Many of these establishments are set up in alleys, and the presence of tables and chairs in a public space can lead to fines. Due to stiff competition among some popular restaurants, the competitors would sometimes sabotage each other through malicious reports. For example, the local police sometimes receive phone calls from someone accusing a store of illegally selling Indonesian cigarettes or hiring runaway workers. This effectively disrupts the operations of the accused business, and such experiences have gradually trained shopkeepers and employees in Little Indonesia to be very sensitive and defensive towards Taiwanese strangers.

Conclusion: Rights and Rhetoric of the City

In this chapter, Little Indonesia serves as a case study for how bottom-up placemaking has created an identifiable, if embattled, space for the marginalized migrant workers. As foreigners in Taiwan who have few guaranteed rights and even fewer avenues for accessing much-needed support, they have made the best use of the gaps in the urban redevelopment process. But their occupation of the dilapidated area near the Taipei Main Station is always under threat from surveillance, inspection of police and immigration officials, and media and public perception. As with so many explicitly marked ethnic neighborhoods, Little Indonesia is a spectacle of multiculturalism that simultaneously maintains a minority way of life while exposing that life to stereotypical oversimplification and government control. As an ethnic gathering place that has existed in the center of Taipei for over a decade, this dilapidated neighborhood is both a testimony to ethnic marginalization in a city and a place with potential for engendering transculturalism.

In examining "migrants' rights to the city," I have studied official activities and grassroots practices of multiculturalism in Taipei during the past decade. Surprisingly, I found that all official multicultural activities have disappeared in Taipei since 2008. There appears to have been no space left for multicultural and bottom-up ways of placemaking since then. Aiming to become a World City, Taipei has apparently turned its back on its largest foreign population. As such, multiculturalism in Taipei exists only as hollow rhetoric. Moreover, from 2008 to 2012, some redevelopment agencies have set up offices in ethnic neighborhoods. One even replaced an Indonesian dance club. The true multicultural landscapes of Taipei have been sacrificed in the pursuit of aesthetic developmentalism with an ever growing number of consumer-oriented, cosmopolitan places.

Notes

1 Citizenship has historically been associated with both positive and negative meanings. Citizenship symbolizes revolution and democracy but also represents conservative values and exclusion. Even legalized citizenship has not been able to deal with ethnicity satisfactorily and has been particularly problematic in terms of former nationality, religion, gender, etc. (Isin & Turner, 2002).

2 According to the statistics of foreign workers in Taiwan in 2011, the Indonesian labor force represents about 28.6 percent (180,006 out of the total population of 630,169).

3 Prior to 1999, the Taipei Main Station became a popular gathering place for many foreign workers. On holidays, workers from other cities of northern Taiwan gathered there to enjoy their precious days off. Some owners of disco pubs in Chungsan (aka. Little Manila) even dispatch free shuttles to transport the workers.

4 The owners can be divided into three types: (1) Indonesian–Taiwanese couples with stores co-operated or operated by Indonesian marriage migrants; (2) overseas Chinese from Southeast Asia (and their foreign spouses); and (3) local Taiwanese in co-operation with their overseas Chinese friends.

5 Taiwan Railways Administration amended their model of governance. They diverted the direct leasehold model into the ROT (Rehabilitation–Operation–Transfer) and selected a branded corporation to rebuild their contracted business. In addition, in order to transform the station into a "national gateway," the space for contracted businesses expanded from being only on the second floor to now include the first floor and underground spaces.

6 As the *Business Times* (2007) reported: "the 'Kingdom,' which was promoted by the developer Radium Life Tech, was sold out in merely 3 months; in the same time the 'Twin Towers' project is also going to bid; the new metro line which links the airport to the Taipei Main Station is estimated to finish in the next 5 years. This means Taipei Main Station will be the first national gateway following the Taoyuan International Airport."

7 In the revised contract, Taiwan Railways Administration released more authority to the turnkey company to reduce the interaction cost of dealing with small tenants.
8 Zhao is the former Minister of Council of Labor Affairs in Taiwan. This speech was originally in Mandarin and delivered at a meeting of the Legislative Yuan. Both text translation and the words noted in the brackets are by the author.
9 As a symbol of department store failures, Jin-Hua Department store was organized more like a marketplace.
10 According to statements by the local police, they patrol for undocumented workers about once a month in Little Indonesia.

References

Bauman, Z. (1989) *Legislators and Interpreters: On Modernity, Post-Modernity and Intellectuals.* Cambridge, UK: Polity Press.
Business Times. (2007) Liao Jian-Han Plays Food and Fashion in Taipei Main Station. March 26, p. A4.
Chatterjee, P. (2006) *The Politics of the Governed: Reflections on Popular Politics in Most of the World*. New York: Columbia University Press.
Chiou, S. W. (2007) The Conflict and Breakthrough between Ghetto Pathology and Network Nodal Points: The Case of the Vietnamese Female Restaurant Owners in Taiwan. *Formosan Education and Society,* 13: 95–120. (In Chinese.)
Ferguson, J. & Gupta, A. (2002) Spatializing States: Toward an Ethnography of Neoliberal Governmentality. *American Ethnologist,* 29(4): 981–1002.
Gordon, C. (1991) Governmental Rationality: An Introduction. In M. Foucault, G. Burchell, C. Gordon, & P. Miller (Eds), *The Foucault Effect: Studies in Governmentality: with two lectures by and an interview with Michel Foucault* (pp. 1–51). Chicago, IL: University of Chicago Press.
Isin, E. F. & Turner, B. S. (2002) Citizenship Studies: An Introduction. In E. F. Isin & B. S. Turner (Eds), *Handbook of Citizenship Studies* (pp.1–10). London: Sage.
Lan, P. C. (2003) Political and Social Geography of Marginal Insiders: Migrant Domestic Workers in Taiwan. *Asian and Pacific Migration Journal,* 12(1–2): 99–125.
Robertson, R. (1995) Glocalization: Time–Space and Homogeneity–Heterogeneity. In M. Featherstone, S. Lash, & R. Robertson (Eds), *Global Modernities* (pp. 25–44). London: Sage.
Scott, J. C. (1990) *Domination and the Arts of Resistance: Hidden Transcripts*. New Haven, CT: Yale University Press.
Wu, T. F. (2002) The Resistance and Adaption of Foreign Workers in Taiwan in the 1990s: Sunday Lives as a Battlefield. *Hong Kong Journal of Social Sciences*, 23, 103–150.
Zhao, Shou-Bo. (1992) *Labor Policy and Labor Issues*. Taipei: China Productivity Center. (In Chinese.)

CHAPTER 20

Regulation and Reception of Public Space in Hong Kong

Kin Wai Michael Siu

Municipal governments generally claim and promise to provide people with inhabitable city space. However, as we review city projects currently in progress, it is readily apparent that many municipal agencies have continuously adopted strategies and plans and sought legislative authority to manage—to control—city space. When undertaking urban development projects, government agencies have followed various forms of operational rationalism while neglecting cultural and human factors, including the continuous transformation of cultures and human needs. In studies of the sociology of everyday life, some sociologists such as Michel de Certeau and Henri Lefebvre point out that everyday life of common or ordinary people offer professionals, including planners, architects, and designers, a new perspective into people's response—reception—to their programmed living environment. These sociologists see reception as a kind of continuous, creative act and art that interacts dynamically with cultural and social changes.

To explore this alternative perspective on urban placemaking, several case studies on the reception of public space were conducted in Hong Kong from 2007 to 2012. The targeted subjects in the case studies were so-called newly arrived persons (NAPs) from the Chinese mainland and local residents.[1] For more than thirty years, due to social and political reasons, a large number of NAPs have entered and lived in Hong Kong. Specifically, the population of NAPs with less than seven years of residence in Hong Kong in 1996 was around 169,000, a figure that had increased to 217,000 by 2006. While the total population of Hong Kong is only about 7 million, the continuously accumulating population of NAPs has had significant cultural, social, economic, and physical impacts on Hong Kong. While the population is mainly made up of people of Chinese origin, and early generations of local residents also originated from the Chinese mainland, NAPs and local residents have cultural backgrounds and life expectations that are similar in some respects but different in others. Borrowing the findings of the case studies, this chapter discusses how city dwellers—users—with different cultural and social backgrounds practice their everyday life in public space, and how they have brought cultural transformations in terms of spatial practices in Hong Kong.

Theorizing Reception

In literary theory, the conception of reception offers a new perspective that can be extended to the users of design. Reception theory was initially a branch of literary studies concerned with the ways in which literary works are received by readers. "Receptor" is a term used in critical theory to mean the person (or group of persons) experiencing a work of art (Cuddon, 1998; Siu, 2001). The term reception has mainly been used to refer to reader-response criticism in general. That is, reading is seen as a creative process in which the text offers guidance as to what is to be produced, and thus cannot itself be the product: "Reading is not a direct 'internalisation,' because it is not a one-way process, [but] a dynamic interaction between text and reader" (Iser, 1978, p. 107).

Although reception theory was originally applied to literature, its arguments provide valuable insights for the study of human behavior in general, including how users interact with designed environments, which helps to understand how design can better fit users (Siu, 2001). Similar to the idea of the incompleteness of text or other formats of discourse, the "creative" design of a place is also never finished. It is incomplete until it is used. The meaning of a place is performed, rather than formulated. In other words, this kind of city users' creation can be considered as an act of production—a creative act, or an art (de Certeau, 1984; Siu, 2001, 2007a). These acts and arts are produced and reproduced at different times and in different environments. Such process has particular relevance to cities, which undergo continuous cultural, social, and physical transformation.

Newly Arrived Persons (NAPs)

In Hong Kong, it is difficult to define the meaning of original residents because many of the earlier generations of residents were not born in the city but came from other regions of China for different reasons in different periods. In fact, "new arrivals" have kept on appearing in Hong Kong for more than a century and have continually made a significant impact on a city with about 1,100 square kilometers in total area, including the continuous land reclamation.

The foundations of contemporary urban development in Hong Kong were laid in the 1950s, prior to which Hong Kong was primarily a trading port for South China. After World War II, and especially after 1949, a major influx of people from the Chinese mainland created both problems as well as opportunities for Hong Kong. In the 1950s, industrial development gave rise to a parallel expansion of urban areas in the city. By the mid-1950s, the territory had a population of 2.5 million. Since then, the population of Hong Kong has continued to increase, mainly due to internal birth growth and steadily increasing numbers of immigrants from the Chinese mainland (Siu, 2001).

The Chinese and British governments signed the Sino-British Joint Declaration on December 19, 1984. In accordance with the "One Country, Two Systems" principle agreed on between the two governments, the Hong Kong Basic Law was implemented in Hong Kong on July 1, 1997. Among all matters related to political and social aspects, one of the key arrangements is that the children of Hong Kong residents have the right to enter and live in Hong Kong. This right applies also to those born in the Chinese mainland if one of the parents is a Hong Kong resident. Since the mid-1980s, with greater economic and social relationships between Hong Kong and mainland China, more Hong Kong residents have married and had children on the mainland. In particular, a large number of low-income older men were unable to find wives in Hong Kong. They instead married mainland Chinese women and applied for their children to enter Hong Kong. Although the wives of these men cannot become residents of Hong Kong automatically, they have higher priority to enter and live in Hong Kong due to concerns about family reunion and the needs of their children.

The large number of NAPs, many of them children and spouses of Hong Kong local residents, have become a cause for concern for many Hong Kong people. Two years after the 1997 Basic Law implementation, due to concerns for social welfare and the economy, the government announced its projection that several millions of NAPs would enter Hong Kong within a very short time, something that Hong Kong's society could not afford. The projection compelled the Hong Kong government to seek interpretation of the Basic Law on June 26, 1999. Since then, residents in the mainland must apply for the One-way Permit (OWP) to enter Hong Kong in an orderly manner.

For the past twenty years or more, the large number of NAPs has had positive and negative economic, social, and cultural impacts and transformations on Hong Kong. Whether the local residents like it or not, due to economic and social considerations, an increasing number of NAPs have entered Hong Kong and have started to have increasing and closer everyday interactions with local residents. Above all, the daily practices of NAPs have significantly influenced the public living environment, bringing change and transformation to Hong Kong.

Case Studies: Sham Shui Po and Tin Shui Wai

To reveal the cultural and social transformations as well as physical changes in the living environment relating to NAPs, several in-depth case studies were conducted from 2007 to 2012. The studies aimed to explore the reception of public space by NAPs from the Chinese mainland and local residents. Sham Shui Po (Kowloon) and Tin Shui Wai (New Territories) districts of Hong Kong were selected as areas for conducting the case studies (Figure 20.1). Demographic data compiled by the Hong Kong Census and Statistics Department (2011) indicate that the two districts have a relatively high percentage of low-income households, as well as single older persons and single-parent families. Each district also has public places typical of Hong Kong, such as market streets, district and local open spaces, and inactive small open areas outside residential buildings. On one hand, the living environment in the two districts significantly reflects the local everyday life. They are densely populated, and different kinds of human interactions can be observed. On the other hand, the two districts are residential areas with a relatively larger population of NAPs. As mentioned above, many NAPs are the children and spouses of local residents with low or no income who participate in the government's Comprehensive Social Security Assistance Scheme (Social Welfare Department, 2011).

Compared to other districts in Hong Kong, people can more easily find affordable living places in Sham Shui Po, an old district in Kowloon of Hong Kong. It consists of many old private buildings and deteriorated public housing buildings. The living environment in the district is notorious for representing the worst quality of life in

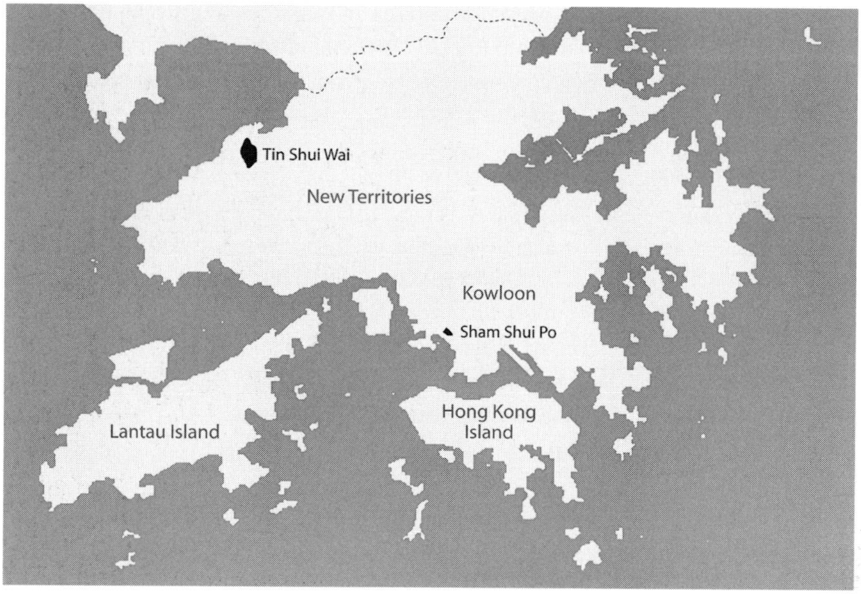

Figure 20.1 Locations of Sham Shui Po and Tin Shui Wai in Hong Kong. Map by Kin Wai Michael Siu.

Hong Kong, such as inhumane cage-houses and packed living partition rooms. Prostitution, crime, and public hygiene problems are also serious in the district. Tin Shui Wai is a new town located in the western part of the New Territories relatively further away from the central business area. It consists of different ages of low-rent public housing estates. Living standards are low and it has been widely regarded as a deprived district since the early 1980s (Chow, 1988). In 2009, it became known as the "City of Sadness" as the mass media widely reported the deprived living conditions and social problems of NAPs and older persons. Due to the relatively lower rent for living places in the district, many low-income people have moved to the district as a stepping-stone, before moving to other districts once their economic status becomes better or they are allocated better public housing flats in other districts. This situation has resulted in Tin Shui Wai becoming one of the districts with the largest number of affordable vacant flats. In turn, many old public housing estates have become home to deprived and single older persons and NAPs. Many NAPs as well as grassroots local residents have chosen to live in these two districts.

The two districts abounding with local human activities provided opportunities for the project team to investigate how different individuals with different interests—NAPs and local residents—act in diverse ways in the public environment. In addition to interviews conducted with planners, designers, and representatives and officers in related departments and organizations, field research was conducted in selected public spaces of the districts to investigate the everyday practices of NAPs and local residents. Unobtrusive observations and direct interviews were the main field research methods that allowed the project team to tap attitudes and behavior and seek descriptive information that would underpin a better understanding of people's behavior in the public living environment. These observations were used to explore how users, or actors, operate in public space.

Reception of Public Space

Strategies vs. Tactics

The case study findings show that, on one hand, the Hong Kong government (including policymakers, planners, designers, executives, and implementers) continues to formulate and adopt *strategies*, and seeks greater authority to control and organize public spaces in the city. The government way of thinking is generally similar to Modernist planning ideas. It views policymakers and professionals as experts, or often as the only experts who can and should change misused city elements, including everyday city space, into efficient facilities, and in turn provide a true urban order under the constructs of industrial society, contemporary cities, and the modern age (Harvey, 1989; Siu, 2003). It also views contemporary plans and designs as the only way to provide a suitable living environment for *different* city users, defined as *average* people with standard needs, and to establish the plans and designs as conditions for everyday happiness and harmony (Harvey, 1996; Hsia, 1995). This way of thinking rejects the possibility that city users are able to manage and find a better way to live together in the environment, including its public spaces, that suits their everyday lives (Sennett, 1970; Siu, 2001).

On the other hand, the findings illustrate that city users continuously seek opportunities under the regulations and strategies promulgated and implemented by policymakers and professionals. However, neither NAPs nor local residents act directly against or challenge these regulations and strategies in public spaces. Rather, they make use of *tactics*, a calculus that does not count on a *propre*—a victory of space *over* time. As de Certeau (1984, p. xix) notes:

> The place of a tactic belongs to the other . . . and tactic insinuates itself into the other's place, fragmentarily, without taking it over in its entirety, without being able to keep it at a distance.

In Sham Shui Po, quite a large number of unlicensed hawkers (including NAPs and local residents) sell daily necessities to earn a living in public space such as streets and small parks. The large number of unlicensed hawkers in the district may be a result of the many residents (including NAPs) receiving money from the government's Comprehensive Social Security Assistance Scheme, who are not allowed to receive any income from part-time jobs or any other means. To make a better living, unlicensed hawking became an alternative. In the past, it was rare to see NAPs conducting hawking activities, but today it has become a lot more common. Some NAPs notice that this local culture—unlicensed business—is a possible way for them to earn more money for a slightly better quality of life. Nevertheless, it is not difficult to identify whether the hawkers are local residents or NAPs by looking at the products being sold: local residents can often be found selling local traditional snacks, while NAPs will be seen selling their own traditional snacks and handicrafts.

These hawkers did not set up their stalls (booths, counters) in places circumscribed as *propre* in response to officers of the General Affairs Teams. The hawkers did not apply for special licenses to sell their goods in permitted areas, or fight for unlicensed hawking to be made a legal activity through legislation. The hawkers did not usually expect to "share authority" or "use a legitimate basis to substitute for the existing legitimate basis" (Habermas, 1995, p. 33). Rather, they insinuated themselves into the areas of control and authority of the General Affairs Teams to safeguard their *survival* (de Certeau & Giard, 1998). They set up small counters that were not eye-catching and did not significantly impede the movement of people. They started doing business before the officers came, and set up their own surveillance teams to notify everyone should the officers arrive.

In the past, to avoid prosecution, the hawkers ran into private buildings and treated them as a "City of Refuge" as the officers were not authorized to enter (Siu, 2001). To deal with such tactics, the government formulated new strategies (introducing stronger legislation and working with the property management companies) to allow the officers to prosecute unlicensed hawkers even in the public areas of private buildings. The new policy caused significant difficulties to the hawkers and was successful mainly due to the support of the property owners, since hawkers often caused hygiene and safety problems. The government also gained support from shop owners (on the ground floor) as unlicensed hawking was a source of competition. To survive, some of the hawkers started to please the shop owners of the buildings by renting small "public areas" in front of the shops. In fact, these types of public areas were not owned by the shop owners and could not be rented legally; they were paid for as an incentive to the shop owners not to complain to the officers about illegal hawkers. While such "unofficial" under-the-table business deals were not legally accepted by the government, such tactics did succeed in making it more difficult for officers to prosecute the hawkers without the support (and formal complaint) of the shop owners.

Similarly, an increasing number of residents (mainly NAPs) gave singing, music, and dance performances in district parks in Tin Shui Wai for their own entertainment, or to make small amounts of money (Figure 20.2). They also omitted obtaining formal permission through skipping such steps as booking a place to perform or applying for a license to perform in a public space. Instead, many of them preferred to perform during a "non-busy" period so their performance would not significantly affect other people. Some of them selected the parks that did not attract many people or to which officers seldom came. They did not ask people to give money, since collecting or begging for money in public space remains illegal in Hong Kong. Instead, they claimed that people gave money voluntarily to help them pay for their musical instruments and the electricity they used.

Therefore, as described by de Certeau (1984), these tactical practitioners did not have a place, or their space "[was] the space of the other" (p. 37). Their tactics were totally dependent on time. They constantly needed to manipulate events to turn them into opportunities, and continually attended to their own ends. As observed by Lefebvre (1984) and Wander (1984), these kinds of tactical acts are an art of everyday life and a radical reorganization of modern life.

Figure 20.2 A group of NAPs practise music at a small corner of a park in Tin Shui Wai during a "non-busy" period. Photograph by Kin Wai Michael Siu.

The Third Realm

The findings in the two case studies here show that it is not appropriate to presuppose that the government (policymakers, planners, designers, management people) and public-space users (NAPs, local residents, other city users) are in a binary relationship of dichotomous opposition. This presupposition of binary opposition is value-laden and would limit the scope for understanding the relationships among government, users, and other related individuals and parties. Instead, by reference to Huang's (1990) theoretical argument in "the Third Realm" (see also Habermas's (1995) theoretical argument in "the Public Sphere"), the use-system of public space is better understood as:

- a formal and official urban policy and implementation system with codified laws and ordinances and government departments;
- informal ways of operating, with well-established customary, traditional, conventional, local, and individual practices; and
- an intermediate realm.

In the third realm, the intermediate realm, the interaction between the government and city users does not always play out in direct confrontation. This is because if interaction is simply a face-to-face struggle and negotiation is simply an enactment of force, then victory must obviously belong to the party of greater force—the *strong*. However, the strong—those holding power and authority—may not win all the time. On the contrary, the *weak* always have ways of not losing (de Certeau, 1984). The trick here is that the weak do not aim to engage in direct battle, or positional war. Rather, they aim to wage guerrilla war (Siu, 2007b).

For example, due to the loud volume of the music and dance performances in the parks of Tin Shui Wai, some serious complaints were raised to the government by local residents. By using the complaints as an excuse, the government started to clear some of the performance places in several district parks and set up additional regulations for the parks. Instead of fighting for a *propre* for performance, NAPs moved their places further away from residential buildings and dialed down the sound volume of their performance. They finished their nighttime "concerts" reasonably early to avoid affecting local residents' sleep. As relative newcomers and belonging to smaller communities in the districts, they also performed in areas of parks where people interested in seeing the performance would not affect other park users—those who had already occupied the places for a long time and taken it for granted that they owned the places.

Moreover, practices in public space have changed due to cross-cultural direct and indirect interactions. For example, in recent years, more local residents have started to accept and appreciate this kind of public performance by NAPs, staying to watch the performances alongside NAPs. A small number of residents even join the NAP groups, or form some similar group, though the groups of NAPs and the local residents do not have much direct interaction. In fact, these kinds of scenes were seldom observed before the 2000s, mainly because of the different cultural and social preferences and practices of local residents and NAPs.

As another example, the Chinese daily practice of sun-drying food and clothes was quite common in Hong Kong until the early 1980s. Following the improvement of living standards in recent years, this practice became less popular except by some village residents who kept up the practice. However, since the arrival of more NAPs in the early 1990s, the practice has become more common again and has started to become a serious issue. As some NAPs dry their things in public space, a ripple effect of cultural and social transformation occurs that results in the practice becoming more popular with local residents (Figure 20.3). Since the early 2000s, after receiving complaints from district councils working to promote district images, the government has introduced more rules and prohibited practices such as sun-drying. Instead of arguing for a *propre* or permitted place, the NAPs and local residents have started to explore new places (for example, places close to some remote construction sites, or places along some waste-water canals at the boundaries of the district) that are not commonly accessed by other people and do not attract the attention of government officers. Instead of applying for permission to sun-dry food and clothes in public space, they prefer to do it tactically by drying their things when officers are not on duty or in places where the responsibilities of different government departments are ambiguous.

The various interactions among the government officers, local residents, and unlicensed hawkers (local residents and NAPs) are also good examples to demon-

Figure 20.3 Local residents dry fish and meat on the gate outside a Lands Department construction site. Photograph by Kin Wai Michael Siu.

strate the interactions between the weak and the strong as well as the transcultural changes of daily practice in public space. Before being formally prosecuted, unlicensed hawkers in Sham Shui Po always tried to avoid direct confrontation with officers and shop owners (local residents) who were likely to think their business would be affected by hawkers. Instead, while ensuring that it was still possible to attract customers, hawkers would operate their business in public places that did not challenge the authority of the officers or influence the business of shop owners. Alternatively, instead of making enemies of the shop owners, the unlicensed hawkers would find another way to please the shop owners, as well as other residents, by including them as "companions" in their interactions with officers.

In fact, these kinds of interactions between the weak and the strong have been shifting due to the cultural, social, and physical transformation of the city. For example, some original residents, whose ancestors were farmers and fishermen in Hong Kong many years ago, conduct their mobile hawking business along the Tin Shui Wai Nullah (Figure 20.4). Many of them sell vegetables they themselves planted. In the past, government officers seldom prosecuted original residents for their unlicensed hawking, largely because the original residents of the New Territories were a group of people the government did not want to confront—particularly the rural communities who were strong enough to influence the effectiveness of governance. However, social, cultural, and political changes in recent years and the increase in the number of unlicensed hawkers have resulted in the need for change. Today, hawking by original residents is not immune to prosecution.

Clearly, because the strong have authority and power, they do not want to see uncertainty and diversity; that is, what they cannot determine and master in a positional war (Siu, 2007b). However, city users—the weak—are not a homogeneous

Figure 20.4 Some original residents of Tin Shui Wai conduct their mobile hawking business along a nullah. Many of them sell vegetables they have grown. Photograph by Kin Wai Michael Siu.

group. For example, while NAPs and local residents collectively partake in drying salt-fish in a park, drying clothes along a footpath, group-dancing in a children's playground, or running a small hairdressing business under a footbridge, they still have their own specific traditions, backgrounds, beliefs, needs, and preferences, and even dreams. These diverse characteristics mean that they may view and practice their everyday lives in city spaces in different ways.

As stated, the strong do not like "uncertainty" generated by the reception of the weak. Thus, in the face of unaccountable and continuously changing city variables, policymakers and professionals prefer to fix them. In particular, the government and planners intend "to transform the uncertainties into readable spaces" for a positional war (de Certeau, 1984, p.36; see also de Certeau & Giard, 1998). For example, once the government departments and the district councils received the local residents' complaints about the noise and crowd problems due to the music and dance performances in the district parks in Tin Shui Wai, they were more than happy to take the complaints as a weapon of rationale to clear the "undesirables" in the parks. They also set up more regulations and warning signs in the parks to further enhance the new controls. This is also the reason it is not easy to find instances of Lynch's (1990) expectation of the openness of space or full-of-flexibility designs either in Hong Kong or in many other cities today.

In some situations, as in the examples mentioned above, the third realm appears not only between the government (considered as the strong) and the NAPs and local residents (considered as the weak), but also between local residents (the strong) and

NAPs (the weak). That is, local residents became the strong (due to their numbers and original well-established/formal practices), while NAPs retained their role of the weak. For example, local residents in Tin Shui Wai seldom shared the same area in a park with NAPs. While local residents were still in the majority, instead of directly competing for space, many NAPs preferred to search for *other* space where they could socialize with their own groups. NAPs sometimes brought their own stools and chairs or ones collected from the street to public spaces to create areas where they could sit and chat with their friends. In other words, the weak do not usually react directly to the force, power, authority, and orders of the strong. Like guerrillas, the weak insinuate themselves into the strong's place to seek opportunities. Furthermore, the weak do not have or maintain their own place. This way of operating makes it difficult for the strong to display their force.

Obviously, all these spatial practices of NAPs and local residents are in a state of flux in Hong Kong today, including the status and roles of NAPs and local residents. For example, after being in Hong Kong more than seven years, NAPs know that they are more likely (and some are more willing) to be called local residents by other people. While they still maintain some original cultural preferences, they themselves have also undergone changes through direct and indirect interactions with (and observations of) the local residents. Some of the NAPs start merging into or integrating with the local communities, and some even prefer to become more detached from their original communities; meanwhile, some of them have held on to their original cultural preferences and lifestyles, and group together to form stronger separate communities.

While NAPs become less of an extreme minority in society, some NAPs have started taking the roles of the strong (or at least not the weak) in that sometimes they do not mind mild confrontation and competition with the local residents (the former strong). The change of spatial use in Tin Shui Wai in the past two years is a good example. NAPs have started *sharing* more places in parks with the local residents. Instead of only using the less popular places, some communities of NAPs move to some popular places to conduct their preferred activities. In the past, most of the plazas and parks in Tin Shui Wai were occupied by the local residents because the number of NAPs in the district was still small. Since the mid-2000s, the overall population of NAPs has increased rapidly such that the use of public space has changed. Today, it is easy to see NAPs and local residents share many public places in the district though they seldom have direct communication and contact (Figure 20.5).

Moreover, in the past, the recreational facilities of the public housing estates were mainly used and occupied by children after school. Adults rarely used such facilities, which stood empty most of the time. This situation has changed in recent years due to demographic, cultural, and social changes. For example, table tennis is a common recreational sports game in mainland China, practiced by many adults. Today, it is common to see housewives (mainly NAPs) playing table tennis in the common space of public housing estates. This cultural and social change seldom causes problems, as housewives play the game during their free time between household chores. For example, they play after taking their children to school and buying food for meals, and before they cook the meals and collect their children from school. Selecting hours of non-peak usage to play also prevents arguments with local residents

Figure 20.5 NAPs and local residents occupy different areas of a plaza outside a shopping center in Tin Shui Wai. Different groups of people there have their own unspoken but well-defined territories that they know where to join and stay. Photograph by Kin Wai Michael Siu.

concerned about conflicts over usage of space and contact between NAPs and the residents' children.

Re-constructing Public Space

Similar to other major cities, Hong Kong has undergone different kinds of transformation. The case study findings as described above illustrate that neither NAPs nor local residents follow or rely on a constant mode of operation as they negotiate for their place in the public realm. They have developed their own ways to adapt and survive, regardless of the regulations and controls imposed in public spaces. Due to the continuing cultural, social, physical, economic, and political transformation in Hong Kong, NAPs and local residents are also required continuously to *re-construct* public space to make it more inhabitable. Some of them *re*-define the meaning of park space as their stage for music and dance performance. Some *re*-territorialize the boundaries of different public space for their own social activities. Some *re*-build the physical environment of street space for their small businesses. Some *re*-establish the public space following rules that avoid confrontation and competition with other users. Some also *re*-order the temporal order of a place, by turning it into an early morning market where both NAPs and local residents benefit from lower prices while avoiding officers' prosecution. All these *re*-s are creative acts and arts based on long-term interactions among different users. These *re*-s are also constantly adapting to the cultural, social, physical, economic, and political transformation of the city. As mentioned before, the NAPs and local residents have continuous cross-cultural interactions. The changing physical environment and results of interactions have also provided further opportunities for the new generation of NAPs and different generations of local residents to have richer and more dynamic transcultural interactions. In sum, all of these aspects are important issues for today's policymakers, planners, and designers; these influential people should be aware of and should understand, respect, and treasure the everyday practice of people when and if they claim that they have good (and perfect) ways to transform Hong Kong into an inhabitable city for all.

Acknowledgments

The author would like to acknowledge the resources extended by the Hong Kong Research Grants Council (General Research Fund; RGC 548310) and The Hong Kong Polytechnic University for the case studies. Part of the findings of this chapter was presented in the Transcultural Cities Symposium at the University of Washington (2011). All field observation photographs were captured by the author and his research assistants, and special thanks goes to Leo Wong and Yihua Huang for their assistance in some of the case study fieldwork, and Pak-hong Wan for his assistance in the drawings.

Note

1 Hong Kong is a Special Administrative Region of China. Thus, the Hong Kong government officially uses the terms "newly arrived persons" and "new arrivals," instead of "immigrants," to describe people who move to Hong Kong as residents. A person continuously living in Hong Kong for more than seven years can apply to become a permanent resident. In this chapter, local resident means a person who has lived in Hong Kong for a long time, that is, a permanent resident.

References

Chow, N. W. S. (1988). *Social adaptation in new towns: A report of a survey on the quality of life of Tuen Mun inhabitants* (Resource Paper No. 12). Hong Kong: University of Hong Kong.

Cuddon, J. A. (Ed.) (1998). *A dictionary of literary terms and literary theory* (4th ed.). Oxford: Blackwell.

de Certeau, M. (1984). *The practice of everyday life*. Los Angeles: University of California Press.

de Certeau, M. & Giard, L. (1998) Ghosts in the city. In M. de Certeau, L. Giard, & P. Mayol (Eds), *The practice of everyday life: Volume 2: Living & cooking* (T. J. Tomasik, Trans.) (pp. 133–144). Minneapolis: University of Minnesota Press.

Habermas, J. (1995). The public sphere (*Öffentlicheit*) (in Chinese). In J. Habermas et al. (Eds), *Socialism* (pp. 29–37). Xianggang: Oxford University Press. (Original work published 1973.)

Harvey, D. (1989). *The condition of postmodernity: An enquiry into the origins of cultural change*. Oxford: Blackwell.

Harvey, D. (1996). On architects, bees, and possible urban worlds. In C. C. Davidson (Ed.), *Anywise* (pp. 216–227). Cambridge, MA: MIT Press.

Hong Kong Census and Statistics Department. (2011). *Social topics studies*. Retrieved January 1, 2011, from: http://www.censtatd.gov.hk/hong_kong_statistics/social_topics_studies/index.jsp

Hsia, C. J. (1995). Making a better-told tale of cities: From Dyos, Kostof to Castells. Paper for the International Conference on Chinese Architectural History: Issues in Methodology (pp. 61–80) (August 7–10). Hong Kong: Department of Architecture, The Chinese University of Hong Kong.

Huang, P. C. C. (1990). *The peasant family and rural development in the Yangzi Delta, 1350–1988*. Stanford, CA: Stanford University Press.

Iser, W. (1978). The *act of readings: A theory of aesthetic response*. London: Routledge and Kegan Paul.

Lefebvre, H. (1984). *Everyday life in the modern world* (S. Rabinovitch, Trans.). New Brunswick, NJ: Transaction Publishers.

Lynch, K. (1990). The openness of open space. In T. Banerjee & M. Southworth (Eds), *City sense and city design: Writings and projects of Kevin Lynch* (pp. 396–412). Cambridge, MA: MIT Press. (Original work published 1965.)

Sennett, R. (1970). *The uses of disorder: Personal identity and city life*. New York: W. W. Norton.

Siu, K. W. M. (2001). *The practice of everyday space: The reception of planned open space in Hong Kong*. Ann Arbor, MI: UMI.

Siu, K. W. M. (2003). Continuous settlement and urban redevelopment on Hong Kong's Chun Yeung Street. *Critical Planning, 10*, 55–71.

Siu, K. W. M. (2007a). Reader's response and reception theories: Discussion on user-oriented design. In W. C. Qian (Ed.), *Transactions of K. C. Wong Education Foundation supported lectures, 28* (pp. 16–26). Shanghai: Shanghai University Press.

Siu, K. W. M. (2007b). Guerrilla wars in everyday public spaces: Reflections and inspirations for designers. *International Journal of Design, 1*(1), 37–56.

Social Welfare Department. (2011). *Comprehensive Social Security Assistance (CSSA) Scheme*. Retrieved March 8, 2011, from: http://www.swd.gov.hk/en/index/site_pubsvc/page_socsecu/sub_comprehens/

Wander, P. (1984). Introduction to the transaction edition. In H. Lefebvre, *Everyday life in the modern world* (pp. vii–xxiii). New Brunswick, NJ: Transaction Publishers.

CHAPTER 21

Peripheralization and Other Roman Stories

Lorenzo Rinelli

> On the one hand the orbis and the urbs, circular, with their extensions and implications (arch, vault); on the other hand the military camp with its strict grid and its two perpendicular axes, cardo and decumanus—a closed space, set apart and fortified.
>
> (Lefebvre, 1992, p. 245)

The social and political questions posed by present-day European migration control policies and their repercussions lead my investigation into the European urban space. Historically, immigrants and migrants tend to be concentrated in the most important urban areas in a receiving country. As cities become home to a greater number of immigrants, urban and immigration policies necessarily overlap. The filtering effect of urban borders, which alternately deflect, tolerate, or promote dwelling, conditions every urban person's existence. For instance, analysis of the architecture of the European Union's border management reveals an interior design that features detention camps for immigrants as ordinary features on the landscapes of the most populated outskirts of European cities.[1] Immigration policies that are adopted to manage international migrations—visa requirements, biometric filtering, militarization of the border—continuously produce anxiety and a population "in excess," while feeding hostility and discrimination. No migrant enters a neutral ideological context when arriving in the EU. Depending on how external controls situate and catalogue migrants individually and collectively, they differently regulate how, and even whether, people are able to enter the EU's urban centers and call them "home" (Balibar, 2003). Once migrants have crossed the frontier of the EU, they come to occupy a precarious place in the city that exists at the margins of integration.

This chapter locates one of the new frontiers of Europe within a fundamental paradox of the city of Rome: an inexorable expansion of buildings together with a mounting rejection and marginalization of an emergent immigrant population that is vital to the city's expansion. There are two reasons to study the city of Rome in relation to the contemporary European policies of migration control. First, the route that goes from East and Central Africa towards Libya via the Sahara Desert has been, at least until summer 2011, the most trafficked and policed in North Africa, and therefore the most dangerous. The main flow of migrants traveling from Sub-Saharan Africa to Europe sails from the Libyan coast and arrives at the southern coast of Italy. Several scholars have made the case for the significance of studying the topic of migration to Italy and its relation to questions of space, time, control, power, and visibility (Parati, 2005; Bullaro, 2010). Rome represents the place where immigrants in Italy can most quickly find points of reference and social coordinates to build new ways of life.

Just as it once was at the apex of its splendor, Rome's vocation is linked to that of the Mediterranean, which leads to the second reason. The new urban plan for the city of Rome explicitly promotes the Mediterranean Sea as a harmonious center of different cultures with Rome as its epicenter, although the history of urban planning in Rome, and particularly fascist ideology, should problematize the Mediterranean as a mechanism that harmoniously regulates the residency of people in Rome's urban space. Officially the plan aims at giving voice to different sounds of the immigrant communities living in the city; in reality, I maintain, it renders their rhythm unintelligible and their essence never a possibility. Questions to be explored in this chapter will be related to the official discourse of multiculturalism set in motion by Rome's town hall and whether it pushes African migrants to the margins of the city, creating distinct places in the city's peripheries. Once at the margins it is important to identify migrants' tactics of resistance and placemaking that support individuals and their communities. It is important to focus on these tactics since they seem to lead to alternative social imaginations beyond marginalization.

Rome—Peace Capital of the Mediterranean

> To talk of the Mediterranean—of its past, present and future—is to move in this disquieting place.
>
> (Chambers, 2008, p. 5)

On March 14, 2008, forty-five years after the last urban plan,[2] the city council approved a new urban plan for the city of Rome. Several successive city administrations took on the difficult task of formulating a new urban plan for Rome. The result is a compromise among groups of private landowners, developers, and city hall to regulate the expansion of the city. Squeezed in the middle is a growing immigrant population which has been left out of the public debate. The introduction to the new plan is solemnly titled "Rome—Peace Capital of the Mediterranean: the new urban plan of Rome" (Morassut, 2003). The introductory document is intended to highlight Rome's crucial role in promoting intercultural and interreligious dialogue featuring social cohesion, as "the richness of contemporary metropolis is over all associated with their standards of social quality" (Morassut, 2003, p. 27). But how does the new plan supposedly promote social quality?

First of all, it does so through environmental protection. It reserves 88,000 hectares of territory to green areas. Second, the protection and preservation of historical Rome is to be extended beyond the classic Roman ruins to include contemporary architecture that is located within the periphery. Third, and most important, the innovation that the urban plan envisions revolves around the concept of the polycentric city: there will be eighteen new urban nodes, or miniature cities, within the city, each with schools, hospitals, and administrative services (Figure 21.1). Lastly, Rome requires the so-called "iron cure," which is the modernization and lengthening of the entire metropolitan rail system.

In the initial document, Roberto Morassut, town-planning councilor in 2003, describes a utopian city that is accessible to its inhabitants, that is filled with social opportunities, and that is modern without forgetting its historical character. But

Figure 21.1 An example of a new urban node in the southern part of Rome. Despite the intentions of the planners, most of the newly constructed urban nodes lack vital services such as transportation, schools, hospitals, or even a well-designed sewage system. The result is the creation of isolated urban dwellings at the margin of the city. Photograph by Lorenzo Rinelli (2010).

what exactly does this document entail? At first sight, the link between Rome and the Mediterranean recalls a distant past perhaps best encapsulated by the Roman imperial tag: *mare nostrum* (our sea). As a matter of fact, the definition of *mare nostrum* appears on the first page of Ceasar's *De bello gallico* (On the War against Gauls), indicating confidence from having acquired full control over the Mediterranean—which led to a change in Rome's foreign policy to be oriented to the North, towards the Atlantic. The idea of *mare nostrum* entailed a totality of control over a sea that ironically was considered by Roman emperors as a mere lake "already conceptualized as a homogenous space of confinement, possession, and colonization" (Pugliese, 2010, p. 11). However, the Mediterranean Sea was never a homogenous space either during the Roman occupation or today when European policies of migration control attempt to purge it of foreign bodies as they traverse the few miles that separate Europe from Africa entering Italian territory.[3]

Multicultural City

Once the African migrant becomes visible in the urbanscape, once she has navigated the surface of the Mediterranean Sea, she comes to be incorporated, more often than not, as a multicultural ingredient that serves to promote the image of a tolerant and multiethnic city. The latter has to be contextualized within the strict relation between market and urban society. Multiculturalism has a major leverage for generating future growth and attracting investment capital and consumers. This phenomenon illustrates a growing relationship between neoliberalism and cosmopolitanism, whereby Rome competes with the other great capitals of Europe, like Paris and London, in their social makeup. It is worth noting here, although briefly, that Rome has only recently approached the discourse of multiculturalism and is doing so in a superficial way. The tardiness of government policies is not necessarily linked to the ethnic composition of its society, which does not share the postcolonial flows of Paris and London; instead it has to do more with the fact that Italian society has not yet come to terms with its colonial past and with its contemporary role as a country of immigration.

Nowadays the immigrant, when visible, stands as a classic example of a fetish where multiculturalism and celebration/construction of difference is the natural outcome of a managerial approach to the city. As Deleuze and Guattari have noted "the dividing line is not between inside and outside but rather is internal to simultaneous signifying chains and successive subjective choices" (1987, p. 178). It implies a fulcrum, a homogenizing principle at its core, a white wall on which to design a rainbow cloth. The idea behind the document that introduces the urban plan is that Rome with its monuments and (imperial) history plays this role. In this sense, the slogan "Rome—Peace Capital of the Mediterranean" signifies a homogeneous space, without conflicts where, therefore, the city's market can flourish and different voices can be distinctly heard. Peace, intended as absence of conflict, is then the necessary chamber to generate harmony. But where can harmony be found other than in an orchestra?

In the following section I show how the documentary entitled *L'Orchestra di Piazza Vittorio* (Ferrente, 2009) exemplifies Rome's organized discourse of multiculturalism. A critical reading of the documentary will give me the opportunity to explore how films, documentaries, and media in general can contribute to reinforce configurations of multiethnic society from above within an ethics of social concern denying the the immigrant equal status with the filmmaker/citizen. Disillusioned with official discourses of multiculturalism, I then move my research toward the discursively external part of the city: the periphery. There, in the southeastern area of Rome, I ultimately recognize the interconnections between Europe's emerging southern border and the urban space of Rome. Like a border, the urban structure does not change spontaneously. Nor does a single powerful agent, such as the state, determine it. The urbanscape changes in ways that are related to forms of inclusion/exclusion that occur through modalities of externalization, or in the case of cityscape, of *peripheralization*.

For Whom Does the Orchestra Play?

The documentary *L'Orchestra di Piazza Vittorio* (the Vittorio Square Orchestra) takes place in the district around Rome's main train station. Today the area is largely populated by immigrants whose activities revolve around a square, Vittorio Emanuele II Square, that is a hub for various cultures, sounds, and odors from all corners of the world (Lakhous, 2006); even Chinese shops are colonizing the entire area (Figure 21.2). This Piazza, its essence, inspired musician Mario Tronco and filmmaker Agostino Ferrente to reunite some of the most extraordinary performers among immigrants, each one unique in origin, instrument, and musical experience, in an orchestra, that plays world music all over the world.

In 2006 a documentary that tells the story of how the project took shape was released and since then screened in many festivals around the world, winning several awards. Watching the documentary, we walk into a sort of harmonious melting pot

Figure 21.2
The area around Vittorio Emanuele square is occupied almost entirely by Chinese shops that have increasingly ousted historical Roman stores. Note the two posters of the radical right wing group, Casa Pound, on the book presentation "La Presa di Roma" [The seizure of Rome] by Claudio Cerasa (2009). The group, headquartered in the area, is very active in promoting anti-immigrant actions. Photograph by Lorenzo Rinelli (2010).

that fits perfectly with the official discourse of Rome as a multicultural place presented with the new urban plan. While the story revolves around the project to revitalize and save an old theatre, the Apollo Theatre, from being transformed into a bingo hall, we have the impression that immigrants exist simply to provide some color to this political mission; they never participate directly in the group's meetings nor are they involved in conversations with the town hall representatives. They remained isolated from the political possibility of this endeavor. They are left aside to do what they are good at doing: playing music.

The two central characters, both Italians and both quite established within the art community, seem to enjoy the practice of fishing for new artists in the area around the square. An attentive reading of these scenes reveals an instinctive hesitancy and incomprehension, if not directly hostile, towards the two Italians. At the beginning of the documentary when the two Italians enter different shops to look for artists, whether a singer or a player of tablas, they do it by intruding into others' spaces with a sort of nosiness and naïveté, "only dimly aware of a certain unease in the air" (Pink Floyd, 1977).

Even when the documentary offers glances of the real life of some of these musicians while they struggle to survive in the city of Rome, the Italian organizers are preoccupied with accomplishing their musical and political project. They panic, for example, because the date of the concert is approaching fast and the Indian player is sent back to India for visa issues; or after several requests for help made by Raul, the Argentinean drummer, because he is going to be evicted from the garage where he lives. The documentary does not rest enough upon the city's urban policies and high rent costs and the consequent marginalization connected with these municipal issues. It seems as though immigrants exist outside of global capitalist social relations and their corollaries of nationalism and racism.

Overall, the documentary does not engage in a dialogue with the immigrants who are part of the orchestra, and without whom the orchestra would not exist. This is possibly because creating a dialogue would highlight tensions that occur within the "organization of difference as a qualitatively homogeneous fetish" (Sharma, 2006). In the same way, the discourse of multiculturalism presented within the urban plan does not imply a reification of separation between immigrants. Instead, the celebration of difference, both within the orchestra as well as between the lines of the urban plan, celebrates sameness within a generic alterity. The fundamental separation exists within hierarchical differences that are aimed at "reducing complex and overlapping relations" (Hardt & Negri, 2001). It suggests the preservation of a firm core around which legal and economic status and place meaning revolve, as the planets do around their sun. As Fernandes brilliantly put it (2007, p. 99):

> The West is cosmopolitan only insofar it understands its culture as the final step in a historical movement. Hence Western cosmopolitanism does not celebrate globalization as flows that breed differences but seeks to discipline the proliferation of differences through colonial-inspired racial and spatial hierarchies.

In the end, a well-liked project such as the *L'Orchestra di Piazza Vittorio*, both despite and because of its politically correct multicultural language, not only fails to question the dominant political organization and creation of differences but also despatializes other ideas of communities and reinforces the homogenization of immigrants' unique experiences of the urban. Instead, true relationships within the urban occur when they are disconnected from the planned idea of difference and are connected to space through practice. Home comes to be defined, then, through experience rather than imposed places and identities. At the margins of the city, it is possible to encounter processes of transculturation that instead of being transcendent, rely on continuous negotiation. Processes of transculturation are indeed linked to subjectification that happen through imaginative reconfigurations of the political in a perennial movement of subjectivity (Fernandes, 2007, p. 133).

Anagnina

In this section, I move into the southern part of Rome where the documentary *Like a Man on Earth* (Segre & Yimer, 2008) in part takes place. It is a work that certainly explores the experience of migration from Africa in Italy while it recovers lost stories of courage and oppression. The documentary produced an original and innovative encounter with the society and city of Rome, along the tradition of Italian Neorealist cinema that "arguably after architecture . . . was the art form most rigorous in confronting the ruinous landscape, charting its provinces, figuring its dimensions, its rifts" (Steimatsky, 2008, p. xii). For Pasolini, who has been a critical lover and attentive observer of the Roman society, the south of the city where city meets fields was the opposite of the northern part; exchanges between people in the north were conducted on the basis of reason, in contrast to life in the south that was fed by passion and life. Today this idea still holds some truth, as this marginal section of the city is again a vital center, for the entire city and for all south-living immigrants outside the municipality of Rome who arrive in the morning at the Anagnina bus station to find a job near the city limits. It is the desire to get into the urban dimension, to be part of the legitimized *polis,* that drives the newcomers here, much as that desire drove Pasolini's *Mamma Roma* and her son together with an entire generation of Italian migrants fifty years ago.[4]

At the beginning of *Like a Man on Earth*, we sense that Dag and his companions have finally reached Europe and the city of Rome, after a long, agonizing odyssey. In the background, gray benches outline the Anagnina bus station, which is at the southeast limit of Rome and was until only a short time ago the unmistakable threshold of the Roman urbanscape (Figure 21.3). Today, the wide open space is a neuralgic yet crucial hub for trade and information sharing among migrant workers who come from the countryside to find work in the city.

The Anagnina bus station is emblematic to the understanding of Rome's peculiarities in relation to the new urban plan. First of all, immigrants live in this area because they cannot afford to pay the rent within Rome proper. The renting market operates together with immigration laws as an effective filter to keep immigrants disconnected from the city without turning away a cheap workforce. It is worth noting here that the station functions as the southeast door of the city as it welcomes the long distance bus routes that arrive from the small villages and towns along the coast south of Rome. These towns are being occupied by newly arrived immigrants who have revitalized the spaces abandoned by Italians and have boosted, if not basically created from scratch, a renting market otherwise sterile along the coastline during winter seasons.

Basically, those who utilize these long routes as means of transportation are of foreign origins. When these buses arrive at the Anagnina Station to drop into the underground metropolitan network, immigrants sit at the margins, genuine

Figure 21.3 A profile of Dagmawi Yimer against a background of gray benches of Anagnina bus station. Source: Segre & Yimer (2008).

strangers to the city, looking for a good bargain or some useful information, before scattering along the ring road that surrounds the Eternal City. Every day, the space of the bus station blooms with vendors selling every kind of object.

These vendors mostly come from the former Soviet block but also from Africa and South Asia. It is a self-sustaining, spontaneous market, still at the beginning of its existence, and does not have the official endorsement of the municipal authority. In fact, the authorities have cyclically challenged their presence, but immigrants have always come back to set up their stands.

This is a transient space, for and by nomads who nonetheless are increasingly involved in organizing a series of social, economic, and cultural activities. They dwell in this space that belongs to them as they intervene in the "distribution of the sensible" (Rancière, 2006) of the political space that otherwise has been sterilized and evacuated. These are border figures that exist in between citizenship and exile, in a space beyond the dichotomy of legitimization and illegitimization, between multiculturalism and complete integration.

Instead of representing a pathological syndrome that can be removed, the nomads, I claim, personify an indication/warning regarding the central paradox of a new urban policy for Rome that permits no alternative to neoliberal multicultural hegemony. There is a widespread consensus among Italians that the urban condition needs to be taken seriously, that a managerially illuminated multicultural approach from above should be implemented to avoid the otherwise inevitable social calamity (see the riots in Paris *banlieus* in 2005 and 2007). As Rancière puts it (1998), consensus, far from representing the absence of conflict or peace, indicates a subterranean permanent violence intrinsic to any form of social purification and political evacuation. The mechanism of externalization related to migration control is indeed an example of consensus wherever it may occur, far from the government geographical location or within the metropolitan areas of Europe. However, it would be erroneous to configure the idea of consensus as a hegemonic and homogenous space, a plaque, where political life occurs separate from personal life. In this space of purity, our lives will be dominated by an overwhelming power and, as Rancière put it, "entrapped in the complementarity of bare life and exception" (2010, p. 11).

Instead, proper politics are all about perforating this space, to produce breaks of elocution, which before were only heard as mere noise. As a matter of fact, the urban border appears only in the moment it is crossed, "for the border is not a thing but, rather, the materialization of authority" that has been contested and subverted (Chambers, 2008, p. 6). The spontaneous meeting point at the station engages the distribution of roles in the community as it contends with exclusion and exists as a transcultural space where practice prevails. In comparison, the well-intended project of the Vittorio Square Orchestra provides us with a pre-assembled, legitimized world and a romantic view of its immigrant subjects. Multiculturalist projects confuse communal distribution of bodies and social concerns with the political as Rancière intended it. He writes (2010, p. 39):

> Political argumentation is at one and the same time, the demonstration of a possible world, in which, the argument could count as an argument, one that is addressed by a subject qualified to argue over an identified object, to an addressee who is required to see the object and to hear the argument that he "normally" has no reason either to see or to hear. It is the construction of a paradoxical world that puts together two separate worlds.

Following the documentary we come across a separate world a few miles from the Anagnina bus station, where the second University of Rome campus, Tor Vergata, is scattered over a vast and desolate industrial area. The neighborhood quarters a huge mall complex and the comforting blue profile of IKEA, plus many clothing factories and home furniture outlets that everyday attract thousands of distracted consumers. One of the university buildings, the former Faculty of Humanities, is a 12-story blue glass building. Here, with neither electricity nor a heating system, lives a community of 250 refugees from the Horn of Africa, including almost 30 children (Figure 21.4). Some of them were born during the exodus through the Sahara Desert and across the Mediterranean Sea. These are people who survived the strenuous journey, crossed many borders, became refugees at the port of Lampedusa, and inexorably crashed against the indifference of the post-political contemporary Europe. After being transferred to identification centers in Sicily, they were set free and, like many others before them, caught the first train to Rome, without any help or directions. This is the last frontier of Europe: its most internal one and probably the most difficult to decipher.

In December 2005, with the support of *Action*, the building was occupied; however, in 2006, the refugees were cleared out. After that, the city hall signed a rent contract with the society that owns the building and in the meantime tried to find solutions for accommodating the refugees. The solutions adopted were always temporary and, according to the refugees, simply inadequate. They complain of the

Figure 21.4 Second university of Rome, Tor Vergata. The glass building used to house the Department of Humanities. Today it is home to 250 refugees from the Horn of Africa who have lived there in precarious conditions since they occupied the vacant building in 2005. Photograph by Lorenzo Rinelli (2010).

size of the rooms where people had to live with five or six others and of the night curfew. Many of those who dwell here would like Italian authorities to clear their biometric data from the Eurodac database that enables countries to help identify asylum applicants and persons who have been apprehended in connection with an irregular crossing of an external border of the EU and therefore facilitates the application of the Dublin Regulation.

According to the Dublin Regulation (2003/343/CE), the state responsible for examining asylum applications is the one through which the asylum seeker first entered the EU. Once officially recognized, that person cannot apply to another state where he/she may have better chances of success and living conditions. The aim of this regulation, which replaces the original Dublin Convention (1990), is to prevent multiple applications and hordes of moving asylum shoppers around the EU and to reinforce the role of the border on those states that geographically confine "problematic" countries.

It is beyond the scope of this analysis to present the consequences of this policy reorganization on the lives of asylum seekers who, more and more, are subject to externalizing practices of interception, forced rejection, and peripheralization. Here it is important to note that the process of externalization operates through the combination of legislation of migration and refugee control together with the new urban plan for the city and its renting market. Muse, one of the residents, recalls that "to find an affordable rent in Rome is a massive challenge considering the current rates. And more, when they [the landlords] hear that you are a foreigner they say that they do not rent anymore" ("Fortress Europe," n.d.). Unfortunately, as I noted above, once inside, asylum seekers cannot freely move within Europe to find the best opportunities for living and working.

In the face of the pattern of exclusion and peripheralization, is there anything conceivable beyond tactics of survival? Is there anything that can be theorized as an innovative form of placemaking? Can a cinematic expression constitute a spatial practice by and for immigrants that not only facilitates mutual understanding but also leaves a mark in the city while engaging the distribution of the sensible?

I Am A Man

I met Dagmawi Yimer, aka Dag, in December 2009, one year after the release of *Like a Man on Earth*. Dag was a student at the Faculty of Law of Addis Ababa University when instances of political oppression and electoral corruption following the general elections of May 2005 ("Ethiopian protesters 'massacred,'" 2006) made him leave his community behind and embark on a long and perilous journey to Rome, where, eventually, he has been granted refugee status. After having survived the violence of the multiple borders all the way through the arid Sahara Desert and the deep Mediterranean Sea, he arrived in Rome. Next, Dag joined the school of Italian language organized by the *Asinitas* cultural association. The school is a meeting space in itself for many African migrants in Rome. Within the safe place of the school, Dag not only learned to speak Italian, but also became literate in the language of video-documentation. He soon decided to gather the memories of others who, like him, had journeyed in exile from Ethiopia to Rome. Memories, gestures, and silences of the people who are represented recall actual difficult challenges that these contemporary migrants had to face along their journey. The kitchen where the interviews take place between Dag and his friends becomes a sort of sanctuary where in Aramaic they can share excruciating memories, knowing that those memories will not be lost. In doing this they become present. They leave a mark in the space and time of the host society that cannot turn its back to its responsibilities in the construction of their memories. It stands for a true placemaking that builds upon a dialectical process of recognition. It reminds us that, "it is in the interrelationship between the individual and society that questions of power and hierarchy, ideology and politics stand most forcefully revealed" (Nichols, 2010, p. 249).

To further underline the political importance of Dag's documentary, it is important to mention that *Like a Man on Earth* is part of a larger project, the Migrants' Memories Archive, which, since 2006, has intended to recuperate both the memory and the dignity of the migratory path of many out of the Horn of Africa.

The project aims to invert the usual process by which the "us" (migration experts, academics, and journalists) speak on the "other's" behalf. It is important to note how the project seems to challenge the very limits of the concept of archive, intended as a "closed" system of storage for memories. In this sense, the active and continuous participation of migrants in building the archive renders the project a "political act" (Segre, 2009, p. 19).

The project is the outcome of conversations between social workers of the Asinitas Association, academic scholars specializing in colonial and postcolonial history of the region of the Horn of Africa, and a group of refugees from the same area. It explores the possibility of filling the lacuna between migrants and the host community. With regard to this social condition, Abdelmalek Sayad (2004) elaborates the idea of *la double absence*, which is the alienation of the emigrant/immigrant. She is condemned to be absent twice, with respect to both her origins and her new society. Marco Carsetti recalls that the idea of building an archive of memories was born after his experience listening to migrants around Vittorio Square. In one of those recordings, Aziz screams his agony out when he says that "we are nothing that exists here, they do not want see us" (Segre, 2009, p. 26).

Through the migrants' active participation in narrating and setting the conditions for the narration, this concept of archive facilitates the spatial and temporal dispersion of the double absence. It succeeds in doing this at the first level by innerving the official discourse around the theme of migration from Africa to Europe. In fact, *Like a Man on Earth* rips open the veil that makes the implementation of the contemporary European policies of migration control in Africa within the urban dimension of Europe invisible and inaudible.

Thus far, conventional academic analyses of migration remain bound to a hierarchy of high over low: the subjects/agents, policies, and data that are associated with states, politicians, and public figures take priority over the lives and experiences of migrants and their interlocutors. Although there are several academic attempts to examine the dynamics of migration from Africa to Europe (Gebrewold-Tochalo, 2007; Pastore, 2007; Huysmans, 2006), scholars tend to either focus on policy analysis and inter-state relations and pay no attention to the human dimension; or they attempt to map the route of the flow, like tracking a herd along geographical lines. Each approach is confined within its proper academic locus.

These hierarchies then commonly filter what can be heard and what can be viewed. We hear politicians arguing for the construction of new walls. We hear the media venting citizens' frustrations about supposed effects of "illegal" migration, such as increasing unemployment, urban insecurity, and cultural contamination. We see boats loaded with desperate people rescued by police authorities. The same is true for what cannot be heard and what must remain invisible. We cannot see a migrant reclaiming her political subjectivity within her new host society and city. In the documentary, Dag himself engages in a conversation with Mr. Franco Frattini, the former EU commissioner in charge of European policies of migration control. Outside the comfort of state media, Mr. Frattini is clearly bothered by Dag's attempt to challenge and reorganize the terms of the discussion simply by presenting his refugee experience as tragic and clamorous as it is banal.

Like a Man on Earth succeeds in more than revealing what has been kept concealed regarding the recent transformations of policies of migration control in Europe and Africa. It addresses the authority of an immigrant to speak within the political realm of the nation-state. It also, perhaps more importantly, questions the multicultural homogenization of a place, that of Europe, that is maintained at the center of a globalized world. But Europe and especially European metropolitan areas are spaces constantly being translated, traversed into a striated space "constantly being reversed, returned to a smooth space" (Deleuze & Guattari, 1987, p. 474). This is, at the end, the essence of transculturalism: a process of performances between two or more communities and individuals within a dialectical zone of tension.

I read Dag's documentary as an effort to reappropriate the capacity to voice his and his companions' experiences in a universal (visual) language that breaks the above-mentioned hierarchy and traverses the space of the city. As Benjamin put it, "the characteristics of the film lie not only in the manner in which man presents himself to mechanical equipment but also in the manner in which, by means of his apparatus, man can represent his environment" (1969, p. 235). The author's capacity for expression reclaims a temporal and spatial mobility both from the society of origin along certain routes and at certain speeds, and also within the city he inhabits. That's why that if we conceive that city as a striated space traversed by memories, sounds, images, and experience, a sort of living and pulsing archive, then *Like a Man on Earth* succeeds to mark a further line on this archive, interacting, translating, and ultimately modifying the surrounding social environment.

Notes

1 www.migreurop.org/IMG/pdf/carte-en.pdf
2 Rome, as capital of the Italian state, has had five other urban plans: 1873, 1883, 1909, 1931, 1962; but only one, that of 1909, has been adopted by the city council. The urban history of Rome confirms its peculiarity in terms of transparency of planning mechanisms (Insolera, 1993).
3 Recently "Interior Minister Roberto Maroni said migrants who have landed on the island of Lampedusa threaten the institutional and social structures of Europe" ("Italy warns Europe over migrants," 2011).
4 In 1962, Pier Paolo Pasolini realized *Mamma Roma,* as a complex critique of the urban development of Rome (Rhodes, 2007). Its central character, played by a superb Anna Magnani, is a prostitute who dreams all her life to leave behind her life on the streets and bring her son with her to have their home *coi signori* (with gentlemen) in one of the new buildings of the expanding southeast periphery of Rome.

References

Balibar, E. (2003). *We, the People of Europe? Reflections on Transnational Citizenship*. Princeton, NJ: Princeton University Press.
Benjamin, W. (1969). *Illuminations: Essays and Reflections* (1st ed.). New York: Schocken.
Bullaro, G. R. (2010). *From Terrone to Extracomunitario: New Manifestations of Racism in Contemporary Italian Cinema: Shifting demographics and changing images in a multi-cultural globalized society*. Leicester: Troubador Publishing.
Cerasa, C. (2009). La Presa di Roma. BUR Biblioteca Universale Rizzoli.
Chambers, I. (2008). *Mediterranean Crossings: The Politics of an Interrupted Modernity*. Durham. NC: Duke University Press.
Deleuze, G. & Guattari, F. (1987). *A Thousand Plateaus: Capitalism and Schizophrenia* (B. Massumi, Trans. and Foreword). Minneapolis: University of Minnesota Press.
Fernandes, J. L. A. (2007). *Challenging Euro-America's Politics of Identity: The Return of the Native* (1st ed.). London: Routledge.
Ferrente, A. (dir.) (2009). *L' Orchestra di Piazza Vittorio*. Documentary. Magnolia Home Entertainment.

Fortress Europe: In fuga dall'Eritrea: il deserto, il mare e i ghetti di Roma. (n.d.). Retrieved from: http://fortresseurope.blogspot.com/2005/12/in-fuga-da-tripoli-verso-i-ghetti-della.html

Gebrewold-Tochalo, B. (2007). *Africa and Fortress Europe: Threats and Opportunities*. Aldergate: Ashgate.

Hardt, M. & Negri, A. (2001). *Empire*. Cambridge, MA: Harvard University Press.

Huysmans, J. (2006). *The Politics of Insecurity: Fear, Migration and Asylum in the EU* (1st ed.). London: Routledge.

Insolera, I. (1993). *Roma moderna: Un secolo di storia urbanistica, 1870–1970 (Piccola biblioteca Einaudi)* (Nuova ed.). Piccola biblioteca Einaudi.

Italy warns Europe over migrants. (2011, February 15).*BBC*. Retrieved from: http://www.bbc.co.uk/news/world-europe-12461866

Lakhous, A. (2006). *Scontro DI Civilta' Per UN Ascensore A Piazza Vittorio*. Imprint unknown.

Lefebvre, H. (1992). *The Production of Space*. Oxford: Wiley-Blackwell.

Morassut, R. (2003). Roma Capitale di Pace del Mediterraneo. Online, available: http://static.repubblica.it/roma/piano_regolatore/presentazione.pdf (accessed February 16, 2011).

Nichols, B. (2010). *Introduction to Documentary*. Bloomington: Indiana University Press.

Parati, G. (2005). *Migration Italy: The Art of Talking Back in a Destination Culture* (1st ed.). Toronto: University of Toronto Press.

Pastore, F. (2007). Recent Developments and Critical Issues. Rome: CeSPI.

Pink Floyd (1977). Sheep (lyrics by Roger Waters). *Animals*. Columbia Records.

Pugliese, J. (2010). *Transmediterranean: Diaspora, Histories, Geopolitical Spaces*. New York: Peter Lang.

Rancière, J. (1998). *Disagreement: Politics and Philosophy*. Minneapolis: University of Minnesota Press.

Rancière, J. (2006). *The Politics of Aesthetics*. London: Continuum.

Rancière, J. (2010). *Dissensus: On Politics and Aesthetics*. (S. Corcoran, Trans.) London: Continuum.

Rhodes, J. D. (2007). *Stupendous, Miserable City: Pasolini's Rome* (1st ed.). Minneapolis: University of Minnesota Press.

Sayad, A. (2004). *The Suffering of the Immigrant*. Cambridge: Polity.

Segre, A. (2009). *Come un uomo sulla terra*. Castel Gandolfo (Roma): Infinito edizioni.

Segre, A. & Yimer, D. (2008). *Like a Man on Earth*. Documentary. Retrieved from: http://likeamanonearth.blogspot.com/

Sharma, N. (2006). *Home Economics: Nationalism and the Making of "Migrant Workers" in Canada* (1st ed.). Toronto: University of Toronto Press.

Steimatsky, N. (2008). *Italian Locations: Reinhabiting the Past in Postwar Cinema*. Minneapolis: University of Minnesota Press.

Index

References in **bold** indicate tables, in *italics* indicate figures, and followed by the letter n indicate notes.